Practical Manual on

BASIC AGRONOMY

(with theory)

2nd Revised Edition

Professor N.R. Das

M.Sc. (Ag.), Ph.D., D.W.P. (CIMMYT-MEXICO), FIBR (INDIA)

Ex-Professor of Agronomy,
Bidhan Chandra Krishi Viswavidyalaya,
Mohanpur, Nadia, West Bengal

Formerly,

Director of Research, Director of Farms and Head, Department of Agronomy;
Faculty of Agriculture, Wheat Agronomist and Officer-in-Charge, All-India Co-ordinated
Wheat Improvement Project (ICAR) and Research Agronomist,
Bidhan Chandra Krishi Viswavidyalaya, West Bengal

SCIENTIFIC
PUBLISHERS

Published by
SCIENTIFIC PUBLISHERS (INDIA)

Jodhpur –

5 A, New Pali Road
P.O. Box 91
Jodhpur - 342 001 INDIA

Delhi –

4806/24, Ansari Road
Daryaganj
New Delhi - 110 002 INDIA

2nd Revised Edition: 2018

ISBN:9789387307018

9 789387 307018

Visit the Scientific Publishers (India) website at
http://www.scientificpub.com
Printed in India

In memory of my teacher late Professor A.T. Sanyal, Professor of Agronomy, the then Kalyani University, West Bengal who inspired me for the field-work activities.

— Author

PREFACE

There is a proverb, *"Practice makes a man perfect"*. From this proverb, it is obvious that confidence which is created out of regular practice anything, ultimately is very helpful for future life and this is true in case of agronomic practice also.

India is a vast country, having different ecosystems, with different plants, animals and human population. Under the prevailing situations, it is quite natural that most of the people of eastern India do not know anything about the plant of '*isophgol*' (*Plantago ovata* Rosk, of *Plantaginaceae* family), though they use it as medicine; on the other hand, most of the people of the western parts of the country, do not know the fibre crop of 'jute' (*Corchorus olitorius / capsularis* L. under *Tiliaceae* family), though they use the jute bag/ sack/carpet etc. But students of agriculture in both the areas, should at least, to be acquainted with some of the information of both the plants, though it is not possible to look at them in their respective areas, due to different ecosystems.

The *'farm'* where agronomy students do the practical class, is a holy place, and from there they can learn many many things which ultimately will be the assets in life. Therefore, agronomy students should visit the farm, of course, with the permission whenever they get time, besides their classes, to increase their knowledge in crops, soils, meteorological aspects or other management practices. Of course, for visiting farms, inquisitiveness or special interest is needed.

For creating such inquisitiveness, the 'farm' should be arranged in such a way, where office-buildings, threshing floors, equipments/tractors/tools sheds, crop museum, irrigation systems, godowns (for seeds and fertilizers), open laboratories, well laid-out fields, good neat and clean beautiful gardens, good pucca road/path, effective machineries, electricity etc. are available.

For good teaching or learning, a good `farm' is essentially needed, with good knowledgeable and work-interested field staffs, along with the interested teachers.

Theoretical knowledge or the background information of the subject, is also required for practical work. With this in view, the present book, entitled, **'Practical Manual on Basic Agronomy'**, has been written in short, as a 'practical text book' in Agronomy subject, based on theoretical background, for thorough knowledge of that subject, after a long teaching experience in the universities. Practical classes, with lesson numbers have been fitted, immediately after theoretical discussions, in different sub-chapters in Chapters, on the same studies. The lesson, has been divided into '*Introduction*', '*Objective*', '*Materials required*' and '*Procedure*'. In some cases, 'assignments' have also been given.

There are 14 chapters in this book and these are: 1. Introduction, 2. Preliminary knowledge for Agronomy studies, 3. Climate and its influence on crop production, 4. Soil environment and its modification for crop production, 5. Tillage and crop production, 6. Seeds, sowing, stand establishment and crop-classification, 7. Manures and fertilizers for crop production, 8. Irrigation for crops and cropping systems, 9. Weeds and weed-management for crop production, 10. Soil moisture management under rainfed conditions, 11. Major crops cultivation (in short), 12. Cropping systems, 13. Harvesting of crops, crop-yields and their storage and 14. Vital information for practical approach. Besides these, there are 170 lessons in it.

For all the students of 4-year-degree courses in all the agricultural universities of the country, this book will be needed. This will be very helpful for other students of agricultural training centres/research institutes of different states of the country also. In different 'Krishi Vignan Kendras (KVK)' of ICAR, this type of book will be required for their training programmes. Even the educated farmers will also be benefited for their upliftment of crop production programmes, with this book.

I thank those persons, who have requested me to, write such a new type of book, for the development of `teaching-in-agriculture' of the country. At the end, I am thankful to my son Partha, daughter Aparna and wife Arati who have helped me directly or indirectly, in writing this book.

N.R. Das

Sabuj Niketan,
Kalyani, West Bengal

LIST OF LESSONS UNDER PRACTICAL CLASSES
(Lessons 1 to 170)

CONTENTS

CHAPTER 1

INTRODUCTION

1.1 AGRICULTURE AND AGRONOMY

1.1.1 Agriculture, Agronomy and Agronomist

'Agriculture' is a vast subject with crops, animals and others (fisheries/forests) having, so many subjects and Agronomy' is one of them, Indian Council of Agricultural Research (ICAR) deals with crops, animals and fishes and their managements.

1.1.1.1 *Agriculture*

'Agriculture' is one of the components of ICAR. It is as old as civilization. The word *'agriculture'* comes from the Latin word *'Agercultura'* which was originated from *'Ager'*, meaning field and *'cultura'*, meaning culture or cultivation of soils to grow plants. Initially agriculture was meant for crop production, rearing of fishes and animals and management of products i.e. pastoral farming (breeding, feeding and management of fishes and animals) and all these were within the scope and contents of Agriculture. Of course, now-a-days agriculture grows into a biological science slowly, along with advancement of other branches of science. Modern agriculture is a very highly developed science, dealing with all aspects of development of crop-plants and their management. It has so many branches. Agronomy is now one of the compulsory subjects in the agricultural faculties of all agricultural universities in the country.

1.1.1.2 *Agronomy*

'Agronomy' being the compulsory subject in *'Agriculture'*, plays a pivotal role in the field of *Agricultural University*. Agronomy is a major branch of agricultural science, which deals with principles and practices of crops, soils and water management.

'Agronomy' is derived from Greek word *'agros'* meaning 'fields', and *'nomos'* meaning 'to manage or knowledge'. Agronomy now, is a vital subject in Agriculture and recently it has assumed newer dimensions and can be defined as a branch of agricultural science that deals with methods, which provide favourable environment to crops for higher productivity (including soil management). Therefore, the top agricultural scientist,

Norman has defined *'agronomy'* as the science of manipulating the crop environment-complex with dual aims of improving agricultural productivity and gaining a degree of understanding of the involved personnel.

1.1.1.3 *Agronomist*

The man/person who deals with agronomy or agronomic aspects, is known as the *'agronomist'*. The agronomists must be acquainted with all aspects of agronomy alongwith other allied sciences connected with agronomy.

1.1.2 Relationship of Agronomy to other sciences

'Agronomy' is related to different allied science subjects like *'Botany'*, *'Zoology'*, *'Entomology'*, *'Chemistry'*, *'Physics'*, *'Statistics'* (Mathematics), *'Soil Science'*, *`Agricultural chemistry'*, *'Pathology'*, *'Engineering'* etc.

For comprehensive developments of crops and soils, all these science subjects are essential. Now-a-days, all the parts of science are directly or indirectly connected with agronomy. Now, agronomy is not a separate subject, it is combination of all, though not in full form. Hence, agriculture vis-à-vis agronomy is standing on different basic subjects of science.

1.1.2.1 *Agronomy vs. Botany*

Agronomy is directly connected with botany, in respect of physiology, taxonomy, ontogeny etc. of the plants.

We, being the agronomists, must know the ontogeny (types of plants, like annual biennial or perennial), the taxonomy (family like *Papilionace* in pulses, etc.), physiology (i.e. changing behaviour under different conditions as low temperature for wheat, high temperature for rice etc.).

1.1.2.2 *Agronomy vs. Zoology / Entomology*

Agronomy is directly connected with zoology and entomology, in respect of environments and physiology of microbes and insects which are directly connected with crops and soils. Entomology is the subject which is directly connected with agronomy, as most of insects destroy the crop plants, under growing conditions and for this danger, entomology is essential.

1.1.2.3 *Agronomy vs Chemistry*

Agronomy is directly related with chemistry, in respect of chemical and soil reactions of fertilizers and plant nutrients within the body of the crops.

We, being the agronomists, seek the help of chemistry in different aspect of changing of soils and plant reactions. Biochemistry is also helpful for production of grains, through photosynthesis. Now-a-days 'Bio-Chemistry' is one of the subjects of agricultural chemistry.

1.1.2.4 *Agronomy vs. Physics*

Agronomy is also related with physics and engineering, in respect of different machineries for land preparations or other aspects. Bio-physics subject is growing gradually for the manufacturing of different machineries/equipments. Meteorological equipments are directly connected with physics and engineering for studies of tractors/ power tillers; tillage implements and now engineering studies in agriculture are essential. Meteorology which is based on physics, is essential for agriculture and ultimately agronomy.

1.1.2.5 *Agronomy vs. Mathematics/Statistics*

Now-a-days Statistics, a branch of Mathematics is a strong subject essentially needed for agronomy and its research. For any research in Agronomy, statistics is essential, because without the statistical analysis, the research in the field is not valid. For, agronomic research, some designs are needed. Presently agronomy and statistics go hand-in-hand in agronomic research activities.

1.1.2.6 *Agronomy vs. Engineering*

'Agricultural Engineering department' is now functioning in the agricultural universities, because of modernization of agriculture, in respect of tillage operations, intercultural operations, harvesting, grain processing operation of crops, food manufacturing industries etc. In these aspects agronomy is directly connected. Now tractors, seed-cum-drills, combined harvesters etc. are used frequently in agronomic operations in the fields, besides irrigation managements.

1.1.3. Scope of Agronomy and the role of Agronomists in relation to crop production in the country

As agronomy deals mainly with crop production and soil management, hence the man behind the activities – agronomist plays a vital role in crop-production programme, both in state and at national levels.

1.1.3.1 *Scope of Agronomy in the country*

Agronomy is a dynamic discipline. With the advancement of knowledge, better understanding of plant / crop environment, agricultural practices are modified or new practices are developed for higher productivity. For example, availability of chemical fertilizers, has necessitated the generation of knowledge on the method, quantity and time of application of fertilizers. Before the arrival of dwarf wheat, tall wheat was fertilized with only 40 kg of N/ha; now in dwarf wheat, it requires 100-120 kg N/ha. Similarly availability of herbicides for the control of weed has led to development of vast knowledge about selection, time and method of application of herbicides.

For any crop, different aspects of fertilization, irrigation, or other crop management practices are required to be updated for all the times to come and for this, research activities are essential. The agronomists are to solve the problems, regarding crops and soils, by experimenting on the field itself. Intensive cropping system is now in vogue, hence research is needed, particularly on minimal tillage, on cropping system, instead of clean cultivation which was done earlier. Here lies the role of agronomists.

1.1.3.2 *Role of 'agronomists' in the country*

In the country, the agronomists have much more vital role. The new technology has to be developed to overcome the effect of moisture-stress under dry land conditions, as major areas of our country are in rainfed conditions.

As new varieties of crops with high yield potential become available, packages of practices have to be developed to exploit their full potential under different situations. Restoration of soil fertility, preparation of soils for sowing seeds with better tillage operations, use of proper seed rate, correct date of sowing of improved crop varieties, proper soil conservation and management of soil moisture and of weeds, are the agronomic practices to make our land and water resources productive.

The agronomist aims at obtaining maximum crop production at minimum cost. He exploits the knowledge developed by basic and allied sciences for higher production of crop.

In a broader sense, agronomist is concerned with production of food, fibre, fodder, fuel and furniture by maintaining soil fertility and production economics. Agronomists, now-a-days, at the national level play a vital role. Their experimental findings are given due importance, for some recommendation in the country. For this purpose, the ICAR Co-ordinated Research Projects on crops and technologies, have been established, for the development of agriculture in the country, and the agronomists play a vital role there.

1.2 PRACTICAL APPROACH FOR SOLVING THE PROBLEMS

In order to solve the problems of the farming community, in different regions of the country, it is necessary to play a vital role further by the agronomists.

1.2.1 Farming communities and their problems

In our country, most of the people involved in agriculture, are poor, illiterate and weak; as a result they can not make progress in life.

They have small amount of lands and feed the huge number of family members, as a result they are in debt. Besides, old methodologies they use in their fields with old varieties of crops. As because they are poor and illiterate, they can not enable higher education to their children and cannot take nutritional food necessary for their health.

1.2.1.1 *Farming communities*

The poor, illiterate and weak persons/people in the farming community exist in different regions of the country, though their problems differ from place to place. The picture of most of the farming communities, is more or less the same in the country, though situations are different.

1.2.1.2 *Farm problems in different ecosystem regions*

The farm problems, change from one ecosystem another ecosystem, because of variations in climatic conditions, land situations, soil types etc. The problems in different ecosystems, are as follows :

(a) Arid ecosystem

This 'arid ecosystem' includes the western Himalayas, western part of Rajasthan, south-eastern parts of Haryana and Punjab, the Kutch peninsular, northern part of Kathiawar peninsular, part of Decan plateau, Karnataka (Raichar and Bellary) and Andhra Pradesh (Anantapur).

This ecosystem is characterized with cold wave and hot arid climate. As a result productivity of crop is very very poor, because of less moisture in the soils. People are also few in number and very poor, and the activities of the farming community are less. Therefore, a dismal picture is seen in this ecosystem in respect of education, health, farming and other business.

At individual levels, moisture conservation techniques should be adopted as far as possible. Proper arrangement should be developed for irrigation.

(b) Semi-arid ecosystem

This semi-arid ecosystem covers the northern plains and central high land of Kathiawar peninsular, Decan plateau and Eastern ghats, covering the part of the Punjab, Haryana, Uttar Pradesh, Rajasthan, Madhya Pradesh, Gujarat, Maharastra, Andhra Pradesh, Karnataka and Tamil Nadu.

This semi-arid ecosystem is characterized with hot and dry summer and mild-to-cool winters. As a result, crop productivity is less, because of less moisture in soils. People are also few in number but higher than that of arid zone. Here also a dismal picture of the farming community is observed because of poor yield of crops.

Emphasis on development of moisture conservation should be given, along with the development of packages and practices of crop cultivation with new varieties.

(c) Sub-humid / semi-wet ecosystem

This sub-humid/semi-wet ecosystem includes the northern Indo-Gangetic Plains, Central high lands of Malava and Bundelkhand, part of Decan plateau, Chhattisgarh and Chhota Nagpur region of eastern plateau, Eastern plains and ghats and western Himalayas, spread over the states of Jammu and Kashmir, Himachal Pradesh, Uttaranchal, Uttar Pradesh, Bihar, Orissa and Andhra Pradesh.

This ecosystem is characterised by mild to hot summer and mild cool to cold winter. Due to climatic conditions, the crop productivity is high but amount of land is small in the hands of the farming community, as a result some of them are poor. The condition of area under this ecosystem is not so bad as in other dry and semi dry ecosystems. Some of the areas are fertile and productivity is high.

Here packages and practices are to be developed with high yielding varieties of crops, along with modern technological developments. Irrigation facilities have to be developed.

(d) Wet / Humid ecosystem

This humid/wet ecosystem comprises of Assam and Bengal plains, eastern Himalayas and north eastern hills, spread over the states of West Bengal, Sikkim, Assam, Arunachal Pradesh, Nagaland, Manipur, Tripura, Mizoram and Meghalaya.

The climate of this ecosystem is characterized with mid-to-hot summers and mild-to-severe winters. In these areas, some are high hills and some are bestowed with plains. The plains of West Bengal and Assam are mostly in the low lying areas, where water stagnation or floods are common.

Of course, land is fertile here, water is in excess, due to high rainfall, with long duration rainy season. Waterlogging and floods are the problems in this area. High human population is another problem. There is good moisture in pre-*kharif* and *kharif* seasons, winter season is dry, hence, most of the land remain vacant/fallow due to lack of irrigation facilities.

Packages of practices should be developed for high yielding varieties of crops, along with finding out the techniques for crop improvement. There is a great scope in improvement of cropping system with rainfed *rabi paira / utera* crop, taking advantage of climate conditions, soil types and land situations.

(e) Coastal ecosystem

The coastal ecosystem covers the western coastal plains, extending from Gujarat to Karnataka. Then this area passes through the states of Orissa, Andhra Pradesh, Tamil Nadu, Kerala, Maharastra and West Bengal.

This ecosystem climate is characterized with tropical or subtropical conditions with hot summer and very mild to mild winters. Because of salinity in soils, lands are not fully utilized with crop, because of low produc-tivity. Development is necessary by growing crops with saline resistant varieties, along with new technologies.

Thus, in different ecosystems, problems are different and under different situations, different solutions should be evolved by the agronomists in these areas.

1.3 DEVELOPMENT OF AGRICULTURE IN THE COUNTRY

After the partition of India in 1947, independent India developed agriculture enormously in different programmes, under the leadership of 'Indian Council of Agricultural Research (ICAR)' which was established in the same year, replacing 'Imperial Council of Agricultural Research' in British rule. The development was mainly on crop improvement (Breeding programmes), improvement of management practices of different crops along with productivity.

1.3.1 Crop improvement

Different high yielding varieties of many crops were evolved by the breeders, associated with pathologists, entomologists and agronomists, etc.

Major works were done in wheat and rice. Some of the wheat varieties have been sent to other countries also for their development.

Wheat breeding programmes were developed in *aestivum*, *durum* and *dicoccum* wheats; breeding programme was done in *multilines*. Some of the wheat varieties evolved in India are as follows :

WH 542, UP 2338, Raj 3077, PBW 343, HD 2687, CPAN 3004, PDW 215 (*durum*), WH 896 (*durum*), PBW 502, PBW 226, Raj 3765, PBW 373, UP 2425, HD 2285, Raj 3777, Raj 3765, PBW 175, PBW 299, K 8804, DL 784-3, HD 2402, HP 1731 (Rajlaxmi), K-9107, HUW 468, HP 1633 (Sonali), HDR 77, K 8962, GW 190, HI-1077, HI-8498 (*durum*), GW 273, Raj 1555 (*durum*), Swati. DL 788-2, MACS 2496, DWR 162, HW971 (*dic*) HUM 318 (dic), DDK 100 (dic) etc.

1.3.2 Improvement of management practices of different crops

Improvement of management practices of different crops for various regions, has been done. Special emphasis has been given to wheat. With the arrival of dwarf wheats, management practices have been worked out specially with different varieties, on method of sowing, sowing date, fertilizers, irrigation depth of seedling, seed rate, weed control, spacing etc. both under irrigated and rainfed conditions.

1.3.3 Overall improvements in crop productivity

Since 1947 (the partition year), crop production of India increased many folds. In 1950-51, production of rice, wheat, *jowar, bajra*, maize, pulses and oilseeds were 20.58, 6.46, 5.50, 2.60, 1.73, 8.41, and 5.16 million tonnes respectively, whereas in 2002-2003 the production of the same crops, were 72.65, 65.10, 7.08, 4.63, 10.30, 11.14 and 15.06 million tonnes respectively. Therefore, it is obvious that production of different crops improved significantly in India, since 1950-51.

CHAPTER 2

PRELIMINARY KNOWLEDGE FOR AGRONOMY STUDIES

2.1 BASIC KNOWLEDGE ABOUT THE LANDS, SOILS, FARMERS, FIELDS, COLLEGE/UNIVERSITY AND ICAR INSTITUTE FARMS

The learners, in B.Sc. (Ag.) Hons. course in agricultural universities, should have the basic knowledge about the lands, soils and other amenities of the farmers' field and that of agricultural college or universities and ICAR institutes or State Govt. farms.

2.1.1 Lands

Land is one of the basic resources for agricultural operations for any country. There are three situations of the land such as (a) high, (b) medium and (c) low, according to water stagnation. In high lands, water after rainfall, is not stagnated, and drained off, whereas in low land after rainfall, water is accumulated on soil surface. In Medium land, rain water is absorbed in soils and no stagnated water is visible on surface, though it is in between high and low land.

In farmers' fields land situations are natural. They grow different crops in different lands under different situations. On the other hand, land situations are manipulated/ created by the machinery both in University/ ICAR/State Govt. farms. There, under different situations, different types of crops are grown according to needs. Rice (transplanted) grows in low lands, maize in high lands and jute is grown in medium lands.

2.1.2 Soils

According to land situations, types of soils also differ. Generally, in high lands, less amount of silt or clays are present in soils, whereas in low land, silts and clays are more. Of course, in medium type of soils, sand, clay and silt are more or less equally present, as a result, more or less all type of crops are possible to be grown in medium lands under different situations.

Normally, high land soils are poor in respect of fertility and sandy in nature; whereas low land soils are rich in nutrients and clayey in nature. However, medium land soils are good for drainage system; nutrient content is good; it is good for soil management also. Therefore, all types of crops are grown in these soils. Water-holding capacity of soil is high in clayey soil, medium in medium land soils and low in high land soils.

2.1.3 Farmers' fields

Different land situations in farmers' fields are different and natural. Under these conditions, they grow different crops at different times under different situations, though they do not get desired yields.

Most of the farmers, of course, level the medium land for growing irrigated crops. During rainy season, high lands are good for some crops like leguminous crops or maize etc. Low lying areas are generally, used by some crops which prefer water (rice/arum). Medium types of lands are used for most of the crops in different seasons, provided irrigation facilities are available.

Naturally, farmers under arid and semi-arid areas, are very poor, but under sub-humid or humid areas, they are rich farmers comparatively, because of their crop productivity. The farmers in coastal areas are also poor because of low crop productivity.

2.1.4 ICAR (Research Institute), University and State Farms

Generally, the farms, unlike that of farmers, under ICAR institutes, or University or State Govt. are quite different. These are as follows :

2.1.4.1 *ICAR Research Institute Farms*

For any institute of ICAR, there is a well-laid-out farm, where particular crops are grown. For these farms, there are provision of field staffs, labourers and machineries. Different types of lands (high, medium and low lying areas) are there, so that different types of crops are grown in different times. Mostly the farms are used for some experiments/trial purpose.

Arrangement for tillage operations, irrigation facilities weeding facilities, facilities for intercultural operations of the crops are always available, along with good threshing floors and godowns.

2.1.4.2 *University Farms*

In most of the agricultural universities, farms are well-laid out with separate farm staff. The students do their practical field work there, batch by batch. The facilities for irrigation, tillage, inter-tillage, weed control devices etc. are available, along with good threshing floors and godown. Arrangement for processing of harvested materials is also there. In those farms research activities also are continued.

2.1.4.3 *State Government Farms*

State Govt. Farms are not located in one place, but in different places, in different district and sub-divisional levels, all over the State. The farm facilities are not like those of ICAR/University farms. Though they do not conduct the vital and basic experiments, they, of course, go for trials. They test the suitability of the new varieties, new methodologies or new packages of practices developed in the universities or elsewhere. Of course, they do not have separate farm staffs specifically, but for entire farming system.

Practical Class: Visit to the farms of the university, ICAR institutes and State Govt. farms and study the land situation, soils and crops there

Lesson 1 : Visit to University Farms (for studies on land situation, soil types, crops grown, area, overall assessment of the farms about the infrastructure facilities and surroundings)

Introduction

Every agricultural university has the individual, well equipped and well laid-out big farms, having different soil types with different land situations. The learners, coming to the university, should visit the farms positively. Generally, the leading Teacher/ Officer/Farm Manager will give the detailed description of the farm, starting from area to all other aspects like soil types, land situations, irrigation system, crops grown, equipment facilities, threshing and storage facilities etc. along with the open field laboratories. The students/learners must take/note down all these records first from the lecture and then sort them out and arrange them in the form of area, surroundings, land situations, soil types, crops grown, required implements, godown, storage facilities, irrigation arrangements, threshing floor, field (open) laboratories etc.

Objective

- o to get the first hand information of the farm,
- o to get the idea of working facilities,
- o to get the idea of crops grown in the farm,
- o to get the idea of land situation and soil types,
- o to get the idea of the surroundings.

Material required

✥ One note book, ✥ One pen/pencil, and ✥ One 15 cm or 30 cm scale.

Procedure

- Note down the lecture fully, giving date and time at farm, standing just below the tree or any shed areas, on different aspects of the farm silently, attentively and politely like disciplined soldiers.
- Arrange the sum and substances of the lectures, for putting into your exercise book/practical exercise book, on these aspects :

Assignments

1. Area of the farm; 2. Surroundings of the farm; 3. Land situation of the farm; 4. Soil types of the farm; 5. Crops grown in the farm; 6. Implements available in the farm – a. Tillage implements, b. Inter-tillage implements, c. Sprayer and duster, d. Sowing and harvesting implements and e. Others (if any); 7. Irrigation arrangement, 8. Godown and threshing arrangement; 9. Plot lay out arrangement (shape, size etc.); 10. Field open laboratories and 11. Other arrangement (if any).

Lesson 2 : Visit to ICAR Research Institute Farms (for studies of land situation, soil types, crops grown, area, overall assessment of the farms about the infrastructure facilities and surroundings).

Introduction; **Objective**; **Material required**; **Procedures**; **Assignments** (all as in lesson 1)

Lesson 3 : Visit to State Govt. Farms (for studies of land situation, soil types, crops grown, area, and overall assessment of farms, infrastructure facilities and surroundings)

Introduction; **Objective**; **Material required**; **Procedures**; **Assignments** (All as in lesson 1)

[**Notes :** Generally, facilities and arrangements are better in both the University and ICAR Research Institute Farms than those of the State Govt. Farms, because of less importance in teaching and research in Govt. farms].

2.2 IDENTIFICATION OF COMMON SEEDS OF FIELD CROPS

In the farms, there are different types of crops, which belong to different families. Some crops give major foods (cereals), some give pulses or *dal* (pulses), some give oils (Oilseeds), some give fibres (fibres), some give sugar/ starch (sugars/rhizome), some give cattle foods (fodder), some crops give tuber (tubers) and so on. All there crops are essential for cattle and human beings for food, fibre, fodder fuel and furnitures. Therefore, the learners must identify the groups of crops.

2.2.1 Cereal crops

(Wheat, rice, maize, *jowar, bajra*, pearl millet, *ragi*, and other millets)

Cereals are the crop-plants, belonging to the grass family *Gramineae* which are grown for their edible carbohydrate starchy seeds, botanically known as *caryopsis*. The word '*ceres*' means the godess of grain e.g. wheat, rice, maize, *jowar*, *ragi*, *bajra* etc. They are grown in different seasons, under different situations/conditions. The seeds of these crops should be identified by the learners from the characters as follows :

Seeds

Cereal Crops	Length (mm)	Breadth (mm)	1000-seed Wt (g)	Shape	Hairy-ness	Colour
Wheat	5-10	2-3	40-50	Creased / grooved	Brush	Whitish / amber
Rice	5-12	2-3	22-25	Pointed end	Rough	Brown
Maize	10-20	5-10	80-90	Blant end	Smooth	Yellow
Jowar	5-12	2-5	40-50	Round	Smooth	Brown
Bajra	5-12	2-5	3-40	Round	Smooth	Brown
Pearl millet	4-10	2-5	25-30	Round	Smooth	Whitish
Ragi (Marua)	4-8	2-4	20-25	Round	Smooth	Brown

2.2.2 Pulse crops

(Gram, lentil, green gram, black gram, grasspea, pea and cowpea)

The edible seeds of legumes, are known as pulses/*dal*. All the pulse crops are under the main family *Leguminoeae* and sub-family *Papilionaceae*. All the leguminous plants fix nitrogen from the atmosphere, through *rhizobium* bacteria present in their roots. Thus the pulse crops fix nitrogen from air and enrich the soil with nitrogen. The cultivation with legumes, is very helpful, as soil fertility is increased.

So, many leguminous crops are there, in the world, but gram, lentil, green gram, black gram, ricebean, grasspea, pea, cowpea etc. are grown as major pulse crops in India. Some of the leguminous crops are used for green manure crops, cover crops, fodder crops or even fibre crops (sunn hemp). The students should identify the seeds of pulse crops, with the special criteria as follows :

Seeds

Pulse Crops	Length (mm)	Breadth (mm)	1000-seed Wt (g)	Shape	Hairyness	Colour
Gram	4-8	3-5	50-60	Pointed end	Smooth	Yellowish
Lentil	3-5	2-3	22-25	Flat oval	Smooth	Redish
Green gram	3-8	2-4	25-30	Round	Smooth	Greenish
Black gram	3-8	2-5	25-30	Round	Smooth	Blackish
Grass pea	4-8	2-5	30-40	Round	Smooth	Brownish
Pea	4-10	2-5	30-40	Round	Smooth	Yellowish
Cow pea	5-8	3-5	50-60	Oval	Smooth	Whitish

2.2.3 Oilseeds crops

(Mustard / rape, groundnut, sunflower, sesame and safflower)

The seeds, from which a slippery viscous liquid or fat like substances are produced, are called the *oilseeds*, and the plants crops are known as the oilseed crops. These crops may be from different families. The major oilseed crops in India are : mustard / rape, sunflower, sesame, safflower, linseed, groundnut, *bheranda / reri*, jatropha etc.

Oilseeds are used as edible, industrial or fuel purposes. The major oil seeds that are used as edible purpose are mustard/rape (*Cruciferae*), ground- nut (*Papilionaceae*), Sunflower (*Compositeae*) and sesame (*Pedaliaceae*) etc.

When oil is extracted from the edible *'oilseeds'*, remaining solid portion, is the oilcake. These oilcakes are the concentrated organic manures, used for cultivation of crops or for feeding to the cattle.

Therefore, the learners should identify the seeds of 'oilseeds' from the following criteria:

Seeds

Oilseed Crops	Length (mm)	Breadth (mm)	1000-seed Wt (g)	Shape	Hairyness	Colour
Mustard	1-2	0.5-1.0	8-10	Round	Smooth	Brown/ yellow
Rape	1-2	0.5-1.0	8-10	Round	Smooth	Brown
Groundnut	5-12	4-5	80-90	Round	Smooth	Brown
Sunflower	4-8	3-5	30-40	Pointed end	Rough	Whitish
Sesame	2-4	1-2	12-25	Glossy	Smooth	Greyey
Safflower	3-5	2-6	10-16	Round	Rough	Brownish

2.2.4 Fibre crops

[Jute (*cap.* and *oli*), mesta, sunnhemp and cotton]

The crops that yield fibres, are called the fibre-yielding crops. Any thread like material, specially those thread like tissues, having sufficient toughness for the use of textile or similar industries, is the fibre. Fibre may be derived from plants, animal or mineral kingdom. The agriculturists are concerned with plant fibres. Plant fibres are of two types i.e bast fibre and lint fibre. Bast fibres are obtained from the stem or bark and lint fibre is obtained from reproductive portion of the plant. Jute, mesta, sunnhemp etc. are the bast fibres, whereas the cotton gives the lint fibres.

Therefore, the learners should identify the fibre crops through seeds. The identifying characters of seeds of fibre crops are as follows :

Notes : cap = *capsularis* and oli = *olitorius.*

Seeds

Fibre Crops	Length (mm)	Breadth (mm)	1000-seed Wt (g)	Shape	Hairyness	Colour
Jute *(Capsularis)*	0.5-1.0	0.3-0.6	5-10	Spherical	Smooth	Brownish
(Olitorius)	0.4-0.8	0.2-0.4	4-8	Triangular	Smooth	Greenish
Mesta	4-8	2-3	30-35	Round	Rough	Brownish
Sunnhemp	3-5	2-6	25-30	Round	Rough	Brownish
Cotton	3-6	2-4	30-40	Spherical	Smooth	Blackish

2.2.5. Sugar crops (Sugarcane, sugar beet)

The crops, that are grown for obtaining sugars/sweet materials, are designated as *sugar crops*. The sugar crops are sugarcane, sugarbeet etc. Sugarcane is cultivated with sugarcane *setts* (small pieces of cane used for planting materials, usually from the top portion of the cane plant). Dry leaves of sugarcane are the *sugarcane trash*. *Bagasse* is the mill residues from the cane sugar industry, consisting of crushed stalks from which the juice has been extracted. Sugar beets is a small *rabi* crop where from sugar is obtained, whereas sugarcane is a big tropical crop. From both of them sugar is obtained from different methods. Yambean is also sugar crop, though percentage of sugar is less. It is the swollen soft root which is edible. Though they yield sugars, they are coming from different families. Sugarcane setts or sugar beet and yambean seeds are to be identified from the following information :

Setts / beets

Sugar crops	Length (mm)	Breadth (mm)	1000-seed Wt (g)	Shape	Hairy-ness	Colour
Sugarcane (Setts)	1-2 m	0.5-1.0 m internode	Sett 0.5-0.6 kg	Elongated	Smooth	Brown/ yellow/ red
Sugarbeet (Beet)	Beet length 8-12 cm	2-6 cm (radius)	Beet 0.5-1.0 kg	Tapering	Smooth	Pinkish/ whitish

2.2.6 Tuber crops (Potato and Sweet potato)

Crops possessing enlarged under ground vegetative stems, roots-like stem, are the tuber crops. Main tuber crops, grown in India, are potato, sweet potato etc. Potato is cultivated with tubers, whereas sweet potato is cultivated with stem cuttings (vine). Of course, now-a-days potato is also cultivated with seeds (TPS-true potato seeds).

Both tubers/seeds of potato and sweet potato should be identified, from the following information.

Tuber / seeds (TPS) / cutting

Tuber Crops	Length (mm)	Breadth (mm)	1000-seed Wt (g)	Shape	Hairyness	Colour
Potato tuber	2.3 cm	1-2 cm	1 tuber 10-15 gm	Round	Smooth	Brownish
Seed	0.4-0.6	0.2-0.3	2-4	Round	Smooth	Brownish
Sweet potato (vine cutting)	0.5-0.8 m	0.3-0.5	9-10 gm	Round	Smooth	Brownish

2.2.7 Fodder/forage crops

(Napier, thinnapier, berseem, lucern, grasspea, pea, ricebean, cowpea and paragrass)

The crops, which are harvested with leaves, stems, flowers or seeds for the animal, livestock feeding (as such or after processing), are known as fodder crops.

The fodder crops are divided into two groups such as (a) Grassy or (b) non-grassy. Among the grassy groups, maize, sorghum, napier, teosinte paragrass etc. are main; whereas in non-grassy group, black gram, rice-bean, cowpea, grasspea, berseem, lucern are the main.

Therefore, they should be identified. The clues are as follows :

Stubble or seed or stem cutting

Fodder Crops	Length (mm)	Breadth (mm)	1000-seed Wt (g)	Shape	Hairyness	Colour
Napier stubbles	2-3m	-	5-7	Round	Smooth	Brownish
Seeds	0.1-0.2	0.001-0.02	2-3	Round	Smooth	Brownish
Thin napier (seeds)	0.1-0.2	0.1-0.2	2.3	Round	Smooth	Brownish
Berseem (seeds)	0.3-0.4	0.1-3	5-7	Round	Smooth	Yellowish
Lucern (seeds)	0.2-0.3	0.1-0.3	5-7	Round	Smooth	Blackish
Grasspea (seeds)	4-8	2-4	30-40	Round	Smooth	Brownish
Ricebean (seeds)	2-3	2-4	20-30	Round	Smooth	Brownish
Cowpea (seeds)	5-8	3-5	50-60	Oval	Smooth	Whitish
Paragrass (stem-cutting)	50-60 cm	-	-	Cylindrical	Smooth	Whitish
Maize (seeds)	10-20	5-8	80-90	Blant end	Smooth	Yellow
Sorghum (seeds)	5-12	2-6	40-50	Round	Smooth	brown

Practical class: Identification of seeds

Lesson 4 : Identification of common seeds / `agronomic seeds' of field crops like (a) cereals, (b) pulses and (c) oilseeds

Introduction

Generally, the size, shape, structure colour etc. of different type of seeds of crops are different. Knowledge about the seed characteristics of crops is highly essential, during different operations like threshing, field observations, intercultural operations, cleaning and drying process, storing testing of viability, sowing and harvesting etc. For all the purposes, the learners should identify the characters of seeds. These lessons (lesson 4 and lesson 5) deal with the methods of identifying the seeds morphologically.

Objective

- o to well equip the learners to identify the different types of seeds morphologically.
- o to utilise the identification marks for different purposes and also for gaining experience about the seeds.

Material required

❖ Different small/medium glass/ebonite/plastic jars/containers, for keeping seeds, ❖ Seeds (about 200 gm) of different categories of crops to be kept in the jars/containers, ❖ Small balance with weights, ❖ Note book, ❖ Pen/pencil, ❖ Small (15 cm or 30 cm) scale.

Procedures

- Display different types of seeds in the containers/jars, with proper labels (with scientific names of the crops).
- The learners will study the seeds minutely, for different character-istics of seeds morphologically, one after another systematically, with the knowledge given in theory.
- The learners will use the scale for measuring the length and breadth of seeds,
- They will use the balance for weighing seeds numbering 100 or 1000.

Assignment

Observations to be recorded

[Under (a) Cereals, (b) Pulses and (c) Oil seeds separately]

- ☐ Measuring length and breadth (cm/mm) of seeds
- ☐ Shape (eye estimation), elongated or bold
- ☐ Size (eye estimation), big or small
- ☐ Colour
- ☐ Hairy-ness
- ☐ Plain or spotted
- ☐ Test weight of seeds (100 or 1000 seeds)
- ☐ Other identifying points (if any).

Lesson 5 : Identification of common seeds/agronomic seeds of field crops like (a) fibres, (b) sugars, (c) tubers and (d) fodders

[Under (a) Fibres, (Jute and mesta) (b) Fodders (Napier and grasspea), (c) Tuber (potato) (d) Fodder (Napier and grasspea)]

Introduction; **Objective**; **Material required** and **Procedures:** All in lesson 4).

Assignment

Observations to be required: (as in lesson 4)

2.3. IDENTIFICATION OF MANURES AND FERTILIZERS

2.3.1 Common organic manures

Now-a-days common organic manures are found in different forms, admixtured of both bulky and concentrated manures, along with chemical fertilizers. We will discuss here only the common orthodox manures, both in bulky organic and concentrated organic manures.

2.3.1.1 *Bulky organic manures/green manures/ leaf manures*

Different types of organic manures are there, having different composi-tions and preparations. These organic manures enrich the soils with respect to nutrients and physical aspects.

Bulky organic manures

Bulky organic manures include (1) farm yard manure (F.Y.M)/farm manure, (2) farm/town compost, (3) night-soil, (4) sludge, (5) green manure and (6) other bulky sources of organic matter. All these manures are bulky in nature and supply (i) plant nutrients in small quantities and (ii) organic matter in large quantities. Among all other organic matters, F.Y.M, compost, and green manure are very important and largely used.

F.Y.M (Farm Yard Manure)

This (F.Y.M) is the decomposed mixture of *'dung'* and *'urine'* of farm animals, along with the litter (bedding materials and left over materials from roughage or fodder, fed to the cattle). On an average well-decomposed F.Y.M contains 0.5% N, 0.2% P_2O_5 and 0.5% K_2O. Of course, this amount varies from country-to-country and place–to-place, depending upon the feeding materials.

Compost

Compost is also the well-decomposed material of village/town/farm refuse. It contains nearly 0.5 of N, 0.15% P_2O_5 and 0.5% K_2O in village compost, though they vary from place-to-place under different circumstances. Town compost, of course, is rich in nutrients. It contains 1.4% N, 1.0% P_2O_5 and 1.4% K_2O, though it also varies from place-to-place.

Composting

Process and methods of decomposition of farm/town/village refuse, is the composting. Composting is largely a biological process in which microorganisms of both types namely *aerobic* (which require air or O_2 for respiration and development) and *anaerobic* (which function in absence of air or free O_2) decompose organic matter and lower the carbon-nitrogen (C/N) ratio of the refuse materials.

Compost may be prepared with weeds, crop-stubbles/residues, *bhusa* or straw, leaves, urine soaked soil, dungs of cattle, goat, excreta of other animals, hedge, clipping (i.e. all plant / animal material) etc. This compost is known as '*village compost*'. The '*town/urban compost*' may be made with night-soil of animals, street refuse, dustbin materials, coming from the houses etc., According to availability of space and material, a trench/pucca tank of 15′ × 5′ × 4′ should be prepared for decomposition.

Green manures

The manure that is obtained in green conditions of the crops, is known as green manure. Green manure is obtained either by decomposition of plants in *situ* (as such in the same place) or by decomposition of green leaves, bringing from out side the fields. For green manure in *situ* generally, *dhainoha* blackgram, sunnhemp, *Pillipesara*, *guar* etc. are used, whereas for leaf-manuring '*Gliricidia*' etc. is used.

Green manuring procedures

Under wet conditions, generally *dhaincha*, blackgram, sunnhemp etc. are sown by the middle of April and after 60 to 65 days after sowing, it is ploughed-under and again after 15 days of first ploughing-under the second time ploughing-under is completed, then it is decomposed well and planting (rice) is possible. This is actually green manuring in *situ*. In some leaves of leguminous plant (*Gliricidia*) are cut and placed in between the planted rice in waterlogged condition. This is *leaf manuring*.

Of course, both advantage and disadvantages are there for green manuring. It adds organic matter to the soil and stimulates the activities of the micro-organisms. It improves soil structures and increases nutrient status in the soils.

Of course, some disadvantages are also there. If green manure crops, that are ploughed-under, are not decomposed properly for want of proper water/moisture, reverse picture will be seen in the field; germination and growth of the succeeding crop will be hampered, as a result the crop will give less yield. In that case termite problem may be seen. Practice of green manuring thus may not be economical because of attack by other disease, pest etc. Hence some times it is a risk.

Leaf manures

Sometimes the leaves of some leguminous shrubs are used. From these plants, leaves or soft twigs are cut and placed over the soils which are turned upside down. But in some cases leaves are put as such, in between transplanted paddy rows. It is a general practice in some areas of Tamil Nadu.

Among the plants, where from leaves or twigs are taken are *Gliricidia* (*Gliricidia maculceata*), dhaincha (*Sesbania speciosa / aculeata*), Karanja (*Pongamia pinnata*) etc.

2.3.1.2 *Concentrated organic manures*

Concentrated organic manures are organic in nature but contain higher percentage of major plant nutrients like nitrogen, phosphate and potash and others, as compared to other bulky organic manures, such as F.Y.M., compost etc. These concentrated manures are made from the raw materials of animals and plants. However, common concentrated organic manures are generally different oilcakes, blood-meals, fish manures, meat-meal, wood waste etc.

Oilcake

Oilcakes can be divided into two groups, such as (i) **edible oilcakes** – suitable for feeding to the cattle and (ii) **non-edible oilcakes** – not suitable for feeding to cattle but suitable for manure. Though some of the edible oilcakes are fed to the cattle, they are used by the cultivators for the soils as manures also.

Oilcakes are generally quick-acting organic manures. Though they are insoluble in water, their nitrogen becomes quickly available to the plants. Nitrogen becomes available by about 7-10 days after its application to the soils. The chemical composition of some of the oilcakes, are is follows :

Average nutrients (N, P_2O_5 and K_2O) of major oil-cakes

Name of oil-cake			Percentage composition		
			N	P_2O_5	K_2O
(A)	**Non-edible oil cakes**				
	(i)	Castor cake	4.3	1.8	1.3
	(ii)	Cotton seed cake, undecorticated	3.9	1.8	1.6
	(iii)	*Karanja* / honge cake	3.9	0.9	1.2
	(iv)	*Mahua* / ippi cake	2.5	0.8	1.8
	(v)	Neem cake	5.2	1.0	1.4
	(vi)	Safflower cake, undecorticated	4.9	1.4	1.2
	(vii)	*Undi / punna* cake	3.6	1.5	2.0
(B)	**Edible oilcakes**				
	(i)	Coconut cake	3.0	1.9	1.8
	(ii)	Cotton seed cake, decorticated	6.4	2.9	2.2
	(iii)	Groundnut cake	7.3	1.5	1.3
	(iv)	Linseed cake	4.9	1.4	1.3
	(v)	Jambo cake	4.9	1.6	1.9
	(vi)	Niger cake	4.7	1.8	1.3
	(vii)	Rape / mustard cake	5.2	1.8	1.2
	(viii)	Safflower cake, decorticated	7.9	2.2	1.9
	(ix)	Sesame / *til* cake	6.2	2.0	1.2

Note : Adopted from **Manures and Fertilizer**, by K.S. Yawalkar, J.P. Agarwal and S. Bokde (1977), Agri-Horti Publishing House, Nagpur.

2.3.2 Common (chemical) fertilizers

Commercial straight fertilizers with nutrients like nitrogen, phosphate and potash are called the common chemical fertilizers. These three elements are the basic needs for all crops. Other secondary and essential macro-nutrients are also needed for plant growth. The basic commercial common straight fertilizers of nitrogen, phosphate and potash are stated below, along with micro-nutrients :

2.3.2.1 *Nitrogenous fertilizers*

Common nitrogenous fertilizers are grouped into four categories such as (a) Nitrate fertilizers, (b) Ammonium fertilizers, (c) Nitrate and ammonium fertilizers, and (d) Amide fertilizers.

(a) Nitrate fertilizers

In these fertilizers, nitrate is combined as NO_3 (nitrate) with other elements. Such fertilizers are, (1) Sodium nitrate/Chilean nitrate ($NaNO_3$) – contains 16% N, (2) Calcium nitrate [$Ca(NO_3)_2$] – 15.5% N etc.

(b) Ammonium fertilizers

In these fertilizers, nitrogen is combined in ammonium (NH_4), form with other elements. Such fertilizers are :

(1) Ammonium sulphate [$(NH_4)_2 SO_4$] – 20% N,

(2) Ammonium phosphate ($NH_4 H_2PO_4$) – 20% N + 20%, P_2O_5 or 16% N + 20% P_2O_5,

(3) Ammonium chloride (NH_4Cl) – 24% N,

(4) Anhydrous ammonia – 82% N,

(5) Ammonia solution – 20% - 25% N.

(c) Nitrate and ammonium fertilizers

These fertilizers contain nitrogen both as nitrate and ammonium. Such fertilizers are :

(1) Ammonium nitrate ($NH_4 NO_3$) – 33% N,

(2) Calcium ammonium nitrate(CAN)- 20% N,

(3) Ammonium sulphate nitrate – 26% N.

(d) Amide fertilizers

These fertilizers contain nitrogen in the form of amide (NH_4 / CN_2). Such fertilizers are :

(1) Urea [$Co (NH_4)_2$] – 40% N,

(2) Calcium cyanamide ($Ca CN_2$) – 21% N.

Generally '*nitrate*' fertilizers are used in aerobic conditions, while '*ammonium*' and '*amide*' fertilizers are used both in aerobic and anaerobic conditions. For top dressing urea fertilizers are preferred.

Sodium nitrate ($NaNO_3$) the white crystal, highly soluble in water, is obtained naturally in Chile, hence it is known as 'Chilean Saltpetre'. This is the oldest and best known nitrate fertilizer. Ammonium sulphate [$(NH_4)_2$ SO_4] is the well known sugar like nitrogenous fertilizer for the cultivators of India. It contains 20-21% N. It is a crystalline salt, stable, soluble in water and stored well. Urea [CO $(NH_4)_2$] is a white crystalline round salt, containing about 44-46% N. It absorbs moisture from the air. Ammonium chloride (NH_4 Cl) is a white crystalline ammonical fertilizer, containing 24-26% nitrogen.

Generally, nitrogenous fertilizers are used for increasing the vegetative parts of the plants/crops, at earlier stage of growth. This is more required by grass-family crops.

In complex fertilizers, generally two or more nutrients of N, P_2O_5 and K_2O are present. Complex fertilizers like Nitro-phosphate/nitric phosphate, Ammonium phosphate sulphate, diammonium phosphate and urea ammonium phosphate are being produced in India.

2.3.2.2 *Phosphatic fertilizers*

The plant nutrient content of all phosphatic fertilizers is expressed as percentage of phosphorus *pentaoxide* (P_2O_5). Phosphorus is an element of which there are several forms. The phosphatic fertilizers can be classified into three groups, depending on the form in which ortho-phosphoric acid or phosphoric acid is combined with calcium. These are as follows :

(1) Phosphatic fertilizers, containing water-soluble phosphoric acid or mono-calcium phosphate

These fertilizers contain phosphoric acid in an available form. These fertilizers are :

(i) Single/ordinary super phosphate – 16-18% P_2O_5,

(ii) Double super phosphate – 32% P_2O_5,

(iii) Triple super phosphate 46-48% P_2O_5, and

(iv) Ammonium phosphate (20% N + 20% P_2O_5).

(2) Phosphatic fertilizers, containing citric acid soluble phosphoric acid or dicalcium phosphate

Such fertilizers are :

(i) Basic slag – (14-18% P_2O_5),

(ii) Dicalcium phosphate – 34-39% P_2O_5,

(iii) Rhenania phosphate – 23-26% P_2O_5 and

(iv) Raw and steamed bone meal (part of P_2O_5 soluble in citric acid).

These types of fertilizers are suitable for the acidic soils, because of low pH.

(3) Phosphatic fertilizers, containing phosphoric acid which is not soluble in water or citric acid or containing insoluble phosphoric acid or tricalcium phosphate.

These fertilizers are :

(i) Rock phosphate (20-40% P_2O_5),

(ii) Raw bone meal – 20-25% P_2O_5 and 3-4 % N,

(iii) Steamed bone meal (22% P_2O_5).

All these fertilizers are well suited for strongly acidic soils or soils with high organic matter.

Phosphorus is good for formation of roots of plants/crops. It is more needed for *leguminous* crops like, lentil, pea, *arhar*, gram etc. In any phosphatic fertilizer 'calcium' is available. Hence productivity of crops will be high, if these phosphatic fertilizers are used in acidic soils, particularly in legumes.

Of course, it is not mobile like other N-fertilizers. Hence placement of P_2O_5 has got some roles in crop production.

Super phosphate is a grayish fertilizer. It is, sometimes, slightly moist with an acidic odour, but it is dry and contains no lumps.

2.3.2.3 *Potassic fertilizers*

Potash, potassic or potassium fertilizers are applied to the soil to supply the crops with potassium (K), one of the essential elements for plant growth. Potassium in fertilizer is expressed as K_2O (potassium oxide) and is commonly referred to as potash. In Indian soils, its presence is high normally, but in some soils, it is deficient. Some crops like tubers (Potato), rhizomes (*amorphophallus* / arum etc.) require more amount of potash.

Potash fertilizers may be classified as (i) fertilizers having K in chloride form and (ii) fertilizers having K in non-chloride form. The most common potassic fertilizer is the '*muriate of potash*' or potassium chloride (KCl), belonging to this group. It is the cheapest and easily available K-fertilizer, and it is extensively used by the cultivators all over the world. Potassic fertilizers, belonging to 'non-chloride form' are (a) sulphate of potash (K_2SO_4), (b) sulphate of potash-magnesia double salt of K and Mg and potassium nitrate (KNO_3). Fertilizers the 'non-chloride form' are costly but for some particular crops like potato, tomato and tobacco it is used for quality improvement.

Muriate of potash/potassium chloride (KCl) is a coarse or fine salt resembling ordinary salt, having no bitter taste, but it is slightly pinkish in colour. It contains about 60% K_2O. Sulphate of potash is the main 'non-chloride' form of potash fertilizer. It contains nearly 48-52% K_2O. Wood ash, coal ash, rice ash etc. also contain some amount of potash, along with N and P_2O_5. Some of the crops, give better yield with addition of higher dose of K_2O, some crops (wheat, rice etc.), of course, do not response to K_2O, as sufficient amount of potash is there in the soils and its requirement is less.

Secondary nutrients

Calcium, magnesium and sulphur are called the secondary nutrients. Calcium occurs in natural soils in the form of carbonate, sulphides and hydroxy form, complex calcium silicates and organic matter. For correcting soil acidity though liming, it is very important. Magnesium, though essential element, has not been observed to be deficient in Indian soils, hence it is less important. Sulphur is not required, generally by the crops under Indian conditions, because some of the sulphur is being incorporated into the soils with other fertilizer like *ammonium sulphate* etc.

2.3.2.4 *Micronutrients*

Out of 16 essential elements for plant growth, seven nutrients are required in much smaller qualities. The nutrients are called the micronutrients.

Out of seven micro-nutrients, 6 nutrients are important, for production view points. It is more important, now because of multiple cropping programmes or introduction of new crops in the areas.

It has been observed that responses of different crops to different micronutrients, are gradually increasing in India. It has been observed that wheat is affected by deficiency of *boron* (B) in Punjab, West Bengal, Rajasthan, Gajarat, Maharastra etc; maize is affected by *boron* in Punjab, cotton in Madhya Pradesh, barley in Rajasthan, Potato in Himachal Pradesh and Paddy in Uttar Pradesh. Deficiency of iron (Fe) affects paddy in Punjab, Haryana and Himachal Pradesh, Cotton in Tamil Nadu, Cu in wheat in Punjab; manganese (Mn) in maize in UP, HP, zinc in wheat in Punjab, HP, Gujarat, maize in Punjab, paddy in Haryana, Punjab, HP, AP, UP, TN, Bihar etc. Molybdenum (Mo) increased the production of berseem in Punjab. Chlorine, of course, affects less in crop production in India.

Practical Class : Identification of common manures and fertilizer

Lesson 6 : Identification of common organic manures

Introduction

The substance, bulky organic in nature, capable of supplying plant nutrients in available form, having low analytical value and having no definite nutrient composition, is the organic manure. The organic manure is obtained from either plants or animals, as a by-product though this manure is low in nitrogen, phosphate and potash, but it is very essential for revitalization of soils. It helps in formation of good soil structure, and also helps to increase water-holding capacity of the soils. It also supplies micro-nutrients to some extent to the plants.

Objective

- O to identify various kinds of organic manures, along with specific nutrient concentrations, for different purposes such as, appearance, handling process, assessment of nutrients etc.
- O to utilise the organic matter for production of crops and building up soil health.

Material required

✣ Samples (about 200 g) of all kinds of organic manures (available locally), ✣ Balance and weights, ✣ Jars of different volumes, ✣ Note book, ✣ Pen / pencil.

Procedure

- Arrange the samples (put into jars or plates) on the bench / table,
- Place the name plates of the organic matter, just near the samples, write the nutrient content of manures on the name plates. [Chart on nutrient (N, P_2O_5 and K_2O %) content placed for ready reference].
- Write down in detail.

Assignment

- ☐ State the nutrients content (N, P_2O_5 and K_2O %) of the following organic manures,
- ☐ See the samples, smell the samples and feel the samples, putting them in between two fingers and write them against all the organic manures like F.Y.M., sewage, sludge rural comost, urban comost, *dhaincha* manures, blood, fish manure, bone meal; paddy husk, mustard cake and sunflower cake.

Chart

Sl. No	Types and name of manures	Nutrient (N, P_2O_5 & K_2O) Per cent		
		N	P_2O_5	K_2O
A.	**Bulky organic manures of plant origin**			
1.	Cowdung (fresh)	0.3 - 0.4	0.1 - 0.2	0.1 - 0.3
2.	Poultry manure (fresh)	1.0 - 1.8	1.4 – 1.8	0.8 – 0.9
3.	Sewage and sludge (dry)	2.0 – 3.5	1.0 – 5.0	0.2 – 0.5
4.	Cattle urine	0.9 – 1.2	Trace	0.5 – 1.0
5.	Farm Yard Manure (F.Y.M.)	0.5 – 1.5	0.4 – 0.8	0.5 – 1.9
6.	Rural compost (dry)	0.4 – 0.8	0.3 – 0.6	0.7 – 1.0
7.	Urban compost (dry)	1.0 – 2.0	1.0 – 1.5	1.5 – 2.0
8.	*Dhaincha* (*Sesbania aculeata*)	0.6 - 0.7	0.15 – 0.20	0.5 – 0.9
(B)	**Manures of animal origin**			
9.	Blood, dried	10.0 – 12.0	1.0 - 1.5	1.0 – 1.5
10.	Fish manure, dry	4.0 – 10.0	3.0 – 4.0	0.3 – 1.5
11.	Bone meal (raw)	3.0 – 4.0	20.0 – 25.0	
(C)	**Crop residues**			
12.	Paddy husk	0.3 – 0.5	0.2 – 0.3	0.3 – 0.5
13.	Straw (paddy)	0.3 – 0.4	0.08 – 0.09	0.60-0.71
14.	Groundnut shell and stem	1.6 – 1.8	0.3 – 0.5	1.1 – 1.7

Lesson 7 : Identification of common chemical fertilizers

Introduction

The materials, that are added to the soils, in order to supply chemical elements, in terms of nutrients (for supplying nutritions to the plants), for production and to improve soil fertility, are called the fertilizers. Therefore, for getting higher yields of crops, fertilizers of different types are added to the soils. Hence the learners must know the different fertilizers and for this they should identify them.

Objective

- O to identify various forms and source of fertilizers,
- O to know their respective nutrient concentration,
- O to see the efficiency of fertilizers,
- O to list the residual effects of fertilizers, when applied to the soils.

Material required

✜ Samples of fertilizers of different types, ✜ Medium jars for keeping fertilizers, ✜ A note book, ✜ Pen/pencil.

Procedure

- Arrange different types of fertilizer samples on the table bench in plates/jars,
- Attach the name plates, giving names of fertilizers and also giving percentage of nutrients (N, P_2O_5 & K_2O),
- Take the colour, odour and appearance of the fertilizers and state according to the statements.

Assignment

- ☐ Give the percentage of N, P_2O_5 & K_2O, of the common NPK-fertilizers and
- ☐ Give, in each case the odour, colour and appearance of the fertilizers, as in the chart, given in details:

Chart : (a) N-fertilizers, (b) P_2O_5 – fertilizer, (c) K_2O – fertilizers.

Hints

(1) Urea : White crystalline granules, completely soluble in water or soil solution, valuable nitrogenous organic fertilizer. Moisture percentage by weight, maximum 1.0; readily absorbs moisture from the air, forming big lump. It contains about 46% nitrogen, spraying is possible with this because of organic in nature.

(2) Ammonium sulphate: Colourless and crystalline stable salt, soluble in water/ soil moisture. Moisture percent by weight maximum 1.0 and can also absorb moisture from the soil. It contains nearly 20% nitrogen. Being sulphate radicle, it also contains about 21% sulphur.

(3) Calcium ammonium nitrate (CAN) : Easy blowing hygroscopic dust free greyish fertilizer. Moisture percentage by weight is maximum 1.0. It contains about 20% N. Because of Ca^{++} content, this fertilizer is suitable for acid soils.

(4) Single super phosphate (SSP) : This fertilizer is the greyish powder, generally dry and contains no lumps but sometimes it is slightly moist with an acid odour. Moisture percentage, by weight 12.0. It is the phosphatic fertilizer having 16% P_2O_5. In double superphosphate, P_2O_5 content is about 32%. Residual effect of SSP is present.

(5) Muriate of potash (MOP) : It is potassic fertilizer, having coarse subtance resembling like ordinary salt, but smells bitter in taste. It is commonly marketed in granular form also available in powder form. It is colourless (of course, some times brown colour); easily soluble in water/soil moisture. Moisture percent by weight maximum 0.5. It contains nearly 60% K_2O.

(6) Suphala (NPK) – 15-15-15 : It is a granular, pink NPK fertilizer, having 15% N, 15% P_2O_5 and 15% K_2O. It is soluble in water/soil moisture. Moisture percent by weight is maximum 1.5.

(7) NPK – 10-26-26 : This is the granular, greyish NPK-fertilizer, having 10% N, 16% P_2O_5 and 16% K_2O. It is soluble in water/soil moisture. Its moisture percent by weight is maximum 1.0.

(8) Ammonium-phosphate-sulphate (20-20-0) : It is also the granular, colourless NP – fertilizer, having 20% N, 20% P_2O_5 and 0% K_2O. Moisture percent by weight is maximum 1.0.

2.4 IDENTIFICATION OF COMMON FIELD CROPS

There are so many field crops in the country, some of which are very important in respect of consumption, utilization, economics, requirements, productivity etc. and some are less important. Different crops are grown in different places, giving importance to those localities for Climatic requirement, traditional consumption, or social demands etc. Jute can be grown easily in West, but not in Rajasthan, again *Isophgol* is grown easily in Rajasthan but not in West Bengal. Some varieties of *diccocum* group of wheat, are still cultivated in some parts of Maharastra for some qualities, preferred by the people in those localities, though productivity is less.

Therefore, for better understanding, only the common crops are grouped together in respect of agronomic/commercial/economic classifications, such as cereals, pulses, oilseeds, fibres, sugars, tubers and fodders. The learners should study the crops under the groups, adhering to the plant height, leaf type, stem conditions, inflorescence, special characters (like hairy-ness), parts used etc. Average identifying data of crops, under the conditions, have been given. These may be recorded from place-to-place.

2.4.1 Cereal crops

(Rice, wheat, maize, barley, *jowar*, *bajra* and oat)

Among the cereal crops, rice, wheat, maize, barley, *jowar, bajra* and oat are the main crops, grown in *rabi*, pre-*kharif* or in *kharif* seasons. Some crops are tall, some are short.

Wheat, maize, barley, oat are grown in *rabi* season, whereas rice, maize, *jowar* and *bajra* are grown in pre-*kharif* / *kharif* season. Maize, *jowar* and *bajra* are tall crops, whereas rice, wheat, barley and oat are short crops.

Maize (C_4 – crop) can be grown in *rabi*, pre-*kharif* and *kharif* seasons, for different purposes. Among all the cereal crops, *wheat* is the `king of all cereals' in respect of nutrient contents. Cereal crops, having carbohydrates and starch, provide major food for the people of the country.

The following characters will help the learners to identify the cereal crop in a better way.

Crops : Cereals

Field crops	Plant height (cm)	Leaf type (B/ N)	Stem solid (S)/ Hollow (H)	Inflorescence (long/short)	Special characteristics	Parts used
Rice	60-80	N	H	Long	No awn	Seeds
Wheat	50-60	N	H	Short	Awn	Seeds
Maize	2-4 m	N	S	Long	Silk & tussel	Seeds/stem
Barley	60-70	N	H	Short	Awn	Seeds
Jowar	2-3 m	N	S	Shot	Sparing inflorescence	Seeds / stem
Bajra	1-2 m	N	S	Long	Compact inflorescence	Seeds/ stem
Oat	70-80	N	H	Long	Star-like flowers	Seeds / stem

Notes : cm = centimeter; B = Broad-leaved, N = Narrow-leaved; H= hollow, S = solid

2.4.2 Pulse crops

(Lentil, green gram, blackgram, pea, *arahar*, gram and grasspea)

Pulses or *dals* give protein to the people of the country. All the pulses are obtained from the main family of *Leguminoceae* and sub-family *Papiliona-ceae*. Major pulses are lentil, greengram (*mung*), black gram/ bengalgram, pigeon pea/red gram/*arhar* / tur, gram, pea and grasspea etc. Some of them are grown in *rabi* (lentil, gram, pea and grasspea) season, and some of them are grown in pre-*kharif*/*kharif* (green gram, blackgram and *arahar*) season. Except, *arhar*, all the pulses are short in nature, although some of them are vine in nature, where length is more with low height.

Production of pulses is low, as a result price is high. Pulses or other leguminous crops, fix atmospheric nitrogen with the help of *rhizobium* bacteria, present in the roots of the plants. So they need high land to grow. Leguminous crops enrich the soil with nitrogen, and provide more organic matter which is the life of soils. The following characters will be very helpful for the learners, to identify the pulse crops.

Crops : Pulses

Field crops	Plant height (cm)	Leaf type (B/N/ m t)	Stem (S/H)	Inflorescence (long/short)	Special characteristics	Parts used
Lentil	30-40	B/m	S	Short	Tendril	Seeds
Green-gram	40-50	B. t	S	Long	star-like inflorescence	Seeds
Black-gram	60-70 l, 40 cm ht.	B. t	S	Long	star-like inflorescence	Seeds /plant
Arhar	1-3	B. t	S	Short	Yellow flower	Seeds / fuel
Gram	30-50	B. m	S	Short	Sack like pod	Seeds
Pea	50-60 l, 40 cm ht.	B. t	S	Short	Long pod	Seeds /pods
Grass- pea	60-70 l, 50 cm ht.	B. m	S	Short	Wider pod	Seeds/pods, plant

Note : t = tri lobed; m = multilobed.

2.4.3 Oilseed crops

(Mustard, rape, groundnut and sesame)

Oilseed crops are obtained from different families in different seasons, though the plants are mostly in the range of of 50-80 cm height.

Recovery percentage of oil from these crops varies from 30 to 40 and all of them give edible oilcakes which are the cattle feeds or concentrated organic manures. Huge amount of stovers of rape, mustard and sesame are used as fuel or crop residues. The groundnut gives the green twigs and leaves (after harvest) which may be used as either fodder or green manure.

Crops : Oilseeds

Field crops	Plant height (cm)	Leaf type (B/N)	Stem (S/H)	Inflorescence (long/short)	Special characteristics	Parts used
Mustard	40-80	B	S	Long	Yellow flower	Oil brown seeds
Rape	30-50	B	S	Long	Yellow flower	Oil from seeds
Groundnut	30-80	B t	S	Nut in soil	Yellow small flower	Oil from seeds
Sesame	40-80	B	S	Short	White flowr	Oil from seeds
Sunflower	70-90	B	S	Head-like	Yellow thalamus	Oil from seeds

2.4.4 Fibre crops

(Jute, cotton, mesta, sunnhemp and flax)

Jute, cotton, mesta, sunnhemp and flax are the main fibre crops in India, of which cotton is the lint fibre and the rest are the bast fibres. All these fibre crops belong different families. All these fibre crops are required for textile industries in the country. Cotton and jute industries are very developed in India. Some of the textile products are being exported to the foreign countries.

All the fibre crops presented are grown in either pre-*kharif* or *kharif* season, except flax which is grown in *rabi* season.

Jute, mesta and sunnhemp are the tall crops, cotton is the medium tall crop, whereas flax is the short crop in height.

In respect of quality of fibre, cotton is the best, followed by sunnhemp and jute, whereas mesta showed the poor quality of fibre. Though production is less but quality is good in flax fibre. The following characters of plants will be very helpful, to identify the fibre crops.

Crops : Fibres

Field crops	Plant height (cm)	Leaf type (B/N)	Stem (S/H)	Inflores-cence (long/short)	Special characteristics	Parts used
Jute	2-3 m	B	H	Short	Bitter/sweety leaf	Bark & stick
Cotton	50-90	B	S	Short	Big flower	Boll & stover
Mesta	2-3 m	B	S	Short	Big flower	Bark & stick
Sunnhemp	2-3 m.	B	S	Short	Yellow flower	Bark & stick
Flax	30-50	B	S	Short	Blue flower	Bark & stick

2.4.5 Sugar crops

[Sugarcane, sugar beet, yambean (*sakalu*) and sweet potato]

Sugar crops may be of high or low content of sugars. Sugarcane, sugar beet, yambean and sweet potato etc. are in this sugar crop group. Out of all these crops, sugarcane and sugar beet showed higher content of sugar and from these two crops (sugarcane and sugarbeet), sugar is produced. Yambean and sweet potato are used as vegetables or they can be eaten as such.

All these crops belong to different families. Sugar industry is highly developed in India with sugarcane. Sugar beet is the sophisticated crop in *rabi* season. Sugarcane area is being reduced due to long duration (1 to 1.5 year).

Sugarcane is a tall crop grown with `setts' in hot weather conditions, but sugarbeet is a small crops, grows in winter season. Of course, yambean and sweet potato are crop of pre-*kharif* / *kharif* season crops, though yambean is a tall vine but sweet potato is a short vine crop. For identification, the following characters are very important.

Crops : Sugars

Field crops	Plant height (cm)	Leaf type (B/N)	Stem (S/H)	Inflorescence (long/short)	Special characteristics	Parts used
Sugarcane	2-3 m	Long. N	H	Very long	Broom like inflorescence	Stem
Sugarbeet	50-70	B	S	Long/ short	Swollen root	Swollen root
Yam bean	Long vine 5-10 m	B	S	Cluster short	Trailing	Swollen root
Sweet potato	Vine 0.5-1.5 m	B	S	Short	Swollen	Swollen root

2.4.6 Tubers / rhizomes

[Potato, elephant-foot-yam / *amorphophallus* (*Ol*) and arum]

Tuber crops and rhizomes are in different families. Potato, the tuber crops, grows underground and it is a good vegetable crop. The plant of potato is short, while *ol,* the rhizome, is tall and short plant, grows under ground. The arum the rhizome are of different types. Some grow in deep water (about 0.8 to 1.5 m), some grow in marshy lands and some in dry conditions. In arum, petiole, rhizomes are eaten as vegetables, along with the runners. *Amorphophallus* (*ol*) is also used as vegetable. In case of potato *small tubers*, in case of *ol, small rhizomes* and in case of arum, *small rhizomes* are used, as 'agronomic seeds' or 'propagules' for propagation or production.

The following characteristics of these crops, may be helpful for identifying the tuber and rhizomes.

Crops : Tuber / rhizome

Field crops	Plant height (cm)	Leaf type (B/N)	Stem (S/H)	Inflorescence (long/short)	Special characteristics	Parts used
Potato	40-50	B	S	Short	White flower	Vegetables
Elephant foot-yam (ol)	0.5-1.5 m	Big B	Soft	Long	Yellow flower	Vegetable
Arum						
(a) High land	50-60	Big B	Soft	Long	Yellow/white	Vegetable
(b) Medium land	50-60	Big B	Soft	Long	Yellow/white	Vegetable
(c) Low land	0.5-1.5m	Big B	Soft	Long	Yellow/white	Vegetable
(d) *Dudhkachu*	50-80	Big B	Soft	Long	Yellow/white	Vegetables

Note : Soft = due to presence of aerenchymatous tissues.

2.4.7 Fodder/forage crops

[**(a) Legume group :** Berseem, lucern grasspea, cowpea, ricebean, blackgram etc. and **(b) Non-legume group :** Napier, thinnapier, teosente and dinanath grass]

Fodders are of two types i.e. (i) *leguminous* and (ii) *non-leguminous* (i.e. grassy). Among the leguminous group, berseem, lucern, grasspea, rice-bean, cowpea and blackgram are the main crops. In other group (i.e. non-legumes), napier, teosinte and thinnapier and dinanath grass are the main crops.

Among the legumes, berseem, lucern and grasspea are grown in *rabi* season, whereas rice bean, blackgram and cowpea are grown in pre-*kharif*/*kharif* season. Among the non-legumes, all crops are under the family *Gramineae*. Napier is very tall plant, followed by teosinte, and thinnapier/dinanath grass.

In legume group, seeds are used for propagation, whereas in napier and teosinte group stubbles are used for propagation, though in those cases, seed may also be used. The following characters may be helpful to identify the fodder crops.

Crops : Fodder

Field crops	Plant height (cm)	Leaf type (B/N)	Stem (S/H)	Inflorescence (long/short)	Special characteristics	Parts used
(a) Legumes group						
Berseem	50-60	B t	S	Short	White flower	Whole plant
Lucern	60-70	B t	S	Short	White flower	Whole plant
Ricebean	Length 25-90	B t	S	Long	Star like inflorescence	Whole plant
Blackgram	Length 25-50	B t	S	Long	Star like inflorescence	Whole plant
Grasspea	Length 80-90	B m	S	Short	Cluster pod	Whole plant
Cowpea	Length 1.0-3.0 m	B t	S	Long	Long pod	Whole plant
(b) Non-legume group						
Napier	2-3 m	N long	S	Long	White inflorescence	Whole plant
Thin-napier	80-90	N short	S	Long	Pink inflorescence	Whole plant
Dinanath grass	80-90	N short	S	Long	Pink inflorescence	Whole plant
Toesente	1.0-3.0 m	N long	S	Long	Silk & tussel	Whole plant

Practical class : Identification of common field crops morphologi-cally in the fields

Lesson 8: Identification of cereal crops like, rice, wheat, maize, barley, *jowar*, *bajra* and oat

Introduction

Huge number of crops are grown in the fields, throughout the year at different seasons, under different conditions/situations, for different purposes. They may be of the same family or different families. Their morphological characteristics, behaviour, nature of growth etc. are different. The learners should know them right in the field at different stages of growth, for different purposes.

Objective

- O to identify the crops morphologically at different stages,
- O to know the uses of the crops,
- O to know the effects (bad/good) on soils,
- O to recognise the crops from growth changes.

Material required

✜ Different small plots for different crops and seasons, ✜ Name plates (medium), with the names of crops, along with the scientific names, and sowing dates, ✜ Note book, ✜ Pen/pencil, ✜ Small scale, ✜ Balance.

Procedure

- Fit the labels / name plates, showing sowing dates, just near the proper crops,
- Mark the plots serially,
- Differentiate the crops by different character at different stages of growth,
- Recheck the crops after removing the plants,
- Put the plates in places again,
- Write in the note book the following (as mentioned in assignment)

Assignment

Observations to be recorded on

- ☐ Serial number, sowing date, observation date,
- ☐ Name of the crops with botanical names,
- ☐ Name of the family to which the crop belongs,
- ☐ Shape, size, colour, hairyness of stem / leaves,
- ☐ Nature of leaf surface, types and margin,
- ☐ Nature of stem (solid / hollow), branching behaviour, colour,
- ☐ Nature of flower, fruit, seed (if any),
- ☐ Plant height (cm or m),
- ☐ Tillers / single plants (per m^2),
- ☐ General appearance and condition of crop,
- ☐ Nature of root system (fibrous / tap),
- ☐ 1000- seed weight,
- ☐ Seed number per ear head / pod / cob / capsule, silique (mustard),

- ☐ Other observations / node in sugarcane, tubers in potato, rootlets in sweet potato, peg in groundnut, cob in maize etc.

Lesson 9: Identification of pulse crops (Lentil, greengram, black-gram, redgram (*arhar*), gram, grasspea and pea).

Lesson 10: Identification of oilseed crops (Mustard/rape, groundnut, sesame and sunflower).

Lesson 11: Identification of fibre crops (Jute, cotton, mesta, sunn-hemp and flax).

Lesson 12: Identification of sugar crops (Sugarcane, sugarbeet, yam- bean and sweet potato).

Lesson 13: Identification of tuber crops and rhizomes [Potato, elephant-foot-yam (*amorphophallus*) and *arum*].

Lesson 14 : Identification of fodder / forage crops : (a) Legume group (*Berseem, lucern, rice-bean, grasspea, cowpea and black-gram*) and (b) Non-legume group (*Napier, thinnapier, teosinte, dinanath grass*).

Note: Less 9 to 14, Introduction, Objective, Material, Procedure and Assignment same as in Lesson 8.

2.5 BASIC KNOWLEDGE ABOUT SOME FARM IMPLEMENTS

Basic knowledge about some farm implements are needed for studies in agronomy in the fields, because the field is the holy place for agronomy. Crop production is connected with agronomy and cultivation of all crops needs some implement. Therefore, the learners should acquire some basic know-ledge of implements required for work in agronomy.

2.5.1 Country / *deshi* plough

It is the common plough used by the farmers in India, though with different forms, shape and size. It is a cultivator rather than a plough, because it simply cuts and loosens the soils but does not invert them. It creates the V-shaped furrows, not covering the seeds. Of course, it is good for puddling and furrow-purposes. It can be used mainly for (i) breaking stubbles. (ii) land preparation, (iii) seed bed preparation, (iv) intercultural operation for some wide-rowed crops (cotton, maize, sugarcane etc.) and (v) sowing purpose (followed by laddering).

It has main parts like (i) body, (ii) shoe, (iii) share, (iv) handle, and (v) beam. Body is the actual part of the plough and it is generally made of wood to which other parts are attached.

Shoe is also main part, made of wood and has different shapes and sizes. It is the semi-round in shape. The inner angle between shoe and the body varies from 110-116°, depending on the plough type used in various operations. The piece of wood used, is 90-

110 cm long, 15-20 cm wide and 10-15 cm thick. The function of the shoe is to cut the slice of the soils, from the main body of the soil. It thus opens the furrow and maintains, its depth uniformly.

Share of the plough is made up of a flat piece of iron, nearly 45-50 cm long, 6-8 cm wide and 1-15 cm thick. One end of the iron piece is sharpened and down to a point of easy penetration into the soil. The share, is generally held on the shoe by means of a U-shaped clamp. The sharpened end of the share projects over the shoe by 5-6 cm. The main function of the share is to penetrate into the soil.

Handle is made up of wood piece, about 45 cm long and is attached to the upper end of the body with nails and by means of a lap joint. It regulates the plough to control bullocks. In case the body and the shoe are made of one piece of wood, a nearly one metre long and 5.7 cm thick piece of wood is fitted to the body. A small piece of wood, about 20 cm in length and 5 cm in diameter, is attached to the upper end of the handle for easy grip.

Beams generally vary from 2.7 to 4.5 metres in length and 7 to 9 cm in cross-section of which one end of it, is fixed to the body by means of a mortice and tenon joint. Generally 3-4 holes are drilled at the upper end of the beam to insert the small iron bars / rods to facilitate the yoking of bullocks and regulating the depth of the furrows.

The depth of this indigenous plough can be adjusted by (i) putting pressure on the handle, (ii) putting pressure on the beam with foot, especially at the corners of the field, (iii) changing the point of hitch on the beam by inserting the pegs in different holes, (iv) changing the angle of the share with the beam.

The work of the *desi* plough begins, when the plough moves forward (by driven bullocks and regulated by plough man), the shoe and the share being inclined, penetrates into the soils and separates the furrow slice from the main body of the soil. The furrow slice becomes V-shaped and the land between the adjacent furrows is sometimes unploughed (dead furrows). Of course, in cross-ploughing it is reduced to some extent. For thorough ploughing with *desi* plough, it needs more or less three-to-four time ploughings followed by 3-4 time ladderings.

Indian farmers complete the work for (i) sowing the arable crops, (ii) intercultural operations in some crops, (iii) puddling for transplantation of crops (rice), (iv) harvesting of crops (potato, groundnut etc.), (v) opening the soils after harvest of crops and (vi) many other purposes.

2.5.2 Mouldboard plough

Mould board plough is the improvement over *desi* plough, particularly for covering seeds, sown in previous furrows. The mould-board plough is used to (i) pulverise the soil thoroughly, (ii) bury the green manure or other residues with the soil thoroughly, (iii) destroy the insects and eggs of insects harbouring in soils by exposing to the sun, (iv) make ridges for some crops, (v) form the bunds and make small plots in the field, for uniform pre-sowing irrigations and (vi) sometimes harvest of ridged crops.

The mould-board plough, having plough bottom, has different 4 main parts such as (i) **frog**, (ii) **share**, (iii) **land slide** and (iv) **mouldboard**, along with other accessories, like (a) **beam** and (b) **handle**.

The (i) *frog,* the central part is made up of one piece of thick steel / cast iron. The various parts like share, land slide and mould board are attached to it with nuts and bolts. The (ii) *share* is the foremost part of the plough that penetrates the soils and makes the passage for the plough bottom. Thus, it cuts the furrows slice and passes on to the mould board. It is made up of steel/ cast iron and is attached to the frog with counter sunk bolts. It consists of 4-main parts i.e. *tip*, *throat*, *wing* and *gunnet*, The (iii) *landslide* is the part which slides along the furrow wall and is on the side opposite to the mould board. The (iv) *mouldboard* is made up of cast iron/steel and is fixed to the frog with sunk bolts. It receives the furrow slices, cut by *share*. The furrow-slice is twisted and inverted, resulting in broken and 'pulversed soil'. The accessories are the *beam* (transmits the bullock power into the implements) and *handle* (to work with and also to lift the plough when necessary) and it is made up of wood or sometimes iron bar.

The mouldboard plough works when it enters the soils, owing to suction. If it is placed on a flat surface, it rests on the tip of the land side, in the rear and with some clearance at the joint of the share and the land side. This clearance is the *vertical suction*. The mould board plough then meets 3 to 5 mm. of suction. Similarly, there is a side clearance of 5 mm. which is created by extending the share towards the land side and this is the *horizontal suction*, which controls the width of the furrows, when the share and its tip wear out, the suctions are reduced and a furrow of desired depth and width can not be opened. The suctions can again be created by replacing the worn out share and sharpening the end of the share.

2.5.3 Harrow

Harrows are the implements which are put into multipurpose uses such as to (i) break the soil crust, formed after shower/irrigation/croppings, (ii) produce the soil mulch on the soil surface after sowing, specially under rainfed conditions, (iii) collect the weeds especially grasses from the fields, (iv) prepare the seed bed, (v) cover the small seeds, along with the pulveris-ation of soils for good tilth.

Of course, there are different types of harrows and these are : (i) *bar harrow*, (ii) *chain harrow*, (iii) *spring-tined harrow*, (iv) *blade harrow*, (v) *disc harrow,* etc.

The *bar harrow* is composed of a triangular wooden or iron frame, seventeen iron pegs, fixed along with 3 arms of the frame. The pegs are, of course, made to points slightly backwards to avoid the accumulation of soils in front and also to avoid the uprooting of the crop plants when used for hoeing.

The *chain harrow* is flexible type harrow, drawn by bullock or tractor. It can be rolled up after easy transport. It is made up of many iron links connected together and spreads. The soil is like a mat. The spike-like projections are provided at the link and they serve the purpose of pegs as in the bar harrow (above).

The *spring-tined harrow*, is composed of a rectangular iron frame to which elliptical C-shaped tines are fixed. Generally, it is not used for hoeing of crop. When there is bad obstruction in front of any type during the operation, it will bend backwards and will slip over the obstruction without throwing the whole implements out of position. Generally, it is used for preparation of soils particularly for pulverization.

The *blade harrow* is the most common hoeing implement, having horizontal blades instead of tines. The implement is made up of a long blade, 130 cm in length and 15 cm. in cross-section. The wooden projection slightly pointing to the front side is fitted on the lower side towards the end of the body to secure the blade. The handle, of course, consists of two wooden bars, joined at the top with a cross-bar.

The *disc harrow* is provided with concave discs of varying diameter and number. They are generally, 40-60 cm. in diameter and are fixed on a square axle at 15-25 cm. Of course, maximum penetration, the angle of a disc to the axle should be 20 degrees. Two gangs of the disc are placed one behind the other on different axles and face in opposite directions. Each disc is provided with a scraper to clean it of the sticking clay/soils during operation. A liver is provided to adjust the depth of the disc. The disc harrows are very effective in cutting and pulverizing the soil, preparing seed beds, besides destroying weeds and breaking the big clods.

2.5.4 Hand tools

The simple tools that are used for daily work or routine work frequently or for farming work in the fields regularly, are actually the handy tools. They can be made locally by the black smiths with wooden parts and partly with steel or cast iron. They are described below in brief:

(a) Spade

The *spade* is a simple handy tool which consists of two parts such as (i) *body* made up of iron in one plate, with a hole in one side, measuring about 20cm × 15cm × 2cm, having the front with sharpened end and the other (ii) *handle* made up of wood, round in shape with the length of about 50-60cm. Of course, different types of spades are available with varying sizes and shapes used for different purposes.

Though the spade is simple, but very important for daily use for different purposes, such as (i) sowing, (ii) soil preparation, (iii) furrow making, (iv) scooping / scrapping, the soils (v) harvesting the crops, (vi) preparation of channels for irrigation, (vi) eartheing-up, (vii) spading the corners of the field after tillage operation with tractor etc.

(b) Khurpi

Khurpi is the local handy tool made up of two parts i.e. (i) *body* and (ii) *the handle*. This body is made up of iron plate (measuring about 10 cm long, 6 cm width, 1 cm thick), sharpened at the front and back pointed nail is attached with handle which is made up of round wood, measuring about 10cm.

Though this is a small and simple hand tool, yet it is very important for its use. It is used for (1) weeding, (2) scrapping soils, (3) soil preparation (in small scale), (4) soil collection (horticultural purposes) and soil scooping (for making holes).

(c) Sickle

It is also a simple handy tool but very important for the farmers, as it is used frequently. Sickle is made up of two parts such as (i) body and (ii) handle. The body is made up of iron plate, measuring but 10 cm long x 6 cm width and 1 cm thick, sharpened at the front, but inner side is serrated and sharp and the one end is pointed. The front portion is semi-lunar in shape. Hand portion is made up of wood, in which the other end of the long (about 10cm) serrated iron bar, is engraved into it.

The essential functions of sickle are: (i) harvesting of some crops, (ii) cutting of other tender leaves/stems etc., (iii) opening of soils with the pointed ends etc.

(d) Hasua

Hasua is just like sickle, but iron blade is plain edged i.e. without teeth but very sharp in the inner side, along with wooden handle, which is straight. *Hasua* is long measuring tool about 30-40cm. in length. Like sickle, it has two parts, (i) *body* and the (ii) *handle.*

It is used for (i) clipping and stripping of sugarcane, (ii) harvesting of jute (sometimes, when stem is woody) and also for (iii) cutting jungles and big shrub-weeds.

(e) Nirani

Nirani is one type of simple but very important tool. It has two parts, one is the *body* and the other is *handle*. The body is a long bending (semi-lunar) sheet which is wide in the front side and sharpened at the foremost end, ending with a sharp pointed edge, while the end portion is long narrow part, put into the wooden-round handle.

It is used for: (i) weeding, (ii) soil stirring (jute and *aus* rice), (iii) sowing some crops in small scale, (iv) scooping the soils etc. In jute this *nirani* is used both for weeding, uprooting (weeds and unwanted plants) and scooping soil for aeration in jute roots.

(f) Hand tine

Hand drawn tine is a simple but important tool. It is made up of two parts such as (i) *body* and (ii) the *handle*. The body is made up of single iron *bar* bent like J-shaped structure, the bending point is sharp and pointed, while the other end of it, is engraved into the long wooden handle. This hand tine is used for making furrows in between the transplanted *kharif* rice rows, under minimal/zero tillage, for sowing wheat. Besides this, it may be used for intercultural operation like mixing of soil with fertilizers, or for weeding and also for sowing purposes.

(g) Hand fork

The *hand fork* consists of two parts such as (i) *multiple tines*, made up of iron and (ii) the *wooden handle*. Several tines are bent like J-shaped structure but joined in single rod, engraved into the wooden long handle. This is used for scratching the soils to keep/

retain moisture, in rice-fallows, under rainfed conditions, scrapping clods/ weeds and also for intercultural operations in arable crop cultivation.

(h) Shovel

Shovel, the simple but important tool has two parts mainly, (i) *body*, big iron rod and (ii) *wooden handle*. The sharp pointed iron rod is engraved into the long wide round wooden structure, the handle.

It is used for scooping soils into the holes or to give pressure into the soils for putting some poles in preparation of fencing etc. sometimes long round single iron rod, pointed at the end, may also work on the soils.

(i) Da

Da, two parts (*body* and *handle*) simple and small tool, is partially semi-lunar in shape at the upper portion. This is made up of iron. The upper portion is wide flat but lower portion is engraved into the short wooden handle. The inner portion of *da* is sharpened and the end portion is pointed.

It is generally used for cutting the materials such as branches of trees, bamboo, ropes for penetration of some material into some objects, its pointed end is also used for urgent work. In the farms, it is used for preparation of bamboo stick/ *bakhari* etc.

2.5.5 Wheel-hoe

Wheel-hoe, though not hand-tool, but is a simple important implement. It (i) removes weeds, (ii) produces soil mulch, (iii) promotes the absorption capacity of soil water, (iv) regulates temperature of the soil in the lower soil layers, (v) prevents the development of cracks on the soils i.e. it improves soil moisture conditions by scratching on the surface of the soil, under rainfed condition. (vi) mixes fertilizers/manures with soils and (vii) breaks the crustation on soil. It is used in other intercultural operations like partial scooping of soil particularly in jute/*aus* crop cultivation etc.

It has 4 parts such as (i) *frame*, (ii) *tines*, (iii) *handle* and (iv) *wheel*. Frame is composed of four iron bars suitably curbed and joined together. All the parts are attached to it. 3-duck foot tynes made of steel and sharp ends are fixed to curved *iron* legs which are further fitted with nuts and bolts to the frame. A double handle, made up of iron/wood is provided to drive the implement, with the wheel.

For working on wheel-hoe, adjust spaces between the tines according to the row spacing of the crops. Dig the spaces between the crop plants to about 2.5cm. depth with the hoe. That should be done a day or two after rain/ irrigation, when the soil is soft and does not affect much resistance. Then push the wheel-hoe between the rows only to avoid touching the crop plants and injuring the plants. While using the wheel hoe, one should avoid the hitting the crop plants.

2.5.6 Seed drill

Seed-drill the sowing implement, is used to sow the crops along with the spacing. It has different parts such as, (i) *frame*, (ii) *wheel*, (iii) *furrow opener*, (iv) *disc*, (v) *seed hopper*, (vi) *soil gatherer*, (vii) *press-wheel*, (viii) *scraper* and (ix) *handle*.

The *frame* is made up of blended in one iron piece. Other parts are fitted with this frame. The *wheel*, a driving wheel is provided at the front to run the machine. The *furrow opener* is like a duck foot tine in wheel-hoe, one furrow opener made up of wheel, is provided in the centre. The *disc*, just like the furrow opener, discs with few grooves, are placed and this is covered by the seed hopper case. Its function is to distribute the seeds with equal spacing. The *seed hopper* is in conical cylindrical shape, made up of iron, is used for containing seeds. Just behind the seed hopper, two flat iron bars are fixed with the frame for collecting soils from side and to cover the furrow and this is the *soil gather*.

Press wheel is made up of round, small wheels, that are joined together and are placed at the rear fo: pressing and for covering the sown lines. The *scraper* is one flat iron piece that is fixed just at the side of the press wheel to scrap the soils from press wheel. The *handle* is made up of two iron/wooden handles that are fixed with the frame for driving the machine.

The learners should adjust the rate of application of seed per unit area (like acre or hectare) can be adjusted with a felt for different crops. The felt is different for different crops. For checking seed rate, just draw the drill for about 5 metres after lifting the tines and judge the dropping of the seed visually, though this process is not measurement for seeds of any crop.

For practising of sowing with this seed drill the learners should set the drill for the seeds at the desired level, where calibration should be done to ensure that the drill is dropping the seeds properly in per unit area. Then test the depth of furrow by adjusting the furrow opener. Of course, it should be done before putting in the seed hopper, if proper, start sowing crop immediately.

2.5.7 Tractor / power tiller

Under total or partial mechanization, tractors or power tillers are used for tillage operation in the fields. There are some advantages of utilization of these machines. Within a very short period, the lands can be tilled for sowing different crops, paving the multiple cropping system, both under rainfed and irrigated condition. Now, it is easy to take 3 or 4 crops a year, under irrigated conditions. Of course, these machineries are dealt with the agricultural engineering section of the agricultural universities of the country. Hence, this should not be discussed here, as it is beyond the scope of the agronomist.

2.5.8 Paddy-weeder

The paddy-weeder or rotary-weeder is a simple implement, for contro-lling weeds or for stirring the soils, under wet or water logged conditions, mostly in transplanted paddy fields. It has three main parts such as (i) *boat*, (ii) *rotary-spike* and (iii) *handl*e.

The boat is a curved one sheet of iron (that floats), attached to the main body rotary *spikes* axis. The rotary spikes, numbering about 18 in two sets, are made of iron but curved and are separated. All these spikes are set on the small wheels which are rotated with the *handle*, made up of iron/wood.

Working principle is this that when the paddy weeder is held with hands and pushed, the boat goes ahead over the surface of water/soil and the spikes moves, turning the weeds (ahead) into muds constantly. Thus, the stirring of soils also continues. Thus, by running the paddy-weeder, under wet conditions particularly in rice field, both weeding and stirring continue. This is good for other waterlogged rice cultivation or other wet-crops cultivation.

2.5.9 Duster

Duster is a simple device for dusting the powdered insecticides, on the crops of insect-infected fields. With the *rotary handle*, the powder-insecticides are dusted though a *mouth-like structure*. The *whole body* of duster is attached to a strong belt which is kept on the shoulder of the operator. Generally, gamaxine, DDT etc. are spread over the crops by the duster.

2.6.10 Sprayer

Sprayer is a simple device for spraying liquid materials over the crops, either for spraying herbicides, insecticides, fungicides or liquid fertilizers in required amounts with water. Pumping pressure is necessary for spraying forcibly. This is visually a cylindrical apparatus made up of brass/ iron/ steel/ plastic, having different capacities. Mainly it has 3 parts, such as (i) *body* (cylindrical), (ii) a *nozzle* and (iii) *belt* to attach with body of the operator. The handle of the nozzle, and the nozzle are made up of brass/steel etc. The belt is of course, made-up of cloth/or cotton/ plastic.

The measured quantity of liquid herbicides, insecticides, fungicides, or fertilizers (solid or liquid) etc. are put into the cylinders and required amount of water is put into it and pumped and sprayed over the crops; the liquid materials then come through the nozzle. The precautionary measures, like wrapping the nostril and mouth are essential, because drifting of liquid is very dangerous for the operator.

2.5.11 Measuring materials

(Balance, scale, tape measuring cylinder etc. for *measuring*)

For work in the fields, in order to get good crop production, some simple tools are needed. Though they are simple, but very essential and important. These are (i) *Balance*, (ii) *Scale*, (iii) *Tape*, (iv) *Measuring cylinder* etc.

2.5.11.1 *Balance for measurement of weight*

Measuring some weights, balances are used. It is of different types. Generally, in the field, spring balances of different capacities, are used under different conditions. Hanging balance is also used for weighing big plants like jute, maize, *arahar* etc. whereas flat balance is used for weighing straw of some small crops, besides grains. For small quantity (mainly for research) 1, 2, 5, 10, 20 kg balances are also used in the fields. Of course, chemical balance, or chenometric balance are kept in the laboratories, for measuring small weight. These are the balances for weighing small amount of soil or plant materials only.

2.5.11.2 *Scale/tape for measurement of length*

Scale is the tool for measuring the height of plants, length, breadth and width of some area/ditch etc. Tape may be used in measurement of more length are as; where ordinary scale is used for measuring small length etc. Of course, special type of scale (1 m long) is used for measuring the height of the crops.

2.5.11.3 *Measuring cylinder for measurement of liquid volume*

Measuring cylinder is made up of glass, graduated in ml (and parts there of). The required quantity of liquid materials (fungicides/insecticides/ herbi-cides/liquid fertilizers), are measured in the cylinder and put into sprayers and from there spraying continues with the addition of water.

Practical Classes : Identification of farm implements and drawing them with labelling

Lesson 15 : Identification of general farm implements and drawing the hand tools

Introduction

Soils in the fields become hard and compact, due to beating action of the rain drops, irrigation practices and subsequent drying, movement of inter-cultivation implements, labourers work etc. at the time of crop growth. After the harvest of the crops, for sowing the next crops, it is very difficult to raise the crops, in this hard-soil conditions. In view of bringing the soils in good tilth, for germination of the seeds and also for establishment of crops, it is very necessary to open the soils with tillage operations. For opening the lands, under uncultivated areas, tillage is also necessary. 'Tillage' is the physical manipulation of soils with tools/implement and also with force, resulting in good tilth for better germination of seeds and subsequently for better growth of the crops. Tillage implements, required for cultivation, may be categorised into several groups, depending upon the purposes for which they are used; these are (i) *primary tillage*, (ii) *secondary tillage*, (iii) *inter-tillage* etc. It is therefore, necessary to identify the common farm implements for tillage operations. To get first-hand information, it is necessary to draw the sketches of hand tools also.

[*Note :* Drawing and labelling should be done as demonstrated].

Objective

- O to know the different tillagers implements, used for different purposes,
- O to know the different parts for different functions,
- O to know the working principles of each implement,
- O to draw the different hand tools used for different purposes in crop cultivators.

Material required

❖ Different tillage implements (tractor, power tiller, wheel-hoe, paddy weeder etc.), ❖ Different hand tools, like spade, *nirani*, *khurpi*, *da*, sickle, *hasua*, *khanta*, shovel, fork, tine etc. ❖ Note book, pen/pencil, eraser, scale (small- 15 cm or 30 cm).

Procedure

- Display the different tillage implements with the names, simply for classification/categorization,
- Display the hand tools (as mentioned above) with the names and functions,
- State the functions of the hand tools.

Assignment

Observations to be recorded

☐ Classify the tillage implements for their functions (such as primary tillage, secondary tillage, inter-tillage),

☐ Draw and label the hand-tools (as mentioned) and give functions.

Lesson 16 : Draw the *desi* and the mouldboard plough and describe the parts and label them.

Introduction

Indigenous or *desi* plough is the original plough of the Indian farmers, though its type/shape varies from state-to-state according to the needs. This type of plough is run with the help of bullocks and it gives 'V'-shaped furrows, and the depth of furrow is not much. This plough can not turn the soil. Generally after criss-cross ploughing, the soils are lavelled and good tilth is made with ladders/plants.

[*Note*: Drawing and labelling should be done as demonstrated].

Deshi plough :

Functions :

❖ It is used for opening the soils, and also for cultivating the soils for preparing the proper seed beds, ❖ It can also be used for intercultural operations in some crops, ❖ For puddling, this *desi* plough is good, though it is not so good for arable soil, as compared to other ploughs (mould board), ❖ For sowing purpose, after proper soil preparation, it is used with different devices like *kera* or *pora* in wheat sowing and ❖ It is sometimes used for harvest of potato or groundnut.

Description of the parts

1. **Body :** It is main part of the plough. It is generally made-up of wood, to which other parts are attached.
2. **Shoe :** It is the lower portion of the body of the plough, attached to the junction between body and beam. It is also made-up of wood, having different shapes

and sizes. Generally, it is semi round in shape. The function of the *shoe* is to cut the slices of the soil.

3. **Share :** It is the lower most sharp end of the plough, made-up of piece of iron. The sharp end goes into the soil and pierces the soil layers. This share is fitted into the *shoe* of the plough.
4. **Beam :** Beam is the long wooden bar, attached to the junction of body and shoe, in one end and with the yoke in another end. Its function is to drag the share and body for tilling the land.
5. **Handle :** It is attached portion of the body of the plough, made-up of wood, but curved in shape, for gripping the plough. This is the upper most portion of the plough, with nails and by means of a lap.
6. **Yoke :** It is the long structure which is attached to the beam and the neck of the bullock. It helps in dragging the body by bullock and it is made-up of woods. Humps of bullocks are very helpful for putting the yoke.

Adjustment : It is done by putting the beam to forward or backward movements, with the help of hoes and sticks.

Working of the plough

- The share and shoe when included, penetrate into soil and separate the furrow slices from the main body of the soil, when the plough moves forward.
- The furrows that are made by *desi* plough, is 'V'-shaped and some lands between adjacent furrows is left unploughed. The uncut space is known as dead furrow. To cover unploughed land, criss-cross ploughing is necessary for 2-3 times, followed by ladderings.

Mouldboard plough

Functions

The mouldboard plough is used -

✥ to pulverise the soil thoroughly, ✥ to bury the green manure/crop residue/manures effectively, ✥ to mix the soils with fertilizers, ✥ to destroy insects, pests as this plough has the curved mould board and opens the egg/ larva etc. and fungus is destroyed by the sun rays, ✥ It is good for making ridges and furrows for crops like potato groundnut etc.

Description of different parts of mould board plough

(1) Plough bottom

The main basal part of the mould board plough that moves along with the surface of soil, is called totally the plough bottom and it consists of four parts such as, *frog*, *share*, *land slide* and *mould board*. Other parts of this plough are *beam* and *handle* (together accessories) yoke and yoke. Short description of the parts is as follows :

(i) Frog

The *frog* is the central part of the plough made-up of one piece of thick steel or cast iron, having various parts like *share, land slide* and *mould-board,* attached with nuts and bolts. It is to some extent curved, 'shoe' shaped structure.

(ii) Share

Share is the front part of the plough that penetrates into the soils and makes the passage for the plough-bottom. Thus, it cuts the furrow slice and passes on to mould board. It is made of steel and very sharp, but attached to frog. Again it consists of 4 parts such as *tip*, *throat*, *wing* and *gunnet.*

a. The *tip* of the share enters the soil first and it supports the plough bottom,
b. The next part is the *throat*, which cuts the furrow slice from the soils,
c. The *wing* of the share is the supporter of plough bottom and
d. The *gunnet* of the share also supports the plough bottom against furrow wall.

(iii) Land slide

Land slide is the part which slides with the furrow wall. This is on the side opposite to the mould board. It provides stability to the plough against the pressure of the furrow slice on the mould board.

(iv) Mouldboard

The mould board is fixed to the frog with sunk bolts and this is made-up of cast iron/steel. It receives the furrow slice, cut by the share. This is very helpful for pulverising and breaking the clods and turning over soils.

(2) Accessories

(a) Beam : It is a long straight wooden piece attached to the plough bottom at one end. The other end is free for connecting it to the yoke. Its main function is to transmit the bullock power and draw the implements.

(b) Handle : It is a major part of the plough and it regulates the direction of the bullock and ploughing. The entric plough can be lifted with whenever necessary. It is also made-up of wood or iron bar. It has a curved grip also.

(c) Yoke : Yoke is the vital part of the plough which is attached with bullocks and the beam. Entire pressure passes through this yoke. It is made-up of wood.

Adjustment of the plough

The plough enters the soil, due to suction (vertical and horizontal suction). This suction may be created by replacing the work out of share and sharpening the end of the share.

Working of plough

- The plough rests on the tip of the share and at the heel of landslide, if placed on flat surface.
- The plough meets 3-5 mm *vertical suction* for easy penetration into the soils, with a side clearance of 5 mm, due to *horizontal suction*.
- Suction is created regularly, turning the soils or other materials into the earlier furrows, covering the seeds sown already.

Lesson 17: Running both the *desi* and mouldboard plough in the field for wheat sowing and record the differences 30 days after sowing and at harvest

Introduction

With the introduction of dwarf wheat, in the country, wheat is grown now in most of the states commercially except the southern states. In some states wheat is sown with mechanical device, but in most of the states wheat is grown traditionally i.e. with *desi* or mouldboard ploughs. Therefore, the learners should go for running both the *desi* or mouldboard ploughs for sowing wheat.

Objective

- o to know the process/technique for running the *desi* or mouldboard plough, for sowing wheat,
- o to gather knowledge/experience of sowing wheat with *desi* and mouldboard plough.

Material required

✤ (a) *Desi* and (b) mouldboard plough, fitted properly, with all parts, ✤ Plough man (if the learner a student), ✤ One pair of bullocks, trained properly, ✤ One ladder (made-up of bamboo/wood), ✤ Note book, pencil/pen, ✤ Scale (15 cm, 30 cm), ✤ Wheat seeds (any variety) for 1000 sq. m. area, ✤ NPK-fertilizers (a) 100 kg N, 50 kg P_2O_5 and 50 kg K_2O/ ha (1/2 of N and full of P_2O_5 & K_2O should be applied in soils before sowing wheat).

Procedure

(a) *Desi* plough

Adjustment of yoke and bullock

- Place the plough between two bullocks so that the beam lies below the lower side of the yoke,
- Insert the peg into the hole of the beam,
- Attach the beam of the plough to the yoke with a chain/rope,
- Adjust the depth of the plough sole, by changing the position of the peg after drawing a test furrow.

Ploughing

- ☐ From one corner, plough the land (suppose point A),
- ☐ Draw the first furrow nearer to border line and reach at the other end B,
- ☐ Turn the furrow at the left as CD parallel to the previous AB at a distance of about 5-6 m. This ABCD should be ploughed and again furrow should be brought by ploughing gradually and after ABCD you begin with EF and so on,
- ☐ This ploughing should be completed in one direction, similarly ploughing should be completed in other direction,
- ☐ After each ploughing, laddering should be done. For laddering, the ladder should be fitted at the end of the bullocks, by ropes attaching ladder and the yoke.

(b) Mouldboard plough

Introduction; Objective; Material required; Mouldboard plough; Procedure (All as in lesson 17).

Assignment

Observations to be recorded

Comparative studies on plant population and growth of wheat both under (a) *desi* and (b) mouldboard plough at harvest.

Item for observations (all average of 10 plants)	*Desi* plough	Mould-board
Total tillers/m^2		
Effective tillers/m^2		
Dead tillers/m^2		
Height of the plant (cm)		
Number of seeds/ear head		
1000-seed weight (g)		
Grain yield of the area (kg)		
Straw yield of the area (kg)		

Note : Area = 1000 sqm

Lesson 18 : Running the wheel-hoe in wheat fields, immediately after application of N-fertilizers, as top-dressing

Introduction

For higher production of crops, the growers apply major (NPK) fertilizers. Immediately after fertilizer application in the arable crops, irrigation is essential to make the fertilizer available to the crops. Hence it is good to mix the soils with fertilizers applied, for better utilization, before irrigation. Hence, wheat is to be fertilized with NPK-fertilizer @ 100 kg N, 50 kg P_2O_5 and 50 kg and K_2O/ha, where 50 kg N is top-dressed, at CRI (crown root initiation) stage, after 21 days of sowing with wheel-hoe.

Objective

- o to mix the fertilizers applied with the soils for better utilization and for higher production of wheat,
- o to make the furrow channel in between the wheat rows, for better soaking of irrigation water in the soils.

Material required

✥ N, fertilizer @ 50 kg N/ha (calculate the amount for 1000 sq.m area), ✥ Pots/bowls for spreading fertilizers – 4, ✥ Balance (spring) – 1 kg, one, ✥ Wheel-hoe (fitted well) – 1, ✥ Arrangements for irrigation, ✥ Plastic container for irrigation (flashing etc.), ✥ Spade –1, ✥ Note book, ✥ Pen / pencil.

Procedure

- Weigh fertilizer (say for nitrogen (N) from urea; for 1000 sq.m. area and put them into the pot/bowl.
- Broadcast the fertilizer (Urea), over the crops uniformly. (Urea should be dry, otherwise it adheres to plants which will be affected by sun rays – scorching),
- Move the wheel-hoe (after application of fertilizers) to-and-fro after putting it in between these crop-rows, by gripping the handle tightly,
- Well-fitted wheel-hoe should be run in between the wheat-rows, for mixing fertilizer added, to the soil and for making furrows for better water passage,
- Irrigate the land immediately after mixing fertilizer with soils,
- Note the date, time and procedure for fertilization, wheel-hoeing and irrigation, in the note book.

Lesson 19: Running the paddy-weeder for removing weeds in trans-planted *kharif* rice

Introduction

Rice, the water-loving crop unlike other crops, is grown under wet conditions. Under these situations 'weed' is a big problem. Nearly 3 weeks after transplantation of rice, stirring and top-dressing with N is needed. Hence, at this stage paddy weeder is run for better production of rice.

Objective :

O to incorporate the weeds into muds (ultimately for manures),

O to mix the fertilizer, applied as top-dressed,

O to stir the soil or to manipulate the soils for better aeration.

Material required

❖ Well fitted paddy weeder – 1, ❖ N-fertilizer (urea) for 1000 sq.m. area @ 50 kg N/ha, ❖ Note book, ❖ Pen/pencil, ❖ Bowl -1.

Procedure

- Broadcast the urea (as per requirement) granules over the crop uniformly,
- Place the paddy weeder in between the paddy rows,
- Grip the handle of the paddy weeder strongly and move the paddy weeder to-and-fro (if weeds are not removed, run the paddy weeder again), so that all weeds are incorporated into the soils and soils are stirred well (Precaution : The paddy weeder should be run by the learners/students under bare-foot conditions. Land should be water logged.
- Note the types of weeds and field conditions before running the paddy weeder and again after running the paddy weeder. Put the date and time of application of fertilizer and running the paddy weeder,

Lesson 20 : Dismantle and mantle the small single-row or jute-seed drill after drawing and labelling each part

Introduction

Jute, the fibre crop is grown in the eastern parts of the country where high rainfall, high temp. and high humidity are present in pre-*kharif* season. This crop is highly affected with weeds, if it is grown by broadcast. The fibre production is high, if it is sown in lines, as it is easier to control weeds in between rows with hoeing. Therefore, single-row seed drill, specially for jute seeds, was manufactured. This is as follows :

[*Note:* Drawing and labelling should be done as demonstrated].

Objective (for using the jute-seed drill)

O to sow the jute crops in lines,

O to regulate seed rate of jute,

O to remove weeds in jute,

O to produce soil mulch,

O to absorb more water by the soils,

O to regulate temperature in the upper layer of soils,

O to sow seeds with uniform depth,

O to prevent cracks of soil surface.

[*Note:* Drawing and labelling should be done as demonstrated]

Lesson 21: Practise the running of sprayer for plant protection measures (weed control) cautiously

Introduction

Weed is a big problem in crop production. Hence, it should be controlled chemically. Some herbicides are used for killing the weeds and for this practice for running the sprayer, is urgently needed. Generally, 2,4-D is used to control broad-leaved weeds @ 0.5 kg ai/ha at 28 DAS, with 600 L of water. The learners should practise running the sprayer cautiously.

Objective

- to control the different weeds in crop fields, for higher crop production.
- to clean the fields.

Material required

✤ A crop field infested with weeds, ✤ Well fitted sprayer-1, ✤ Measuring glass cylinder-1, ✤ Herbicides (to be used), ✤ A long piece of cloth (to cover the nostril and mouth), to avoid drifting, ✤ Required amount of water (fresh), ✤ Note book, ✤ Pen/pencil

Procedure

- Put the required amount of herbicide of particular brand, into the measuring cylinder, (as per calculation)
- Put herbicides into the sprayer,
- Put fresh required amount of water into the sprayer,
- Put pressure on water in the sprayer,
- Test the nozzle for spraying,
- Put the cloth over the nostril and mouth,
- Spray the herbicides, over the plants and weeds uniformly,
- After use wash the sprayer.

2.6 IDENTIFICATION OF WEEDS AND OTHER ENEMIES OF CROPS

Production of crops is very risky. It passes through ups and downs and also through so many enemies; one of the enemies is the weed and the others are pests, diseases etc. Crop protection, therefore, is a must for getting higher yields of crops. Weeds, pests, diseases, birds and animals are the enemies of the crop production and these are discussed.

2.6.1. Weeds

The unwanted plant in the crop-field, is the *'weed'*. Jute crop may be the weed in rice field, again rice may be the weed in jute, though both of them are the crops. Weeds are of different types. They may have broad leaves, narrow leaves or neither broad nor narrow – a typical type of leaves and this weed is the sedge. All these weeds rob nutrients, water, air, light etc. given to the crops and as a result the crop production is reduced. Therefore, for higher production, crops should be free from all weeds.

2.6.1.1 *Broad-leaved weeds*

The weeds, having broad leaves, are the broad-leaved weeds. All of them are in *dicot* group. These groups of weeds grow in all the seasons and damage the crops. Sunberry (*Physalis minima*) is a broad-leaved weed and it grows all the year round. Generally, 2,4-D is used to kill the broad-leaved weeds, but not always. Hand weeding is the best to kill the broad-leaved weeds, though not possible commercially.

2.6.1.2 *Narrow-leaved weeds*

The weeds, having narrow leaves, are the narrow-leaved weeds. All of them are in *monocot* group. These groups of weeds also grow in all the seasons and damage the crops heavily. Barnyard grass/*shyamaghash* (*Echinochloa crusgalli*) grows in rice field in all the seasons (*aus, aman, boro*). Generally, selective herbicides are used to kill them.

2.6.1.3 *Sedges*

'Sedge' group of weeds, is neither broad nor narrow-leaved weeds. They have solid stem-like structure of leaves, though *monocot* (*Cyperus rotundus*) – a sedge, is difficult to control. It has got *nuts*, below the soil surface.

2.6.2 Other enemies of the crops

Other than weeds, crops are to pass through so many enemies like diseases, pests, birds, animals etc. For higher crop production, all these should be controlled, though very difficult to do all the times.

They are controlled gradually, rather than control fully at one time.

2.6.2.1 *Diseases*

'Disease' is one of the most dangerous enemy of crops. Different types of diseases affect different crops in different seasons. For getting higher crop production, the diseases should be controlled. Of course, it is not possible to control disease fully, rather partially it can be controlled.

Recently through breeding programmes, disease has to some extent, been controlled. Janak, HP 1209, UP 262, HD 2270, UP 2338, WH 291, K 8020, HP 1633 etc. are rust-free varieties of wheat. The day is ahead when varieties of other crops will be produced, through breeding programmes, free of diseases.

2.6.2.2 *Pest (Insect/spider/aphid/jassids)*

'Pest' is also one of the most dangerous enemy of crops. Different types of pests are observed in different crops in respective seasons. For getting higher yields the pests should be controlled. Though it is not possible to control pests fully, partially it can be controlled.

By high-tech-breeding programmes, both diseases and pests are being controlled recently, in cotton and mustard.

2.6.2.3 Visiting enemies (*Birds/jackle/cattle/monkey/pig etc.*)

Other than pests and diseases, the crops should be protected from the attack by the birds and animals also, as the crop production is hampered due to their damages, caused by them.

Birds : It has been observed that starting from sowing of wheat to harvest, different birds attack the crop at different stages of growth. Immediately after sowing, the crows eat away the germinated wheat seeds. At later stage, sparrow, moyna etc. eat away the wheat seeds at harvest, while parrots take away the earheads of wheat, at dough stage. As a result of the birds attack, the wheat production is reduced to about 20-30%, though not always. For protection of wheat, men should be kept in the fields, to drive away the birds.

Jackle : The jackles, the cunning animal, attack the sugarcane crop, before its harvest. As the juice of sugarcane is very sweet, they chew the nodes and internodes of sugarcane, for the juice and as a result, the crop is partially damaged and production of sugarcane is reduced. For protection of sugarcane or maize, men should be posted with `sound-making-arrange-ments', in the fields.

Cattle : Maximum damaged is done by the cattle when they graze the crop-fields partly (some time fully). Beside the cattle, goats, buffaloes etc. also damage the crops.

For these attacks of crops, the fields should be protected. For this protection, fencing is needed sometimes. In the vast field, of course, time-to-time careful watch is needed for their presence. Besides this, in the village, a committee may be constituted to check the arrival of the cattle and to punish the keepers of the animals, for the damaging of the crop.

Monkey : The monkey, one of the swiftest animals, may sometimes damage the crops particularly, in the fields of pea, cucumber, etc. Some fruits they eat, some fruits they destroy, besides damaging the crop-fields. They come to the fields, in a group, big or small, and run away after damage. Men should be there to drive them away. Of course, it is very difficult to check them. On their arrival, keep a visit and make a strong sound to keep them away.

Pig : Pigs are the destroyer of potato crop. As potatoes are grown in the soils, they visit the fields and eat away the potatoes by removing the soils from the ridges prepared, for potato cultivation. Alert vigilance is needed for them. If needed, the owner of the pigs should be punished for the damage of crop.

2.6.3 Types of weeds

Different types of weeds are observed in different seasons in respective crops. The learners and growers should identify them and destroy them for higher crop-production.

2.6.3.1 *Broad-leaved weeds in rabi/pre-kharif/kharif seasons*

Broad-leaved weeds are observed in different seasons in different crops. They rob of the nutrients, water, light air etc. and reduce the crop yields. They should be destroyed by any means. For this, it is necessary to identify them.

***Rabi* / winter season**

In *rabi* season, the following main broad leaved weeds are observed in different crops :

(1) Pig weed (*Chenopodium album* L.), (2) Anagallis (*Anagallis arvensis* L.), (3) Bengal vetch (*Vicia benghalensis* L.), (4) White sweet clover (*Melilotus alba* Medik), (5) Yellow sweet clover (*Melilotus indica* All), (6) Sweet clover (*Melilotus parviflora* Desf), (7) Fumitory (*Fumaria afficiralis* L.), (8) Ting vetch [*Vicia hirsuta* (L) Gary], (9) Common vetch (*Vicia indica* L.), (10) Sunberry (*Physalis minima* L.), (11) Yellow vetchling (*Lathyrus aphaca* L.), (12) Launaea (*Launaea pinnatifida* Cass), (13) Nicotiana weed (*Nicotiana plumbaginifolia* L.), (14) Golden dock (*Rumex maritimus* L.), (15) Night shade (*Solanum nigrum* L.) etc.

Summer or Pre-*kharif* or *kharif* seasons (rainy)

In pre-*kharif* and *kharif* seasons, the following main broad leaved weeds are observed in different crops :

(1) *Jangli amli* (*Phyllanthus niruri* Hook), (2) *Laldudhi* (*Euphorbia hirta* L.), (3) Croton (*Croton sperciflora* L.), (4) Sunberry (*Physalis minima* L.), (5) Cassia (*Cassia sophera / tora* L.), (6) Eclipta (*Eclipta alba* L.), (7) Commelina (*Commelina benghalensis* L.), (8) Alternanthera (*Alternanthera sessilis* L./*amoena Voss*), (9) Swamp cabbage/water spinach [*Ipomoea reptans* (L.) Poir], (10) Mikania (*Mikania scandens* Hoog / *cordata* Robinson), (11) Beggar weed (*Desmodium tortuosum* D.C.), (12) Parthenium (*Parthenium hysteophorus* L.), (13) Lantana (*Lantana camara* L.), (14) Tridex (*Tridex procumbens* L.), (15) Boerhavia (*Boerhaavia diffusa* L./*purpurea* L.), (16) *Kakranda* (*Blumea lacera* D.C.), (17) Bellyacha bush (*Jatropha* gossypifolia L.), (18) Wild tobacco (*Nicotiana plumbaginifolia* L.), (19) Cocklebur (*Xanthium strumarium* L.), (20) Trailing indigo (*Indigofera endecaphylla* Jacq), (21) Indian sorrel (*Oxalis corniculata* L.), (22) Mexican prickly-poppy (*Agremone mexicana* L.), (23) Malva (*Malva parviflora* L.), (24) Sensitive plant (*Mimosa pudica* L.), (25) Cud weed (*Gnaphalium indicum* L.), (26) Goat weed (*Ageratum conyzoides* L.), (27) Marsilea (*Marsilea qudrifolia* L.), (28) Pig weed (*Chenopodium album* L.) etc.

2.6.3.2 *Narrow-leaved weeds in rabi/pre-kharif/kharif seasons*

Different types of narrow-leaved weeds are observed in different seasons in raspetive crops. They rob of the nutrients, water, air and light, provided to the crops and as a result the crops yields reduced. Hence, it is necessary and to are identify destroy them.

***Rabi* / winter season**

In *rabi* season, the following narrow-leaved weeds are observed in different crops : (1) Bermuda grass (*Cynodon dactylon* Pers.), (2) Nut grass (*Cyperus rotundus* L.), (3) Cannary grass (*Phalaris minor* Retz.), (4) Wild oat (*Avena fatua* L. / *ludoviciana Durien*), (5) Darnel grass (*Lolium temulentum* L.), (6) Barn yard grass [*Echinochloa crusgalli* (L.) Beauv.] etc.

Out of all these narrow-leaved weeds, cannary grass, wild oat and dernel grass resemble to wheat plant and these obnoxious weeds are observed in wheat fields. All these 3 weeds have been noticed in wheat fields recently, after the arrival of the dwarf wheats.

Pre-*kharif* / Summer and *kharif* / rainy seasons

In pre-*kharif* / *kharif* seasons, the following narrow-leaved weeds are observed in different crops :

(1) Nut grass (*Cyperus rotundus / esculentus* L.), (2) Barn yard grass [*Echinochloa crusgalli* (Beauv)], (3) Digitaria (*Digitaria sanguinalis* Scope.), (4) Bermuda grass [*Cynodom dactylon* (L.) Pers], (5) Gamba grass (*Andropogon gayanus* Kunth.), (6) Goose grass [*Eleusine indica* (L.) Gaertn.], (7) Brachiaria [*Brachiaria ramosa* (L.) Stapf], (8) Hillo grass (*Paspalum conjugatum* Berg.)], (9) Sporobolus [*Sporobolus diander* (Retz.) Beauv.], (10) Cat tail millet [*Setaria glauca* (L.) Beauv], (11) Cogon grass [*Imperata cylindrica* (L.) Beauv.], (12) Rice grass (*Leersia hexandra* Sw.), (13) Scirpus (*Scirpus erectus* L.), (14) Brachiaria [*Brachiaria mutica* (Forsk) Stapf], (15) Barnyard grass (*Echinochloa crusgalli / hispidula* Retz), etc.

All the narrow-leaved weeds are very difficult to control. Specific herbicides are used for specific weeds in respective seasons.

2.6.3.3 *Sedges in kharif/pre-kharif/rabi seasons*

Some of the sedges are very difficult to control, as a result they grow profusely at the cost of crops. Some of sedges have been placed in narrow leaved weeds, because leaves are narrow in nature.

Though number of sedges is less, they are stated below, for their eradication from the fields: Nutgrass (*Cyperus rotundus* L.) is very obnoxious weed and it grows all-the-year-around in most of the crops provided land is not water logged. In waterlogged areas, other types of *mutha* are present, that is Water-mutha (*Cyperus iria* L.) scirpus (*Scirpus erectus* L.) are very obnoxious weeds in waterlogged rice. Asphodelus (*Asphodelus tenuifofolius* L.) is very obnoxious weed in wheat also, if grown after rice.

Practical Class : Identification of weeds in the fields

Lesson 22: Identification of different broad and narrow-leaved weeds, with scientific names, in *rabi* season

Introduction

Generally, 'weed' is a menace for production of crops, as it robs of all types of nutrients, water, light etc., depriving the crops, as a result the productivity of crop is less. It is therefore, urgent need to eradicate or destroy them. For that it is necessary to identify different types of broad and narrow- leaved weeds in crops of different seasons. For identification, the learners should know their characters, families, nature etc. along with the names, including scientific names.

Objectives

O to remove, destroy or eradicate the weeds in crops,

O to utilize weeds in different forms,

O to utilize weeds as support base of some crops, under *paira* / utera crops conditions.

Material required

✣ Lense (powerful) – 1, ✣ Scale (small 15 cm or 30 cm), ✣ Balance – 1 kg, ✣ Long scale (1m) – 1, ✣ Note book, ✣ Pen/pencil, ✣ Field with weeds (broad and narrow-leaved) at flowering conditions.

Procedures

- Observe the weeds in detail at flowering,
- Note the types of roots, shoots, leaves and inflorescence,
- Note the speciality, if any.

Scientific names, families and natures of some broad and narrow leaved weeds in winter season

English names	Scientific names	Families	Natures
(A) Broad leaved weeds			
(1) Bengal vetch	*Vicia benghalensis* L.	Papilionaceae	An.
(2) White-sweet clover	*Meililotus* alba *Medik.*	Papilionaceae	An.
(3) Sweet clover	*Melilotus parvi* flora Desf	Papilionaceae	An.
(4) Yellow sweet clover	*Melilotus indica* All.	Papilionaceae	An.
(5) Anagalis	*Anagallis arvensis* L.	Primulaceae	An.
(A) Broad leaved weeds			
(6) Sunberry	*Physalis minima* L.	Solanaceae	An.
(7) Yellow vetchling	*Lathyrus aphaca* L.	Papilionaceae	An.
(8) Launaea	*Launaea pinnatifida* Cass	Compositae	An.
(9) Golden dock	*Rumex maritimus* L.	Polygonaceae	An.
(10) Nicotiana weed	*Nicotiana plumbaginifolia* L.	Solanaceae	An.
(11) Pig weed	*Chenopodium album* L.	Chenopodiaceae	An.
(12) Prickly poppy	*Argemone mexicana* L.	Papaveraceae	An.
(B) Narrow-leaved-weeds and sedges			
(1) Bermuda grass	*Cynodon dactylon* (L) Pers.	Gramineae	An / Pe
(2) Nut grass (Sed.)	*Cyperus rotundus* L.	Cyperaceae	An / Pe

(3)	Wild oat	*Avena fatua* L. / *A.* ludo *viciana* Durien	Gramineae	An
(4)	Darnel grass	*Lolium temulentum* L.	Gramineae	An.
(5)	Cannary grass (Fala - ghash)	*Phalaris minor* Retz.	Gramineae	An.

Note : An = Annual, Pe = Perennial, Sed = Sedge.

Assignment (Identification of local weeds (broad and narrow) in *rabi*/winter season

Local/English name	Scientific name	Speciality of			
		Root	Stem	Leaf	Inflorescences
1.					
2.					
3.					
4.					
5.					
6.					
7.					
8.					
9.					
10.					

Lesson 23 : Identification of some broad and narrow-leaved weeds, with scientific names, in pre-*kharif* season

Introduction; Objective; Material required; Procedures (All as in lesson 22).

Scientific names, families and natures of some broad and narrow leaved weeds in pre-*kharif* season

	English names (1)	Scientific names (2)	Families (3)	Natures (4)
(A)	**Broad-leaved weeds**			
(1)	Sunberry	*Physalis minima* L.	Solanaceae	An.
(2)	Sickle-senna	*Cassia sophera / tora* L.	Caesalpiniaceae	An.
(3)	*Bhuiamla*	*Phyllanthus niruri* Hook	Euphorbiaceae	An.
(4)	*Dudhia*	*Euphorbia hirta* L.	Euphorbiaceae	An.
(5)	Black-night-shade	*Solanum nigrum* L.	Solanaceae	An.

(6)	Croton weed	*Croton sperci* flora L.	Euphorbiaceae	An.
(7)	Corchorus weed	*Corchorus acquitangular* L.	Tiliaceae	An/Pe
(8)	Eclipta	*Eclipla alba* L.	Compositae	An.
(9)	Commelina	*Commelina benghalensis* L.	Commelinaceae	An / Pe
(10)	Water spinach	*Alternanthera sessilis* Voss	Amaranthaceae	An / Pe
(11)	Parthenium	*Parthenium hysterophorus* L.	Compositae	Pe
(12)	Lantana	*Lantana camara* L.	Verbanaceae	Pe
(B)	**Narrow-leaved weeds / sedges**			
(1)	Barnyard grass	*Echinochloa crusgalli* Beauv.	Gramineae	An.
(2)	Nut grass (Sed.)	*Cyperus rotundus* L.	Cyperaceae	Pe
(3)	Bermuda grass	*Cynodon dactylon Pers*	Gramineae	Pe
(4)	Goose grass	*Eleusine indica* Gaerth	Gramineae	Pe
(5)	Cogon grass	*Imperata cylindrica* Beauv.	Gramineae	Pe

Assignment (Identification of local broad and narrow-leaved weeds in pre-*kharif* season)

Local/English name	Scientific name	Specialty of			
		Root	Stem	Leaf	Inflorescences
1.					
2.					
3.					
4.					
5.					
6.					
7.					
8.					
9.					
10.					

Lesson 24 : Identification of some broad and narrow leaved, weeds with scientific names, in *kharif* season, along with family and nature

Introduction; Objective; Material required; Procedures (All as in lesson 22).

Scientific names, families and nature of some broad and narrow leaved weeds in *kharif* season

	Common / English names	Scientific names	Families	Nature
(A)	**Broad-leaved weeds**			
(1)	*Makoi / Banbegun*	*Solanum nigrum* L.	Solanaceae	An.
(2)	Sunberry	*Physalis minima* L.	Solanaceae	An.
(3)	Commelina	*Commelina benghalensis* L.	Commelinaceae	An.
(4)	Horse purslane	*Trianthema monogyna* L.	Aizoaceae	An.
(5)	Alternanthera	*Alternanthera amoena* Voss.	Amanthaceae	An./Pe
(6)	Wild brinjal	*Solanum khasianum* L.	Solanaceae	An.
(7)	Swamp cabbage	*Ipomoea reptans* Poir	Convovulaceae	Pe
(8)	Moon flower	*Ipomoea alba* Fosk	Convolvulaceae	Pe
(9)	Mikania	*Mikania scandens* Robinson	Compositae	An.
(10)	Beggar weed	*Desmodium tortuasum* (Sw) D.C.	Papilionaceae	An.
(11)	Parthenium	*Parthenium hysterophorus* L.	Compositae	Pe
(12)	Lantana	*Lantana camara*	Verbenaceae	Pe
(B)	**Narrow-leaved weeds / sedges**			
(1)	Nut grass	*Cyperus rotundus* L.	Cyperaceae	Pe.
(2)	Barnyard grass	*Echinochloa crusgalli* Beauv.	Gramineae	An.
(3)	Digitaria	*Digitaria sanguinalis* Scope.	Gramineae	An.
(4)	Bermuda grass	*Cynodon dactylon* (L) Pers	Gramineae	Pe
(5)	Goose grass	*Eleusine indica*	Gramineae	Pe
(6)	Hilo grass	*Paspalum conjugatum, Berg.*	Gramineae	Pe.

Assignment (identification of local broad and narrow-leaved weeds in *kharif* season)

Local/English name	Scientific name	Speciality of			
		Root	Stem	Leaf	Inflorescences
1.					
2.					
3.					
4.					

5.					
6.					
7.					
8.					
9.					
10.					

Lesson 25 : Identification of sedges with scientific names, family and nature, in different seasons

Introduction; Objective; Material required; Procedure (All as in lesson 22).

Sedges in different seasons

	Common / English name	Scientific names	Families	Nature
(a)	**Winter / *rabi* season**			
(1)	*Nut grass*	*Cyperus rotundus* Beauv.	Cyperaceae	Pe
(2)	Asphodalus	*Asphodallus tenuifolius*	Cyperaceae	An.
(b)	**Summer / pre-*kharif* season**			
(1)	Nut grass	*Cyperus rotundus* Beauv.	Cyperaceae	Pe
(2)	Water-nut grass	*Cyperus iria* Beauv.	Cyperaceae	An.
(c)	**Rainy / wet (*kharif*) season**			
(1)	Nut grass	*Cyperus rotundus* Beauv.	Cyperaceae	Pe.
(2)	Water-nut-grass	*Cyperus iria* Beauv.	Cyperaceae	An.

Lesson 26: Identification of obnoxious weeds of wheat crop (wild oat, *phalaris* and *temulentum*), with scientific names, family and nature.

Introduction

Wheat, the second cereal production of India, is very important for food production of the country. If, by any means, wheat production is less, there would be a hue and cry in the country for food. In view of this, it is very important not to disturb the production rate of wheat. Weed a menace some times, disturbs the productivity of wheat, if it is obnoxious. There are some obnoxious weeds in wheat fields, that may disturb the productivity of wheat, because they are very dangerous, in view of their appearance. They are exactly like the wheat plants and hence very difficult to identify and control them, before flowering.

Objective; Material required; Procedure (All as in lesson 22).

Obnoxious weeds in wheat fields

Common name	Scientific name	Family	Nature
(1) *Phalaghash / Gudidanda*	*Phalaris minor* Retz.	Gramineae	An.
(2) *Bunojai / Banjai*	*Avena fatua* / ludoviciana Durien	Gramineae	An.
(3) *Karat ghash / Machni*	*Lolium temulentum* L.	Gramineae	An.

■■■■

CHAPTER 3

CLIMATE AND ITS INFLUENCE ON CROP PRODUCTION

3.1. CLIMATE

Climatic conditions are very important for growth, development and yields of all crops. Climatic conditions vary from season to season. Wheat is grown in winter, whereas jute is grown in high hot summer under huge hum-idity. Hence effects of climate are observed in any biological phenomenon.

3.1.1 Elements of water

Climatic conditions are the long term effects of whether conditions, whereas short time conditions of different parameters, such as rainfall, temperature, humidity, sunshine etc. are combined known as wheather. That means weather is the combination of rainfall, temperature, humidity sunshine etc. of a particular place, giving variations into seasons, like *rabi*, pre-*kharif* and *kharif*. Rainfall, temperature, humidity etc. are the elements of weather.

3.1.1.1 *Rainfall*

Rainfall is very important for growth and yield of any crop, as it helps in dissolving of nutrients into solution, making easily available to the crops, besides, it is the physiological balancing material of the plants.

The place with high rainfall, gives better crop yield, than that under low rainfall areas. Rainfall of a particular area changes the face of development of any biological affair, establishing the economic status either good or bad. Rainfall is essential for all crops particularly the water loving crops.

3.1.1.2 *Temperature*

Like rainfall, temperature is also important for crop production. High temperature with high rainfall, gives better yields in some crop, but high temperature with low rainfall is very detrimental for crop production. Low temperature with more soil moisture, is good for some crops (wheat, barley etc.), whereas low temperature without soil moisture is also very detrimental for the crops. Some crops, of course, are grown with high temperature but with low rainfall, though yields are less.

3.1.1.3 *Humidity*

For some crops, humidity is very essential. Jute mesta, sunnhemp, *dhaincha* etc. are grown well in high rainfall, high temperature, along with high humidity. Where rainfall is high, humidity will be naturally high. Of course, for some crops (wheat), humidity is very detrimental at a latter stage of growth.

3.1.1.4 *Solar radiation*

Solar radiation is also very important for some crops. Productivity of rice in rainy season is less, whereas it is high during summer (*boro*) season when solar radiation is high. Solar radiation, helps in increasing the photosyn-thates or rather, increases the efficiency of photosynthesis. Of course, it varies from crop-to-crop. Under the same solar radiation, yield of rice is less, whereas yield of jute does not change at all, during pre-*kharif* season.

3.1.1.5 *Cyclone/storm/tarnado*

Cyclone, storm, tornado which occur occasionally etc. are the effects of weather abnormalities. All these hamper the productivity of crops heavily.

Storm, cyclone tornado etc. are the intensities of storm. In our country storm may occur regularly, cyclone occurs occasionally and tornado is not seen in our country.

3.1.1.6 *Snow/hails/ice/dew*

Snow, hails and ice are very detrimental for crop productivity, though dew is better for it. Winter crops suffer due to snow or ice. Hails (lump of ice pieces at normal temperature during raining), are sometimes observed in some places. Crops are heavily damaged by the hail storms, along with rainfall.

3.1.1.7 *Monsoons in India*

The term 'monsoon' is generally used in the context of cyclic reversal of the pressure and wind system within a year, in association with sequential changes in the cloud, rains and weather patterns, mainly in the tropics. In India two types of monsoons are observed i.e. (a) South-west and (b) North-east monsoons.

(a) South-west monsoon

Among the two types of monsoons, South-west monsoon is very important, because, it affects the entire area of the country and entire economic position depends on this monsoon, for better conditions of the crop. It starts from Kerala in May-end and comes toward bay of Bengal at about earlier part of June and at latter stages go to different places like, U.P., J&K, Punjab, Delhi, MP, etc. Monsoon is closed in September-October.

(b) North-east monsoon

Other monsoon is the north-east monsoon. And it affects only the southern states of the country and as a result two times of rainfalls were observed there in Tamil Nadu, Kerala, Karnataka and Andhra Pradesh and as a result plants grow better, throughout the year. Of course, by its influence, in other parts of the country, rainfall is rare in winter season though, South-east monsoon prevails.

3.1.1.8 *Abnormalities in weather*

Of course, sometimes abnormal weather conditions are observed in some places where rainfall is withdrawn fully, or partly, as a result drought is observed and on the other hand, if excess rainfall is there, flood is observed.

(a) Flood : Due to abnormal weather conditions, excess rainfall brings the disaster and as a result flood in some places take place. Floods, in some places, destroy the crops, damage the animals, uproot the trees, remove the people from their houses, showing the miserable conditions in the places.

Of course, though disaster is there, some good things are also there with flood. Flood though brings the tears on the eyes of flood-affected people, but it improves the land with better soil fertility sometimes, because of washing of enriched soils but some soil fertility is destroyed due to deposition of sands or saline waters.

(b) Drought : Drought in some places, is due to abnormal weather conditions. Due to drought, no crop is possible to be grown, because of less or no rainfall, along with high temperature. Even drinking water is not available and as a result both human beings, animals, birds etc. flee-away from that drought affected areas to the safer places. Of course, less rainfall areas are, generally, highly affected with drought.

(c) Storms/hailstorm/tornado: Storms, hailstoms, '*andhi*', '*Kalbai-sakhi*' etc. are all due to weather vagaries. Due to storm, cyclone, hail-storms etc., crops are partly/totally destroyed; men, animals, birds, etc. die instantly and people become penniless and ultimately the place become shabby. Of course, tornado is a strong storm which does not take place normally in India, if it takes place, entire area is destroyed.

3.1.1.9 *Weather forecasting (for crop improvements)*

As weather is very important for crop production, weather forecasting sections have been opened up recently in most of the state departments of agriculture. They forecast weather before 2/3 days of occurrence, for the farming community, in order to do their agricultural operations (intercultural operations) properly in time. Recently in most of the agricultural universities and the Co-ordinated Weather Forecasting Centres have opened up different centres for the farming community in the vicinity of the agricultural university, for improvement of intercultural operations of the crops.

Generally, 'bulletin' on weather of that locality, is brought out by the, university or co-ordinated 'weather forecasting centres', giving advice to the farmers to go for particular operations, for particular crops in different seasons. This is very vital service to the farmers, for improvement of crop production ultimately.

3.1.1.10 *Weather instruments/apparatus*

Following are the instruments/apparatus, for measuring the particular 'weather parameter' in 'weather forecasting centres'.

	Instruments	For weather parameters
1.	Maximum thermometer	Highest temperature attained during the day
2.	Minimum thermometer	Lowest temperature reached on a day
3.	Soil thermometer	Soil temperature
4.	Thermocouples	Soil temperature
5.	Thermograph	Air temperature measurement (with graph)
6.	Barograph	Atmospheric pressure
7.	Wind vane	Wind direction
8.	Anemometer	Velocity of wind
9.	Hygrometer (Psychrometer)	Relative humidity
10.	Assman's psychrometer	Relative humidity in the crop canopy
11.	Hygrograph	*Recording humidity continuously through graph*
12.	Open-pan-evaporimeter	Evaporation measurement
13.	Sunken screen-can-evaporime-ter /screen-can-evaporimeter	Evaporation or Evapo-transpiration under field conditions
14.	Ordinary rain-guage	Amount of rainfall measurement
15.	Self-recording rain-guage	Amount of rainfall measurement automatically
16.	Net-radio-meter	Net radiation measurement
17.	Spectro-photometer	Wavelength of light measurement
18.	Heliograph	Amount of bright sunshine-hours measurement

3.1.1.11 *Meteorological (weather) symbols used all over the world*

	Meteorological elements	International symbols
1.	Fog	≋
2.	Dew	_o_
3.	Snow	✳
4.	Rain	•
5.	Continuous slight rain	••
6.	Continuous heavy rain	o o o o
7.	Drizzle	,
8.	Shower	▽
9.	Hail	△
10.	Thunder storm	⛈
11.	Lightning	ϟ
12.	Cloudy	●
13.	Partly cloudy	○

Practical Class : Identification of meteorological instruments and symbols

Lesson 27 : Identify the meteorological major instruments with diagrams

Introduction

There are so many weather parameters like temperature, rainfall, humidity etc. And these parameters are measured by some instruments at different times under different situations, for forecasting weather in particular area, in view of proper improvement of packages of practices for crop growth, processing and harvesting for higher crop production. These instruments should be acquainted by the learners.

Objective

O to develop the idea for weather conditions,

O to know the instruments for the purpose it is used,

O to utilize these instruments for proper place,

Material required

- Important instruments (for weather forecasting), ✤ Note book,
- Pen/pencil, ✤ Scale (small, 15 cm, 30 cm)

Identify (as displayed in the laboratory) the following major meteo-rological instruments with diagrams

(1) Maximum thermometer,
(2) Minimum thermometer,
(3) Soil thermometer,
(4) Wind Vane,
(5) Open-pan-evaporimeter,
(6) Ordinary rain-gauge,
(7) Hygrometer/Psychrometer and
(8) Light meter.

Lesson 28. Show the meteorological international symbols of (a) Dew, (b) Rain, (c) Fog, (d) Snow, (e) Drizzle, (f) Shower, (g) Contin-uous slight rain, (h) Continuous heavy rain, (i) Hails, (j) Cloudy weather, (k) Thunder storm, (l) Lightning, (m) Partly cloudy, with drawings.

Introduction

Crop production is directly connected with meteorological parameters. But conditions of parameters partially or fully, are symbolized with some symbols, for better understanding in a nutshell for entire people of the world. Therefore, international symbols are used for better understanding in a short period of time, all over the world.

Objective

O to understand the conditions of meteorological parameters, for the centres of the world,

O to express the conditions of meteorological parameter to others, all over the world.

Material required

✤ Note book, ✤ Scale (small, 15 cm, 30 cm), ✤ Pen/pencil.

Procedure : (as shown and sec. 3.1.1.11).

Assignment :

International meteorological symbols (to be placed by the students).

	Meteorological items	International symbols
(a)	Fog	
(b)	Dew	
(c)	Snow	
(d)	Rain	
(e)	Continuous slight rain	
(f)	Continuous heavy rain	
(g)	Drizzle	
(h)	Shower	
(i)	Hail	
(j)	Thunder storm	
(k)	Lightning	
(l)	Cloudy	
(m)	Partly cloudy	

3.2 INFLUENCE OF WEATHER ON CROPS

Different crops are grown in different seasons, under different environments in changes of elements of weather. Influence of weather is very important, on growth and production of crops. Different crops, whether cereals, pulses, oilseeds, fibres and others, grow differently with variations of temperature, rainfall, humidity etc. If these variations coincide with particular amount, the productivity is high.

3.2.1 Cereal crops (*Rabi*/pre-*kharif*/*kharif* seasons)

There are different cereal crops which grow in different seasons, for the variations of temperature, rainfall, humidity etc., though they are under the same family,

Gramineae

(a) Rabi : Wheat, oat, barley etc. are the winter/*rabi* crops. During the winter season, temperature is low. Without low temperature these crops cannot be grown. Of course, with particular temperature, rainfall humidity etc., particular crops grows in a better way. Wheat production will be very good, provided temperature is higher at germination, lower at tillering and higher at flowering and maturity, if proper adequate soil moisture is available. Rain is good for wheat at earlier stage but not at maturity stage, similarly barley and oat. Maize, though a tropical crop, may be grown in winter

also, provided adequate soil moisture is available. *Boro* paddy also can be grown with good yield in winter, provided water stagnation is there with higher temperature at later (at flowering) stage of growth.

(b) Pre-*kharif* : During pre-*kharif* / summer season, *aus* rice, cat-tail millet (*Sitaria italica*), maize, *jowar, bajra, ragi,* other millets etc. are grown with high temperature, small but frequent rainfall and high humidity etc. Though this type of weather is available only in eastern part of the country, but not in other parts, generally.

(c) Kharif : With high rainfall, high temperature and high humidity, transplanted *kharif* rice, *ragi*, maize, *jowar*, bajra, millets etc. are grown. These crops are the main basic grain food crops in the country. If rainfall is adequate and temperature is moderate then there is a bumper crop production in the country and for this, economic condition in the country is also good.

3.2.2 Pulse crops (*Rabi*/pre-*kharif*/*kharif*)

Different types of pulse crops are grown in different seasons, because of variations of temperature, rainfall, humidity etc. Pulse crops, being in the *Papilionaceae* family, fix atmospheric nitrogen with the help of *rhizobium* bacteria which harbour in the nodules in the roots of plants. Therefore, in waterlogged conditions, pulse can not be grown.

(a) Rabi : Among the *rabi* pulses, pea, lentil, grasspea, gram etc. are the main crops. Generally, all these crops are grown in arable condition, but grasspea, gram etc. could be grown as rainfed *rabi paira / utera* crops also, where tillage is omitted. During winter/*rabi* season, in most of the areas, rainfall is less, temperature is low and humidity is also low. Under these conditions, all these crops grow well provided soil moisture is available.

(b) Pre-*kharif* : In summer or pre-*kharif*, season, greengram, pigeonpea / red gram (*arahar*), cowpea etc. can be grown, as temperature is high, alongwith moderate rainfall and high humidity. All these crops can be grown with low moisture in the soil. Even they can be grown as rainfed crops, provided frequent but less rainfall is there.

(c) Kharif : In high lands, greengram, pigeonpea, cowpea etc. can be grown without irrigation, because of high temperature, high rainfall and high humidity.

3.2.3 Oilseed crops (*Rabi*/pre-*kharif*/*kharif*)

Different oilseed crops are grown in different seasons, under different climatic conditions. Oilseed crops are coming from different families.

(a) Rabi : In *rabi* season, mustard/rape, safflower, niger linseed etc. are grown. All these crops require low temperature, less rainfall and less humidity, provided adequate soil moisture is there. Generally, all these crops require less amount of moisture and hence, they could be grown even under rainfed condition. It has been observed that mustard and linseed can be grown as *paira / utera* crops after *kharif* transplanted rice, without tillage operation, with minimum cost.

(b) Pre-kharif : During pre-*kharif* season, sesame, groundnut, *bherenda,* jatropha etc. could be grown, because during pre-*kharif* season, particularly in eastern part of the

country, temperature, rainfall and humidity are high. All these crops require this type of weather conditions.

(c) Kharif : In other parts of the country, of course, sesame, groundnut *bherenda*, jatropha etc. can be grown in *kharif* season, when temperature, rainfall and humidity are high.

3.2.4 Fibre crops (*Rabi*/pre-*kharif*/*kharif*)

Different fibre crops are also grown in respective seasons, under different climatic conditions. Fibre crops also come from different families. Fibre crops are of 2 types i.e. bast fibre and lint fibre. Cotton is the lint/boll fibre, whereas jute, mesta, sunnhemp, flax etc. are the bast (from bark of stem) fibre crops.

(a) Rabi : During *rabi* flax, sunnhemp (for seed) are grown. In *rabi*/winter season, mostly in all the places, temperature, rainfall and humidity are low. Sunnhemp may be grown as *paira / utera* crops also after transplanted *kharif* rice, without tillage for seeds.

(b) Pre-*kharif* : In pre-*kharif*/summer season, particularly in eastern part of the country, jute, mesta, sunnhemp etc. are grown for fibre, when temperature, rainfall and humidity are high. All these crops, of course, are grown in eastern part, as a rainfed crops, because of frequent rainfall there, though amount is not so high as in *kharif* season.

(c) *Kharif* : In *kharif* season, temperature is moderate, rainfall high and humidity is also high. Under these conditions jute (for seed and fibre), mesta (for seed and fibre) and sunnhemp (for seed and fibre) are grown in eastern parts, whereas in northern, western and southern parts of the country, cotton is grown in some places.

3.2.5 Sugar/starch/rhizomes/tuber crops (*Rabi*/pre-*kharif*/*kharif*)

Sugar, tuber, rhizome crops etc. are grown in different seasons, under different situations.

(a) *Rabi* : During *rabi*/winter season, temperature rainfall and humidity are low. Under this condition, sugar crop like sugarbeet, tuber crop like potato etc. are grown in winter/*rabi* season, though they come from different families.

(b) Pre-*kharif* : Pre-*kharif* season is represented by high temperature, moderate rainfall and high humidity, particularly in eastern parts. Under this conditions, *amorphophallus* (ol), sugarcane, arum, asparagus, different types of vegetables etc. are grown.

(c) *Kharif* : In other parts of the country during *kharif* season, sugarcane, arum, *amorphophallus*, some vegetables, arum, asparagus etc. are grown.

3.2.6 Fodder/forage/hay crops (*Rabi*/pre-*kharif*/*kharif*)

Fodder/forage/silage crops are grown in different seasons under different climatic conditions, though they belong to different families.

(a) *Rabi* : During winter/*rabi* season, in most of the parts of the country, berseem, lucern, oat, barley etc. are grown, as temperature, rainfall and humidity are less, provided adequate moisture is present in the soils.

(b) Pre-*kharif* : In pre-*kharif*/summer season, temp. is high, rainfall moderate and humidity is high. Under these conditions, napier, thinnapier (dinanath grass), ricebean, blackgram, cowpea, horsegram etc. are grown, both under rainfed and irrigated conditions, for fodder/forage/hay crops etc., particularly in eastern part. In other parts of the country, all these crops can be grown, provided adequate irrigation arrangements are there.

(c) *Kharif*: Fodder crops like napier, thinnapier, teosinle, cowpea, blackgram, rice-bean etc. can be grown in *kharif* season, where tmpeature is moderate, rainfall high and humidity is high, in most of the places, in India.

3.2.7 Green manuring crops (*Rabi*/pre-*kharif*/*kharif*)

Green manuring crops, for improvement of fertility of soils and increase of food production, are grown in different seasons under different situations and conditions.

Green manuring crops are grown in the fields and then they are ploughed under and decomposed *in situ* i.e. in the same place. For decomposition of green manure, a good amount of water is needed.

(a) *Rabi* : In *rabi*/winter, generally temperature is low, along with low rainfall and low humidity and hence it is very difficult to decompose the green manuring crops, if platned in *rabi*. Generally grasspea may be used as green manuring crop, but that also will not be economical.

(b) *Pre-kharif* : Most of the green manuring crops are grown in eastern India, during pre-*kharif* season, for the use in *kharif* crops. These green manuring crops are mainly *dhancha*, sunnhemp, blackgram, cowpea, rcebean etc. All these crops are grown under rainfed conditions, particularly in eastern India.

(c) *Kharif* : In *kharif*/rainy season, some crops like *guar*, sunnhemp, blackgram, cowpea etc. are grown for the *rabi* season crops like wheat etc. During *kharif* season, temperature is moderate, rainfall high and humidity is also high.

Practical Class : Growing and ploughing-under *dhaincha*, grasspea and *guar* as G.M. *in situ*

Lesson 29 : Growing of *dhaincha* and ploughing-under it, in pre-*kharif* season (elaborate fully)

Introduction

Continuous cropping with high cropping intensity, deteriorates soil fertility, along with breakdown of soil structure. As a result, crop productivity is decreased. To improve soil structure and soil fertility, some crops (specially leguminous crops) are grown as such, in the fields; these crops are then ploughed-under and incorporated into the soil, for improving soil structure and soil fertility and thereafter the next crop is grown. In this lesson (lesson 29) *dhaincha* (*Sesbania aculeata*) is grown in pre-*kharif*, where as grasspea (*Lathyrus sativus*) is grown in *rabi* (lesson 30) and *guar* (*Cyamopsis tetragonotoba*) in *kharif* seasons (lesson 31).

Objective

- O to improve the soil structure,
- O to increase the soil fertility, (increase of N, P_2O_5, K_2O and other nutrients),
- O to add micronutrients into the soils.

Material required

(a) Growing of *dhaincha in situ*

✣ Plough the land (say, 1000 sqm area) with moisture,

✣ *Dhaincha* seeds @ 30 kg/ha, ✣ Container – 1, ✣ Bullocks (2) with 1 plough man and *desi* plough. ✣ Ladder – 1.

(b) Plough-under *dhaincha* at 45 days after sowing

✣ Bullocks (2) with 1 ploughman and *desi* plough, ✣ Ladder – 1 or Plank (Plank is better).

Procedure

(a) Growing of *dhaincha in situ*

- One-time ploughing the land, with *desi* plough,
- Broadcast *dhaincha* seeds (as per calculation) uniformly,
- Laddering two times, length and cross-wise,

(b) Plough-under *dhaincha* at 45 days after sowing

- Fit the two bullocks (without plough) with the plank/ladder, at the end of the bullocks,
- Run the bullocks and ladder/plank, over the tender *dhaincha,* for ploughing-under (water should be present during this operation),
- Keep this conditions (laying down the plants of *dhaincha* in water for 7 days, as such,
- After 7 days, the land should be ploughed with *desi* plough, again,
- Laddering should be done twice length and cross-wise,

- After 7 days again, land should be ploughed two times. After one ploughing, basal dose of N, P_2O_5 and K_2O should be applied @ 50 kg N, 50 K P_2O_5 and 50 kg K_2O/ha (for transplanted rice).
- Now, the land is ready for *kharif* rice transplantation.

Lesson 30 : Growing of grasspea and ploughing-under it, in *rabi* season (elaborate fully i.e. write the procedures for G.M.)

Introduction and **Objective** (both as in lesson 29).

Material required

(a) Grasspea growing (as in lesson 29)

(b) Ploughing-under (as in lesson 29) (Of course, instead of wet conditions, it will be in dry condition)

Procedure

(a) **Grasspea growing** (as in lesson 29)

(b) **Grasspea ploughing – under** (as in lesson 29) (Of course, instead of wet condition it should be in dry condition)

Seven days after ploughing-under, it should be thoroughly ploughed.

Now the land is ready for direct-seeded rice (*aus*) or other arable crops.

Assignment

Write the details of the procedure of –

(a) Growing of grasspea and (b) Ploughing - under it

Lesson 31 : Growing of *guar* for G.M. (green manure) and ploughing – under it, in *kharif* season (elaborate it fully), Write the procedure

Introduction; Objective; Materials needed : All as in lesson 29.

Procedure

(a) **Growing *guar in situ*** : (as in lesson 29)

(b) **Ploughing-under the *guar* as GM** : (as in lesson 29).

[Note : This should be in dry condition and instead of *dhaincha* or grasspea it should be *guar* in all respect].

Assignment

Write in detail about the procedure of

(a) Growing of *guar*

(b) Ploughing – under *guar in situ*.

[Note : (i) No fertilizer should be given for the green manure crop,

(ii) Only 7 days will be the rest period for further ploughing in dry condition both in pre-*kharif* and *kharif* seasons.

3.3 INFLUENCE OF WEATHER ON WEEDS

Weeds are also the plants and they are affected with the variation of elements of weather. As a result, they behave differently in different seasons. Particular weeds grown in particular season, associated sometimes, with particular crops behave differently. They can be destroyed also, in some cases, by changes in the weather conditions.

3.3.1 Broad-leaved weeds (*Rabi*/pre-*kharif*/*kharif*)

Broad-leaved weeds are found throughout the year i.e. in different seasons.

(a) *Rabi season* : In the *rabi* season, the following broad leaved weeds are observed to damage the crops : sunberry, launaea, eclipta, *amaranth*, golden doc, *melilotus* spp., *vicia* spp. anagalis, pig weed/*bathua*, *nicotiana* etc. Of course, under rainfed and *paira* condtions, Bengal vetch/*hetka*, *melilotus*, *Vicia* spp. etc. grow well. In *rabi*, in general, temperature rainfall and humidity are low. Vetch is associated with grasspea crop.

(b) Pre-*kharif* and (c) *Kharif* seasons : In summer/pre-*kharif* and *kharif* / rainy season, when temperature is high along with low high rainfall and low/high humidity, the following weeds grow well :

Jangli amli, *dudhia*, croton, sunberry, eclipta, *kanchra, senchisag,* mikania, beggar weed, parthenium, *cassia* spp. lantana, tridex (*Tridex procumbens*) *punarnava* (*Boerhaavia diffusa*), blumera (*Blumera lacera*) *bheranda*, jatropha, wild tobacco, burweed (*Xanthium strumarium*), *sushni* (*Oxalis corniculata*), mexican poppy (*Argemone mexicana*), sensitive plant (*Mimosa pudica), cudweed (Gnaphalium indica), goat weed (Ageratum conyzoides), marsilia (Marsilea quadrifolia*). During pre-*kharif* seasons/ *kharif*, these weeds affect the yields of jute and *aus* rice (direct-seeded rice). Marsilia is associated with transplanted *kharif* rice, under waterlogged conditions.

3.3.2 Narrow-leaved weeds (*Rabi*/pre-*kharif* /*kharif*)

(a) *Rabi* / winter season : In *rabi* season, when temperature, ranfall and humidity are low in most of the areas, the following weeds grow in a better way, provided soil moisture is available in adequate amount : *Chiria bajara* (*Phalaris minor*), wild oat (*Avena fatua* / *ludoviciana*), dernal grass (*Lolium temulentum*), bermuda grass (*Cynodon dactylon*) etc. *Phalaris,* wild oat, *lolium* etc. are the obnoxious weeds and are associated with wheat.

(b) Pre-*kharif*/summer and (c) *Kharif*/rainy seasons : In pre-*kharif*/ summer and *kharif*/rainy seasons, temperature, rainfall and humidity are high. Under these conditions the following narrow leaved weeds grow well: Barnyard grass (*Echinochloa crussgali*), bermuda grass, gamba grass (*Andropogon gayanus*), hilograss (*Paspalum conjugatum*), cogon grass (*Imperata cylindica*), rice grass (*Leersia hexandra*), *scirpus* (*Scirpus erectus*), *brachiaria* (*Brachiaria mutica*), *goose* grass (*Eleusine indica*), *digitaria*

(*Digitaria sanguinalis*), cat-tail millet weed (*Setaria glauca*) etc. Barnyard grass is associated with rice crop.

3.3.3 Sedges (*Rabi*/pre-*kharif* /*kharif*)

(a) *Rabi season :* During *rabi* season, temperature, rainfall and humidity are less. Under these conditions, *Mutha* (*Cyperus rotundus*), *Piaji* (*Asphodelus tenufolius*) etc. weeds, under sedge group, grow well. *Piaji* is associated with wheat crop in some low lying areas of eastern India.

(b) Pre-*kharif* and (c) *Kharif* seasons : During pre-*kharif* and *kharif* seasons, temperature, rainfall and humidity are generally high. Under these conditions, *mutha*, *jalmutha* (*Cyperus iria*), *guria* (*Scirpus erectus*) etc. are grown well. *Guria* is associated with transplanted *kharif* rice.

Practical Class : Studies on weed behaviour in seasons

Lesson 32 : Observations, on weed behaviour of obnoxious weeds of @ *Phalaris minor*, (b) *Avena fatua* and (c) *Lolium temulentum* in wheat in *rabi* season

Introduction

Wheat, being the important cereal crop in India, controlling weeds, particularly obnoxious weeds like *phalaris*, wild oat and *lolium* etc., is very important for higher production. For controlling these obnoxious weeds, it is necessary to eradicate them fully, with some clues. From the behaviour of these weeds, the clues are expected to come out. These weeds may be controlled chemically.

Objective

O to find out the clues for controlling them chemically,

O to find out management practices for wheat crop.

Material required

❖ Wheat fields with these weeds, ❖ Weed counting materials (quadrate etc.), ❖ Note book, ❖ Pen / pencil

Procedure

- Select the wheat plot with the obnoxious weeds.
- Record the observations on number of weeds/sq.m at different dates of the season.

Assignment

Observations on number of weeds/ sq.m of *Phalaris*, *Avena* and *Lolium* to be recorded with wheat at 30, 60, 90 and 120 DAS

Items for observations		Obnoxious weed (with wheat)				Remarks
		Phalaris	*Avena*	*Lolium*	*Wheat*	
(i) Tillers/m^2	30 DAS					
	60 DAS					
	90 DAS					
	120 DAS					
(ii) Height (cm)	30 DAS					
	60 DAS					
	90 DAS					
	120 DAS					
(iii) Dry weight (of 10 plants) g	30 DAS					
	60 DAS					
	90 DAS					
	120 DAS					

Lesson 33: Study on the growth behaviour of *Physalis minima* in (a) *rabi*, (b) pre-*kharif* and (c) *kharif* seasons, with the associated crops

Introduction

Physalis minima (sunberry weed) is an annual but all-season weed. It shows different behaviour different seasons, under different circumstances. This weed is very difficult to control with 2,4-D herbicides, though it is broad-leaved weed. Therefore, it is necessary to identify the week points of this weed in different seasons, so that it can be controlled properly or to utilize properly for other purposes. The learner should take record on growth behaviour of this obnoxious weed. Associated crops of this weed, are wheat in *rabi*, jute in pre-*kharif* and mesta in *kharif* seasons.

Objective

- O to identify this (*Physalis minima*) obnoxious weed in different seasons,
- O to record the growth behaviour of this weed, for utilizing and controlling it, from the or experience.

Material required

- Long scale (1m) – 1,
- Pen / pencil,
- Note book,
- Small scale (15 cm and 30 cm) - 1

Procedure

- Select the fields with this weed in different season (*rabi*/pre-*kharif*/*kharif*)
- Balance (1 kg) - 1
- Take the observations in different seasons with different crops,

Assignment

Observations to be recorded, on weed in different season (*rabi*/pre-*kharif*/*kharif*) with the associated crop.

Sl. No.	Observations	Seasons			Remarks
		Rabi	Pre-*kharif*	*Kharif*	
1)	Height of the plant (cm)				
2)	Number of branches/plant				
3)	Number of internode/plant				
4)	Length of node (cm)				
5)	Total green biomass (gm)/plant				
6)	Time of flowering				
7)	Time of seed setting				
8)	Association with crop				

[Note : Basic information from studies (at Kalyani, W.B.)]

(i) In *rabi* season with wheat, the weed is short but with bigger leaves,

(ii) In pre-*kharif* season with green gram, weed is medium in size, shape and height

(iii) In *kharif* season with jute/mesta (for seeds), the weed is lanky with less number of branches and leaves.

■■■■

CHAPTER 4

SOIL ENVIRONMENT AND ITS MODIFICATIONS FOR CROP PRODUCTION

Soil environment is very important for better germination, growth and yield of crops, along with aerial environment. Of course, aerial environment is less amenable for modification, except some adjustments like maintaining plant population and reduction of weed competition for light, space, nutrients etc. Therefore, soil environment has been discussed in brief in respect of biological, physical and chemical aspects.

4.1 SOIL ENVIRONMENTS

Soil environment is very important for crop production and that can be studied under (i) biological, (ii) physical and (iii) chemical environments. For higher productivity, higher soil fertility is also needed.

In one hand, soil fertility is the status or the inherent capacity of the soil to supply nutrients to plants in adequate amounts and in suitable proportion; on the other hand soil productivity is the capacity of the soil to produce crops with specific systems of management and is expressed in terms of yields. Of course, all productive soils are fertile, but all fertile soils need not be productive, for some reasons or otherwise.

4.1.1 Biological environments

Soil biological environment is composed of several organisms, belonging to both plant and animal kingdom, resulting to organic matter, Organic matter, produced, due to decomposition of plants and animals, is the main source of natural nutrients in the soils.

4.1.1.1 *Organic matter*

Soil organic matter is very important for productivity of crops. The soil organic matter is obtained from the decomposition of both plants and animal materials by the microbe or soil micro-organisms micro-flora/fauna, (i.e. bacteria, fungi, actinomycetes, algae, etc.) or macro-organism (micro-fauna, protozoa, nematodes etc.). Soil micro-organisms secrete several enzymes such as urease, sulphates, phosphatases, cellulase, amylase and proteinase. These enzymes are utilized for the breakdown of complex compounds present in plant or animal residues. In legumes, atmospheric nitrogen is fixed, by the *rhizobium* bacteria through nodules in plants.

4.1.1.2 *Soil microbes/insects*

Based on the oxygen requirement, microbes are divided into 3 groups, such as (i) aerobic (O_2), (ii) anaerobic (no -O_2) and (iii) facultative (O_2-high or low). Soil organisms may be divided into 2 groups i.e. (i) flora and (ii) fauna. In flora again, it is divided into (a) macroflora (roots of plants) and (b) micro flora (bacteria, fungi, actinomycetes, algae, etc.). Fauna is divided into (a) macrofauna (earthworm, mole, nuts, etc.) and (b) microfauna (protozoa, nematodes, etc.). All the organisms act with good moisture, proper temperature and oxygen, to produce good organic matter.

4.1.2 Soil physical environments

The soil physical environment is, of course, controlled by soil physical characters like (i) texture, (ii) structure, (iii) aeration, (iv) water (moisture), (v) mechanical resistance and (vi) soil depth.

4.1.2.1 *Soil texture*

Soil texture is actually the relative proportion of (i) sand, (ii) silt and (iii) clay. Normally fragments of rock, larger than 2 cm are stones, between 2 cm to 2 mm are gravel, smaller than 2 mm in diameter is the fine earth. Sand, silt and clay together constitute the fine earth.

Sand : The sand particles, having 0.2 to 2.0 mm diameter are (i) *coarse sand*, and 0.02 to 0.2 mm diameter is the (ii) *fine sand*. Because of large size of sand particles, macropores exist between them, which facilitate free movement of air and water.

Silt : The size of the silt particles is in the range of 0.02 to 0.002 mm in diameter. Of course, unlike sands, silt particles adhere to the film of clay, exhibiting some plasticity, cohesion, adhesion and adsorption. These silt particles, of course, can hold more amount of water than sand, but less than clay.

Clay : The fraction of clay is the most important constituent of soils that controls most of the soil physical and chemical properties. The particle size is less than 0.002 mm in diameter. They can absorb water or nutrients around them and retain them, as a result the land is fertile.

Textural classification : Soils with different proportion of sand, silt and clay, are classified as follows : (i) Sandy, (ii) loamy sand, (iii) sandy loam, (iv) loam, (v) silt loam, (vi) silt, (vii) sandy clay loam, (viii) clay loam, (ix) silty clay loam, (x) sandy clay, (xi) silty clay and (xii) clay.

4.1.2.2 *Soil structure*

Sand, silt and clay, the primary soil particles, are usually grouped together in the form of aggregates. The arrangement of primary soil particles and their aggregates into certain defined patterns, is called 'soil structure'. Natural aggregates are called the 'fragments', while 'clod' is used for an artificially formed soil mass. Thus, soil aggregates are those which resist disruptive forces such as water, wind etc.

Soil structure is the important soil property that influences the soil environment through its effect on the amount and size of pore-space, water holding capacity, availability of nutrients, development of microorganisms etc. The changed soil environment affects germination of seeds, roots of plants, growth and yields of crops.

The fine clay particles i.e. colloids, block or group together due to cohesion and form a cluster. The amount and types of colloidal clay influences aggregate formation. Organic components such as cellulose substances, on decomposition of organic matter, release sticky materials like mucus or mucilage. Shape and size of soil structure are (i) platy, (ii) prismatic/ columnar, (iii) blocky and (iv) spherical/spheroidal.

4.1.2.3 *Soil air / aeration*

In the soil aggregates, pores are there. In those pores, air is there. Therefore, porous soil is always better for aeration. Solid materials, water and air in aggregates are continually changing, as the soil undergoes wetting, drying, swelling, or shrinking, due to tillage operations, compaction, aggrega-tion or dispersion. Specially, water and air always compete for occupation of the same pore space. If pore is filled up with water, there is no air in soil, but if pore is filled up with air, there is no water.

Composition of soil air differs from atmospheric air. The carbon dioxide (CO_2) content and relative humidity of soil air are higher and oxygen (O_2) is less compared to atmospheric air. When oxygen content of the soil air is limited, the growth of most crops is highly affected and stops when oxygen content reaches to a minimal level (below 2%).

Soil air can be controlled mainly by providing good drainage facilities, tillage operations or by scooping the soils time-to-time for respiration of the roots (as in case of jute intercultural operation – *nirani*).

4.1.2.4 *Soil moisture/water*

Soil moisture or soil water is very important characteristic for production of crops. The moisture is held around soil particles in thin film, due to adhesion and cohesion. Of course, it is held with greater force (i.e. 10,000 bars) near the particle surface. Soil water, which is held around the aggregates, has major influence on modifying the soil environment. It minimizes the soil tempera-ture from extremes. Water helps in making nutrients available to the plants. Biological activities, including mineralization increase, due to presence of water. Soil water is controlled by rainfall, irrigation and drainage. Rainfall received is effectively used by the plants itself.

Temperature of the soil has considerable influence on several aspects of growth, especially in early stages of growth as in wheat. Optimum soil temperature for germination for different crops are : winter cereals 15 to 18°C, maize 32°C, sorghum, sugarcane and cotton 30°C and mustard 18-23°C.

Mineralization, nitrogen fixation in legumes and other microbial activities are all temperature related. Nutrient availability is also higher at optimum temperature.

4.1.2.5 *Mechanical resistance*

Plants strike roots into soil for both for uptake of water, nutrients etc. in one hand and mechanical support in another. Tap roots of dicots go deep into the soil straight, producing primary, secondary and tertiary roots. These tap roots are always better than fibrous roots for anchorage of plants, against some storms. For anchorage, clay soil or silty soils are better than sandy or loamy soils. Fibrous root are obtained from monocot plants and these do not go deep into the soil but go laterally. Of course, for binding the soils, these fibrous roots are better.

4.1.2.6. *Soil depth*

In the soil profile, soil depth is very important, in terms of anchorage of the plants. If some stones are there in the lower profile, anchorage for plant roots will be less. If soil depth is high, good anchorage of roots of dicot plants are high. Generally, sandy soils are bad for anchorage of root, whereas other soils specially heavy soils, anchorage of root of plants is good.

4.1.3 Soil chemical environments

Soil chemical environments are very important for higher production of crops. It is the dynamic reactions that maintain dilute solution of nutrient elements which are indispensable for continual plant growth. Of course, transformation of nutrients and their availability in soils depend on soil pH, clay minerals, cation and anion exchange capacity.

4.1.3.1 *Soil pH and soil major nutrients (Nitrogen, phosphate, potash, organic matter and micronutrients)*

Crop productivity depends on soil pH and soil nutrients like nitrogen, phosphate, potash, organic matter, micronutrients, etc.

Soil pH : 'pH' is defined as the negative logarithm of hydrogen ion concentration/activity which indicates its acidity or alkalinity. The pH of soils influences the nutrient availability, soil physical conditions and plant growth, ultimately yields of crops. Decomposition of organic matter is the slowest at pH below 6 and the fastest between 6 to 8. Availability of Fe, Al, Mn, Ca, Zn are influenced by pH of soils, through its effect on solubility of the nutrient compounds. Availability of phosphorus is also dependent on soil pH. If it is in acidic side, its availability is decreased.

4.1.3.2 *Soil major nutrients* (Nitrogen, phosphate, potash and organic matter)

Nitrogen : Plant nutrients present in minerals and organic matter are in non-available forms; weathering of minerals and mineralization of organic matter make the nutrients available to the plants, with the availability of water.

The nutrients present in the soils, are subjected to physical, chemical and biological changes or transformations. In mineralization, organic matter releases ammonium (NH_4) and nitrate (NO_3) and these are available to plants. Mineralization, of course, is temperature dependent process and rate of mineralization is more at higher temperature. Fertilizer materials, added to the soils are transformed and several losses occur during this process. Immediately after application of urea, it is hydrolysed in presence of urease and forms ammonium carbonate.

$$CO\begin{matrix} NH_2 \\ NH_2 \end{matrix} \quad H_2O \xrightarrow{Urase} (NH_4)_2CO_3$$

Ammonium carbonate [$(NH_4)_2CO_3$] is an unstable compound and decomposes into ammonium (NH_4) and carbondioxide (CO_2). Ammonium is adsorbed on the clay complex, as it is positively charged. This fertilizer is used by rice, arum, paragrass etc. (water loving plants), as such immediately. In arable soils, NH_4^+ is converted into NO_3 which is available to the arable (dry area) crops.

Phosphate : Phosphorus is present in the soils in organic and inorganic forms. Inorganic phosphorus is more than organic phosphorus. Minerali-zation of organic matter, immensely releases phosphorus in available form.

Normally phosphorus in soils or in soil solution depends on rate of decomposition of organic matter and rate of reaction with inorganic fraction. Organic and inorganic phosphorus is in equilibrium with phosphorus in soil solution, thus :

$$\text{Organic P} \rightleftharpoons \text{Solution P} \rightleftharpoons \text{Insoluble inorganic P.}$$

Fixation of phosphorus is by adsorption, isomorphous substitution and double decomposition. Fixation is influenced by soil pH, nature and amount of clay, free oxides of Fe and Al, calcium carbonate and organic matter. Acidic soils fix more phosphates than neutral, alkaline and calcareous soils. At pH 2 to 5, fixation is mainly by formation of insoluble Fe and Al phosphates, as Fe and Al are more active in acid soils. Soluble phosphates (P_2O_5) are first converted into tricalcium phosphate and subsequently to carbonate apatite.

Potassium : Water soluble potassium is present in the soil solution as K^+ ion. Exchangeable K is present on soil clay complex which is adsorbed. Fixed potassium is bound between the units of clay minerals. The three forms viz. fixed, exchangeable and water soluble K are dynamic equilibrium in soil. Fixed K $\rightleftharpoons$ Exchangeable K $\rightleftharpoons$ Water soluble K, when fertilizer is added, K in the solution is increased. Some amount of K goes to exchangeable form and in due course even to fixed form.

Organic matter : A variety of bacteria are involved in breaking down complex organic matter to mineral form of nutrients. During this process, nitrogen (N), phosphate (P_2O_5), potash (K_2O), sulphur (S) and several other micronutrients are also made available.

The source of nitrogen ultimately, is from atmosphere. Immediate and important source of nitrogen is soil organic matter. Plant/animal material and microorganisms, etc. are the source of soil organic matter. Plant material contain nitrogenous and non-nitrogenous compounds. The nitrogenous compounds are: (i) amino acid, (ii) amino sugars (iii) nucleo-proteins, (iv) nucleic acids, (v) amides, (vi) purins, (vii) pyrimedines, (viii) alkaloids etc. The non-nitrogenous compounds are: (i) cellulose, (ii) hemicellulose,

(iii) starch, (iv) sugars, (v) gum, (vi) lignin, (vii) tannin, (viii) fats, (ix) oil, (x) waxes, (xi) resins and (xii) pigments.

Among them, cellulose and hemicellulose are more abundant and constitute about 30 to 60 percent of dry matter. The dry organic material are decomposed by cellulose decomposing bacteria viz. *Cytophaga, Sporocyto-phaga, Angiococcus, Polyangium* etc.

The proteins present in organic matter are hydrolysed into polypeptides which are subsequently broken down into amino acids by the activities of *Bacillus* and *Pseudomonas* bacteria. *Amino acids* are then further broken down to ammonium by the process known as *ammonification*. Ammonium thus released, is first attacked by *nitrobactor* and that is converted into nitrate (NO_3). Nitrate is attacked by *nitrosomonas* bacteria and at the end the nitrate into nitrite (NO_2).

Materials obtained from cereal-straw, stubbles of rice, wheat, sorghum and other millets, contain high carbon: nitrogen ratio (C : N). As sufficient quantity of carbon material is available, bacteria multiply using the mineralized nitrogen. There is a temporary lock up of nitrogen due to consumption by bacteria, which is made available after decomposition of dead microorganisms. In the process, aeration (O_2) is highly necessary. For this process, O_2 is needed, at a pH range of 6.5 to 7.5 and temperature range between 30 and 35°C.

Micronutrients : During decomposition and mineralization of organic matters, some bacteria and fungi are capable of acting on it and release different micronutrients such as Fe, Mn, Zn, Bo, Mo, etc. These micro-nutrients, though in small amount, meet up the need of crop production. Of course, in particular place, particular micronutrients may be needed, due to high production of crops. That should be replenished by organic matter. Deficiency of micronutrients may be acute in multiple cropping programes.

Practical class : Soil fertility assessment by visual observations of crops (rough estimation)

Lesson 34 : Growth records of wheat, in general, for assessing the soil fertility in *rabi* with '1 – 10' scale

Introduction

In a bigger plots, soil fertility may be assessed roughly, by visual observations, by observing the patches of poor growth of any crop. Of course, it requires some experience of crop growth, in general. Suppose, in our study, a big general field of wheat is available. At a first glance, you will find the poor patches of growth of wheat. From there, you can roughly assess the poor soil fertility of wheat crop by comparing the patches of good growth of crop by some observations.

Objective

- to assess the soil fertility,
- to find out the difference of parameter, like height, tillers, population etc.

Material required

✥ A vast general field of wheat, ✥ A note book, ✥ Pen/pencil, ✥ Long scale (one metre), ✥ Some pegs (made of bamboo), ✥ Note book, ✥ Pen/pencil

Procedure

- Go inside the wheat field.
- Put the pegs in 1 m x 1 m area both in poor patches and good patches, at least 4 times in each case.
- Record the (i) number of population, per m^2 area, in both cases,
- Take observation in (ii) number of tillers/m^2, in both the cases,
- Record the observations of (iii) height of the plant (10 plants and average them) in both the cases.

Assignment (Observation to be recorded)

Items of observation	Wheat crop									
	Poor patch					Good patch				
	1	2	3	4	Mean	1	2	3	4	Mean
i) Plant population/m^2										
ii) Number of tillers/m^2										
iii) Plant height (10 plants)										

Lesson 35 : Growth and visual observations of *aus* rice for assessing soil fertility in pre-*kharif* season

Introduction and Objective : (as in lesson 34; but instead of wheat crop, it should be *aus* rice (i.e. direct-seeded rainfed rice).

Material required and Procedure : (As in lesson 34).

Assignment [Observations to be recorded (45 DAS)]

Items of observation	Aus rice (direct seeded rice)									
	Poor patch					Good patch				
	1	2	3	4	Mean	1	2	3	4	Mean
(1) Population/m^2										
(2) Number tillers/m^2										
(3) Plant height (cm) (10 plant average)										

Lesson 36. Grow wheat crop in (a) one plot without organic matter and in (b) the other plot with organic mater (FYM, well decom-posed) @ 3 tonnes/ ha and record the following at 40 days intervals :

(i) Number of tillers/m^2, (ii) Height of the plants (cm, from 5 plants) and (iii) Leaf-rolling (at the tip of the leaves) (1-10 scale)

Introduction :

'Soil fertility' may also be assessed by the application of manures/ fertilizers, from the observations of some growth parameters. The learners must know the difference in observations (as above) of growth parameters by addition of manures in wheat.

Objective :

O to observe the difference of parameters,

O to show the overall growth of the crop (wheat),

O to assess the soil fertility of that soil,

O to get the crop in good condition.

Material required :

✤ A well-ploughed land, ✤ Seeds of wheat for 1000 m^2 area, ✤ Well decomposed manures (FYM) @ 3 tonne/ha (calculate the amount, needed for 1000 sqm, ✤ Bullocks (2) with ploughman and plough, ✤ Ladder, ✤ A note book, ✤ Pen/pencil, ✤ Bowls for seeds and manures

Procedure :

- Sow the seeds of wheat in one plot, @ 100 kg/ha (500 m^2)
- Sow another plot (500 m^2 area) with same amount of seeds but with the application of FYM @ 3 t/ha (calculate the amount of F.Y.M.) for 1000 m^2 and divide it by two),
- Record the observations of wheat, on (i) Number of tillers/m^2, (ii) Height of the plant (10-plant average) in cm, (iii) Leaf-rolling at the tip of leaves (10 plants average) at different stages of growh.

Assignment (Learner should record the data)

Item of observation	Treatment		Difference
	No FYM	FYM @ 3 tonnes/ha	
At 40 days after sowing			
i) Number of tillers/m^2			
ii) Plant height (cm)			
iii) Leaf-rolling (at the tip of leaves) (1-10 scale)			
At 80 days after sowing			
i) Number of tillers/m^2			
ii) Plant height (cm)			
iii) Leaf-rolling (at the tip of leaves) (1-10 scale)			
At 120 days after sowing			
i) Number of tillers/m^2			
ii) Plant height (cm)			
iii) Leaf-rolling (at the tip of leaves) (1-10 scale)			

Note : For assessing soil fertility, test crop should be cereals.

4.2 MODIFICATION OF SOIL ENVIRONMENTS

For higher production of crops, soil environments may be modified by (i) biological aspect, (ii) soil physical aspect and (iii) soil chemical aspects.

All these aspects work simultaneously for mineralization/ decomposition of organic matter by bacteria and fungi, release of nutrients by some bacteria and fungi and ultimately plants uptake nutrients physiologically, for higher production of crops.

4.2.1. Biological aspects (by addition of organic matter)

When some organic matter is added to the soil, either through crop- residues or other organic matter addition, the particular bacteria and fungus attack the material and prepare `humus', from particular nutrients and these are obtained after mineralization. Therefore, organic matter either from plant material or animal products improves the soil nutrients.

Nitrogen from atmosphere, is fixed in legumes by *rhizobium* bacteria, harbouring in the nodules, provided some nutrients such as phosphate (P_2O_5) is available in the soils having organic matter.

4.2.2 Soil physical aspects (by addition of organic matter)

Soil physical aspects is very important for higher production of crops. With the addition of sand in heavy clay, a good combination may be created. With the addition of organic matter, that soil gives better yield. Besides silt and clay soils are full of nutrients and with the addition of organic matters nutrient availability further increase, giving higher production of crops. Hence addition of organic matter improves aggregates/structure and nutrient availability.

4.2.3 Soil chemical aspects (by addition of organic matter/chemical aspect (fertilizer etc.)

In soils, there is a big laboratory where continuous changes are going on with the rise of pH, availability of some nutrients increase; as a result crop production is higher. If in soil Fe and Al is there and pH is low, then phosphorus will be fixed and they will be unavailable to crops. If organic matter is added to the soils, more or less all the nutrients will be available, giving higher crop yields.

Practical Class : Addition of organic manures and inorganic fertilizers

Lesson 37 : Application of well-decomposed organic manure @ 2 t/ha to rice crop transplantation

Introduction

Organic (bulk) manure is very important for higher crop production and also for maintaining proper soil structure. Under multiple cropping system, the addition of organic matter, is very important for maintaining soil health. Here the learners are to add bulky organic manure (well decomposed compost) to the transplanted *kharif* rice crop, under wet conditions, @ 2 t/ha. The addition of organic matter will improve the structure and soil fertility.

Objective

- O to increase soil fertility,
- O to improve soil structure,
- O to increase crop production,
- O to show the residual effect,
- O to add micronutrients in addition to N, P_2O_5 and K_2O.

Material required

✤ A field in low lying/medium areas, having water sources or filled-up with rain water, ✤ One pair of bullocks, ✤ One ploughman, ✤ One plough, ✤ Bulky organic manure (well-decomposed compost) @ 2 tonne/ha (calculated the amount for 1000 sqm area), ✤ Some buckets, ✤ One spade, ✤ Ladder / plank – 1, ✤ Spring balance 1 or 2 kg capacity

Procedure :

- Give two ploughings/puddlings (assuming adequate water in the field),
- Before third ploughing, add the calculated amount of bulky organic manure to soil,
- Plough (3rd ploughing) again after spreading the organic matter / compost uniformly
- Give two ladderings, length and cross-wise. Now the land is ready for transplantation.

Lesson 38 : Apply urea fertilizer @ 50 kg N/ha in wheat as top-dressing and irrigate the field

Introduction

Application of nitrogenous fertilizer increases the production specially of cereal crops. The N-fertilizer splitted into two part i.e. *basal* and *top-dressing*. Generally, basal fertilizer is given at sowing and top-dressing is given at CRI stage, i.e. at 3 weeks after sowing wheat but just before irrigation, for higher production. In case of wheat N is applied @ 100 kg N/ha and ½ the amount (i.e. 50 kg N/ha) is applied, as top dressing.

Objective

O to get more yield of wheat,

O to get higher number/amount of yield-components in wheat,

O to increase the growth of wheat

Material required

❖ A field with wheat crop, ❖ Amount (to be calculated @ 50 kg/ha) for 1000 m^2 area) of urea fertilizer, ❖ Bowls – 2, ❖ Sacks – 2, ❖ Spring balance, 1 or 2 kg capacity (for field use), ❖ Spade – 1, ❖ Irrigation arrangements, ❖ Note book, ❖ Pen/pencil

Procedure

- Weigh the calculated amount of urea fertilizer from the sack,
- Take the fertilizer in the bowls for spreading over the wheat crop, uniformly, at CRI stage (i.e. 3 week after sowing),
- Mix the fertilizer with the soils, with the help of wheel-hoe/tine,
- Irrigate the land, giving 6-7 cm of water depth,
- Keep the spade ready for making easy flow of water in the field and
- Note the date and time of application of fertilizer in wheat

Lesson 39 : Recording observation, on average basis of @ height (cm), (b) tillers/sqm^2 of wheat crops, where wheat was grown in lessons 37 and 38, 3 weeks after application of organic manures and fertilizers and compare them with comments

Introduction : (as in lessons 37 and 38); **Objective :** (as in lessons 37 and 38); **Material required :** (as in lessons 37 and 38); **Procedures :** (as in lessons 37 and 38).

Assignment (Observations to be recorded)

	Items of observations	Bulky manure (3 t/ha)	N from Urea (50 kg/ha)
42	**DAAMF (days after application of manure and fertilizer)**		
	a) Height (cm) of wheat (cm)		
	b) Number of tillers/sqm		
63	**DAAMF**		
	a) Height of wheat (cm)		
	b) Number of tillers (sqm)		
84	**DAAMF**		
	a) Height of wheat (cm)		
	b) Number of tillers/sqm		
105	**DAAMF**		
	a) Height of wheat (cm)		
	b) Number of tillers/sqm		

4.3 SOIL CONSERVATION

Soil, which is the life line of a nation, should be protected/conserved against erosion and degradation.

4.3.1. Soil protection against erosion (by mulching/cover crops/soil binding crops/slope adjustment)

Soils should be protected against erosion by mulching, cultivation of cover or by binding crops and by adjustment of slopes (to some extent), for higher production of crops. Selection of proper crops is also good for soil protection.

Mulching

Soil erosion may be checked by mulching with rice or wheat straws, leaves of some crops like banana etc. Of course mulching materials can not be procured everywhere easily, because in places where erosion takes place, all types of crop can not be grown. Mulching may be done with polythene sheet also, but it is very costly.

Cover crops

Soil erosion may be controlled, to some extent, with cover crops. Cover crops may be grown in terrace places of hills/hillocks in rainy season. Among the cover crops in rainy season, groundnut, cowpea, ricebean, black gram etc. (all legumes) are main. In winter season, sunnhemp for seed may also be grown both as crop and cover crop. This crop (sunnhemp for seeds) may control weeds also.

Soil binding crops

In order to check erosion partially, soil binding crops may be grown. Crops under *gramineae* family such as thinnapier, napier, maize, *Eleusine coracana*, and weed (like *Eleusine indica*, *Spolrobolus diander* etc. may be grown. The crops, weeds etc. under grass group, give fibrous roots which bind the soil. Other grasses may also be grown, for binding soil, in order to check erosion.

Minimising slopes

Soil erosion takes place mainly due to higher slopes of the land in hills/hillocks. Soil slopes may be minimised to some extent, by growing some big crops under *Gramineae* family. Slopes may be minimised by growing some tall grassy crops like phragmites (*Phragmites comunis*), arundo (*Arundo sativa*), thatch (*Saccharum spontaneum / Imperata cylindrical). Motra, sital pati* (*Clinogyne dichotoma*) etc. To check slopes, *dhaincha*, small bamboo, grandiflora (*Sesbania grandiflora*) etc. under *Papilionaceae*, can also be grown, other than dressing the tops for terracing the areas. All these practices will check wind erosion also to some extent.

4.3.2. Soil protection against degradation (by addition of organic matter and chemicals)

Soils, may be degraded due to so many reasons. The degraded soils should be protected against degradation by addition of organic matter or by adding chemicals.

Addition of organic matter : Addition of organic matter, improves the physical condition of the soils, making unavailable nutrients available to the crops. With the formation of *humus* from organic matter, soil aggregates are improved, as a result availability of nutrients is increased.

Addition of chemicals : In acidic soils, dolomite or calcium carbonate may be used. In fertilization, calcium ammonium nitrate (CAN) may also be used in acidic soils.

4.3.3. Soil protection against soil sickness/exhaustion (by growing legumes/ addition of organic matter)

Sometimes soil protection is needed for some soils which are sick/exhausted by any means.

Growing of legumes : It has been observed that after growing some crops (thinnapier or linseed), the land becomes sick/exhausted. In order to improve the yields of the next crops the land should be developed. For this, cultivation of legumes or addition of organic matter are urgently needed.

Among the crops, blackgram, cowpea, ricebean, green gram etc. are good. To check, to some extent, for the exhaustion or soil sickness, add organic matter in rainy season, from any source. In low lying areas, *dhaincha* or sunnhemp or blackgram may be green-manured for the next rice crop.

Addition of organic matter/crop residues : Organic matter from any source should be added to the arable soil before rainy season starts. In low lying areas green-manuring crops like, *dhaincha*, sunnhemp, blackgram may be grown; the green-manure crops may be ploughed-under *in situ,* in order to improve soil fertility and crop productivity. FYM or other well decomposed materials may be added to arable soils before rainfall.

Practical Class : Checking of soil erosion, rectification of soil degradation and correction of soil sickness

Lesson 40 : Checking of 'soil erosion' by different methods

Introduction

'Soil erosion' is very detrimental for any crop production, because of poor soil fertility. It happens, generally, in sloppy areas. The soil erosion may be minimized by application of some foreign materials or by some other means. The learners must know that erosion can be checked, for higher crop production.

Objective

- O to check soil erosion,
- O to improve soil fertility,
- O to improve productivity

Material required

❖ Rice/wheat straw (for mulching), ❖ Leaves of plants (for mulching), ❖ Organic matter (soil mulching), ❖ By growing cover crops (like ricebean, cowpea, blackgram or some soil binding grasses).

Procedure

- Spread rice/wheat straw as mulch @ 3 tonnes/ha on soils,
- Spread leaves (same amount) on soil,
- Spread bulky organic manure both for addition of nutrients, absorption/ retention of soil moisture and also for checking erosion (for growth of cover crops)
- Grow some crops like ricebean, cowpea, blackgram and some soil binding grasses in different seasons.

Lesson 41. Rectification of soil degradation by different methods

Introduction

There are some soils which are degraded in nature, due to some reasons. Some soils are acidic in nature, some are alkaline. Some soils are saline, some are calcareous. Again some soils are gaseous etc. In all these soils, it is very difficult to cultivate crops under normal conditions. They are to be normalised with some technique, so that crops can be grown. If the soil is acidic in nature, application of lime stone is needed.

Objective

- O to know first, the type of degradation,
- O to rectify accordingly,
- O to grow some crops particularly with legumes,
- O to grow other crops,
- O to identify soil (if the soil is acidic in nature).

Material required

❖ Limestone / dolomite, ❖ Sacks – 4, ❖ Bowls – 4, ❖ Bullocks (2) with ploughman and plough, ❖ Ladder, ❖ Spring balance 1-2 kg capacity (for the field), ❖ Note book, ❖ Pen/pencil

Procedures

- Plough the land twice, followed by ladderings,
- Before the third ploughing, spread uniformly the powder of limestone/dolomite (as calculated for 1000 sqm area according to pH of soil),
- Again plough twice, length and cross-wise followed by ladderings,
- Sow some legume crops, say blackgram/cowpea, in that land, and
- Grow other crops (legumes), onwards, as the degraded soils have been rectified.

(Note : Other degraded soils should be rectified by other means).

Lesson 42 : Correction of soil-sickness by different legumes crops

Introduction

Sometimes some soils are sick due to some reasons like, continuous croppings, crops or some other reasons like allelopathic effect etc. If it is highly necessary to correct the sickness, for higher production of crops, the leguminous crops are good for correction of sickness. Among the crops, blackgram, ricebean, cowpea, *guar*, sunnhemp, *dhaincha* etc. are good, according to land situations and conditions.

Objective

- O to improve the structure of the soils,
- O to increase nutrients for crops,
- O to revitalize the cropping systems,
- O to correct the sickness of the soils

Material required

✣ Seeds of some leguminous crop (say, black gram), ✣ Bullocks (2) with ploughman and plough, ✣ Ladder – 1, ✣ Superphosphate @ 60 kg/ha, ✣ Bowls – 4, ✣ Note book, ✣ Pen / pencil, ✣ Balance – 1 (1-2 kg capacity).

Procedure

- Weigh the seeds of blackgram @ 30 kg/ha, for 1000 sqm area,
- Spread the seeds over the land and SSP uniformly, after 2nd ploughing,
- Ladderings (2) should be given, and
- Record the date and time with the area, amount of seeds etc.

(Note : Moisture at the time of sowing, should be ensured for germination of seeds).

■■■■

CHAPTER 5

TILLAGE AND CROP PRODUCTION

For crop production, tillage is highly necessary, irrespective of rainfed or irrigated agriculture. It improves germination of seeds, growth and development of crops, finally giving higher yields.

Of course, now-a-days, in some special situations, tillage is avoided. Tillage has been referred to its type, tilth, objectives, implements and along with new concept and 'special tillage'.

5.1 TYPES OF TILLAGE

Generally, tillage operations are grouped into two types, based on the time (with reference to crop) at which they are carried out. They are : (i) preparatory or primary tillage/cultivation which is done before the sowing of the crops and (ii) secondary tillage, i.e. for sowing the crops.

5.1.1 Preparatory or primary tillage and secondary tillage

'Ploughing' is opening of the compacted soil with the help of different ploughs or power tillers or tractors etc. Generally, ploughing is done to open the hard soil and to separate the top soil from lower layers. In addition, ploughing should ensure inversion of soil, uprooting weeds and stubbles and less cloddy soil surface. Optimum time of ploughing of course, depends on soil moisture. When the soil is too dry, it is very difficult to open the soil, again when ploughed-under wet conditions, the soil sticks to the plough; this is not good. Optimum moisture is the best time for ploughing.

5.1.2. Secondary tillage

Tillage operations, after primary tillage, for lighter and finer further operations performed, are known as secondary tillage operations. After ploughing, the fields are left with large clods with some weeds and stubbles and all those weeds are partially uprooted. Harrowing is done to a shallow depth to crush the clods and to uproot the remaining weeds and stubbles. For finer preparations, having good tilth, soil-clod crushing should be done with disc harrow, cultivators, blade harrows, ladderings etc. Thus the field is made ready for sowing. Generally, sowing operations are also included in secondary tillage. The ordinary farmers operate both preparatory and secondary tillage operations simultaneously for sowing crops, provided adequate moisture in the soil is available.

5.1.3 Special tillage

In some special cases, special type of tillage operations are done. Rice is grown in wet conditions and it is a water loving crop. Therefore, soil structure is broken down for `puddling' which is a special type of tillage. Generally sugarcane, groundnut, maize, potato etc., are cultivated with special tillage, called 'earthing-up' either light or heavy. Heavy earthing-up is done in sugarcane and light in groundnut.

5.1.4 Inter-tillage

Inter-tillage operations are also done above the soils, in some arable and wet land, in between sowing to harvesting of crops. Hoeing or wheel-hoeing or tining is done in arable soils, whereas `stirring' is done under wet conditions in the paddy lands. All these are called as inter-tillage.

Inter-tillage operations are done generally for (i) mixing fertilizers with soils, before irrigation, (ii) controlling weeds partially, (iii) penetrating air into soils for respiration of the roots of plants, (iv) exposing excess water/ moisture in wet/marshy lands and also for (v) breaking crustation of the soils after rains/irrigation in the crop fields (particularly in wheat).

5.2 GOOD TILTH

For (i) good germination of seeds, (ii) growth and development of crops, (iii) root formation properly, (iv) retention of moisture in the field and (v) good yields; good tilth is needed for higher production of crops. Good tilth can be obtained by breaking clods or by using planks/ladders in the fields.

5.2.1 By breaking clods/ploughing cross-wise

Good tilth of soils may be obtained by breaking clods with the help of ploughs, *itamugar* (a hammer like structure, made up of wood) or by using harrows, ladder or planks.

(a) Breaking clods : In order to get good tilth of soil, breaking of clods is necessary with the help of *itamugar*. Generally, big clods are observed in clayey soils, no clods are observed in sandy soil. Clod formation is related to moisture content of soils and different proportions of silt, clay and sand.

(b) Ploughing length- and cross-wise: Ploughing the soils several times in lengthwise and crosswise, diminishes the amount and size of the clods, to some extent. After this laddering/planking are necessary for getting good tilth. It has already been stated that ploughing alone minimizes the clods for good tilth only in loamy soils, but not in other soils, though partially clods may be reduced. Clods may be reduced to some extent by cross-ploughing the land.

5.2.2 By using harrow/ladders/plank

Soil of 'good tilth' may be obtained by using *harrows* or *ladders* or *planks* etc.

(a) Harrows : For any big clods, harrows are the best, because the teeth of the harrows grab the big clods piercing the body and minimize the size of clods. Thus, the small clods are further broken by planking. In clayey soil, big clods are observed but not in sandy soils.

(b) Ladders/planks : For good tilth of the soils, ladders and planks have a good role to play. After breaking down the big clods, small clods are observed. These small clods are broken further into pieces with the help of ladders or planks, giving good tilth of the soils.

5.2.3 Rough tillage for water-loving plants (rice)

There are some water-loving plants/crops for which rough tillage is necessary. Rice under transplanted conditions, requires more water stagnation from transplantation to harvest. But the same crop under rainfed conditions, can be grown under direct-seeded situations. In that case, tillage operations should be rough, instead of good tilth, as it (rice) is a water-loving crop, having aerenchymatous tissue, on the leaves and stems. Similar crops are arum, para-grass etc., rough tillage is necessary for stagnation of rainwater. Even transplanted rice, requires less ploughing in *kharif* season, provided clayey soils are there.

5.3 OBJECTIVE OF TILLAGE

The basic objective of tillage is to prepare a good seedbed, where seeds germinate properly, grow crops vigorously and produce optimum yields of crops heavily and also to maintain pollution-free environment activities in the soils, though there are so many objectives. For convenience, all the objectives of tillage are discussed as, (i) Biological, (ii) Physical and (iii) Chemical activities.

5.3.1 Biological objectives of tillage

The biological objectives of tillage are as follows : (a) to control weeds under different situations/conditions, (b) to control rodents/rats/vermins etc., (c) to control soil-borne pathogens and soil-living insects by exposing them to sun rays, if soil is turned-up by tillage, (d) to improve distribution, nutritions and growth of roots by reducing penetration resistance of soil, promoting root respiration and providing accessibility of roots to the moist zone in the soils, (e) to promote acceleration of fixation of nitrogen in leguminous crops by *rhizobium* bacteria inside the soils (O_2 enhanced the nitrogen fixation by bacteria and respiration higher), (f) to provide an optimum habitat that encourages early and uniform emergence and establishment of seedlings (g) to provide greater space, due to fine tilth, for the underground development of storage-roots and stems, (h) to encourage growth and activities of soil inhabitating beneficial flora/fauna, including symbiotic bacteria, (i) to provide conditions, suitable for necessary field operations, and (j) to provide better but total underground environments, for a prolonged period of growth and yields of crop-plants and better care and management of the soils, even when there is no crop in the fields.

5.3.2. Physical objectives of tillage

The physical tillage objectives are as under: (a) to cut, heave and shatter the dense soils, at desirable depth and to break the clods and crust to a desirable extent, for suitable seedbed/planting field and to burry in or cover and pack the seeds/propagules with soils, (b) to redistribute the constituents, particularly soil particles, organic matters, micro-organisms, moisture and air, (c) to improve the capacity of soils in receiving rain or irrigation water, retaining and releasing moisture for crops or to increase percolation or drying excess water, (d) to incorporate crop-residues, green manures and other organic manures, fertilizers, soil amendments and conditioners and agro-chemicals and also to remove shift roots, stubbles, root stocks and stumps, (e) to maintain proper structural conditions, improving soil aggregates, (f) to prepare the land-surface suitable for easy, early and uniform movement of irrigation and drainage water, (g) to increase soil aeration, particularly in the non-capillary space, (h) to provide anchorage to the crops by adding soils around the crop-base, (i) to prepare soil-mulch or to preserve the particle-mulch to conserve soil moisture and to make an impervious soil layer in wetland paddy, to check loss of water through percolation.

5.3.3 Chemical objectives of tillage

The different chemical objectives of tillage are as follows : (a) to accelerate the weathering process of the soils, where from new/modern soils, come up after a long period of genesis, (b) to convert and chemically modify the applied fertilizers and manures into simple forms, in order to make available to the crops, (c) to expedite the reclamation of problematic soils by cation exchange process, (d) to improve the availability of plant nutrients by enhancing the rate of decomposition of organic matter and also to mineralize the soil nutrients, (e) to expel gases from the prolonged reduced soil conditions and also (f) to detoxify soils from any erratic use of agro-chemicals.

Practical Class : Practising ploughing and bringing soils into good tilth

Lesson 43 : Practise ploughing with mouldboard plough in wheat sowing

Introduction

Ploughing is very much essential in raising crops, particularly for establishment of stand, development of roots and anchorage of plants with the soils. Generally, 3-4 ploughings are needed for raising irrigated wheat, in order to bring soil into good tilth.

Objective :

- O to bring the soil into good tilth,
- O to absorb water into the soils,

- O to establish good crops with ploughing,
- O to help in better germination of seeds,
- O to uproot and destroy the weeds,
- O to kill the larvae of insects and fugi (by exposing the soil with ploughing)

Material required

✣ Bullock – 1 pair, ✣ Ploughman –1, ✣ Mouldboard plough – 1, ✣ Ladder – 1, ✣ Seeds @ 100 kg/ha, for 1000 m^2 area, ✣ Fertilizers @ 50 kg N, 50 kg. P_2O_5 and 50 kg K_2O/ha, for 1000 m^2 area, ✣ Bowls – 4, ✣ Balance – 1, ✣ Sacks – 4, ✣ Note book, ✣ Pen/pencil

Procedures :

- Attach the bullocks with yoke,
- Fit the plough in between two bullocks, keeping the beam over the yoke,
- Adjust the depth of ploughing, by dragging the plough over the yoke,
- Stand behind the plough, holding the grip of the plough,
- Now practise, cautiously trail the land, turn right and after about 10 m away turn the plough again, keeping a land unploughed, but gradually that will be ploughed.
- Spread the fertilizers (as per calculation) just after two ploughings, by broadcasting.
- Give the ladderings after each ploughing.
- Sow wheat seeds (as per calculation) just behind the plough-sole, so that seeds are covered by the soils from the next-row ploughing.
- Do not apply ladder, because before mouldboard ploughing, which turn the soil and cover the seeds of the earlier row (If you feel proper tilth has not been brought about, then another ploughing with laddering should be applied, before putting seeds i.e. before 3rd ploughing. Remember proper moisture should be ensured for proper germination of seeds).

Lesson 44 : Use the ladder/plank with mouldboard plough for bringing good tilth in soils, for wheat sowing

Introduction

During ploughing in most of the soils, clods will be there. It will be less in optimum soil moisture in alluvial soils, but in some clayey or clayey loam soils, under no or less moisture conditions, clods will be higher in number and bigger in size. Therefore, it is very necessary to break the soils, for bringing good tilth, in order to have good germination of seeds of any crop.

Wheat which requires good tilth, should be ploughed at optimum soil moisture conditions, particularly in alluvial soils. For this, after each ploughing, laddering/planking is needed. Generally ladder/plank is attached with rope at the yoke and at the rear end of the bullocks, (out side the legs of the bullocks). The ploughman stands on the ladder/plank with a rope attached to the middle of the yoke.

Objective

- O to break the clods,
- O to get finally good tilth/pulverization,
- O to fit the wheat seeds for proper germination,
- O to get the soil moisture preserved.

Material required

❖ Unploughed land – 1, ❖ Ladder –1, or Plank – 1, ❖ Bullocks – 2, ❖ Ploughman – 1, ❖ Plough (mould board) – 1, ❖ *Itamugur* – 1, ❖ Spade – 1, ❖ Basal dose of N, P_2O_5 and K_2O fertilizers @ 50, 50 and 50 kg/ ha, for 1000 sqm area, ❖ Wheat seeds amount, @ 100 kg/ha for 1000 m^2 area (to be calculated), ❖ Bowls – 3, ❖ Balance–1 (spring balance, 1 or 2 kg capacity), ❖ Note book, ❖ Pen/pencil

Procedures

- Plough the land, under optimum moisture, with mouldboard plough,
- After first plough, set the ladder/plank with ropes just behind the legs of the bullocks, so that the ladder does not touch the legs of the bullock,
- The rope should be attached with the yoke, from two sides of the ladder,
- The plough man or the learner will stand on the mid portion of the ladder/plank, by holding another rope attached with the middle of the yoke,
- The bullocks will be set across the ploughed lines, so that clods are broken,
- Direct the bullock to move on, thus the most of clods will be crushed,
- If some big clods are still there, then those should be broken by *itamugur* (a drum-like structure made of wood with a long handle), manually,
- If some portion is unploughed then, spade it (particularly in the corners of the field),
- Run the ladder/plank, after each ploughing,
- After two ploughings, spread uniformly the basal dose of fertilizers, as calculated,
- After the third ploughing, sow the wheat seeds with the mouldboard plough (as per calculated amount of seeds), and
- Then after sowing wheat, setting of ladder is not necessary for final laddering/planking [because of covering of seeds with soils automatically by mouldboard plough].

5.4 TILLAGE IMPLEMENTS

For tillage operations, different tillage implements at different stages of growth of crops are needed. Major implements are needed at and before the sowing of crops. Some are needed in between sowing and harvesting of crops. The following tillage implements are described in brief. Big implements under mechanized farming are not discussed here.

5.4.1 Country/*desi* plough

Country/*desi* plough, though old, is still prevalent in most of the parts of the country. Major tillage operations are done with the *desi* plough, with one pair of bullocks and a ploughman. Of course, *desi* plough is good for puddling the land under wet conditions. For normal ploughing, there are some untilled ridges in fields, for which several time ploughings are necessary (length-wise and cross-wise). This country ploughs differ from place-to-place, because of types of soils and crops.

5.4.2 Mouldboard plough

Mouldboard plough is the improved plough over the *desi* one, and it turns the soils, covering all plant materials including organic matters beneath the soils. Of course, it is good for sowing wheat, where seeds are put into the furrows and next ploughing covers the seeds with soils. This plough is very good for incorporation of manures/other organic plant-residues etc. It can be run by a plough man with one pair of bullocks or it may be attached to the tractor.

5.4.3 Clod breaking agents (Harrow/ladder/plank/*itamugur*)

Harrow, ladder, plank etc. are used for levelling the soils after ploughing and also for crushing the smaller clods for good tilth which is good for germination of seeds, growth and higher yields of crops.

Harrow : 'Harrow' is a device attached to the tractor for levelling the soils after sowing seeds. This is also used for crushing the bigger/smaller clods before sowing. Harrow is used by the tractor, immediately after using cultivators which break the big clods by piercing through them. Of course, optimum moisture in soils is the best for proper use of harrows. Harrows are generally made of iron rods, but some bamboo harrows are also available, though less. For good tilth, it is very necessary.

Ladder and '*itamugur*' : '*Ladder*' is made up of strong and hard bamboo nearly 3 m in length, 30-40 cm width; in between supporting poles/*bakhars* are used. The two ends of the ladder is a little bit tapering. The ladder is attached to the end of the ploughman who is running the plough with bullocks. Under proper moisture ('*joo*') condition in soils, the ladder works well. Most of the farmers in the country use this ladder for levelling the soils and crushing the clods. Sometimes *itamugur* is used for crushing the clods. It is an wooden cylindrical hamper like small implement, with long handle, to break the clods for fine soil-tilth.

Plank : '*Plank*' actually the sliced and rough wooden log, about 3 metres long, 30-40 cm breadth and 8-12 cm width, having heavy weight. It is also put behind the ploughman of bullock drawn plough. It also grabs the clods and breaks into pieces and it also levels the soils after sowing operation. As because, it is weighty, it is better than ladder in some heavy soils. It is good for levelling the puddled soils.

5.4.4 Hand tools (Spade, sickle, tine, *nirani, khurpi*)

There are so many hand-tools which are required for different operations before and after sowing crops. Though they are smaller in nature but for works, they are essentially needed.

Spades : These are of different types for different purposes. Spade is an iron sheet, having wooden handle. The sheet may be narrow or broad, long or short. Again handles may be long or short. Some times the sheet is bent and short handle is used. Long handled spades are used for light earthing-up in potato in earlier stage of growth; whereas at latter stage bent spades with short handle is used.

Sickle : '*Sickle*' is the semilunar serrated fine bending iron bar (12 to 18 cm), with an wooden handle. For some works (harvesting of small crops, weeding in rice field at latter stage), sickle is essentially needed. Due to serration, the stems of crops cut well. For other purposes (such as cutting weeds), it is also used.

Tine **:** It is the sharp end-bent-iron rod with an wooden handle. It is nearly 12 to 24 cm length. It is generally used for sowing wheat in untilled soils or other works like tining the soils, after application of manures/ fertilizers or for breaking the crustation. This tine may also be used in soil stirring in wet paddy soils.

Nirani *:* It is a small tool made of iron, bending like semilunar at the sharp-wide-end with an wooden hand. It is about 10 to 12 cm in length. Generally, it is used for hand weeding in direct-seeded (*aus*) rice or jute. It is good for scooping the soils for respiration of roots of jute crop at earlier stage of growth.

Khurpi *:* It is also a small tool, made up of long and flat iron-sheet with sharp end, having an wooden handle. It is about 10 to 12 cm in length. It is used by pushing ahead, for scrapping soils/weeds etc. It may be used for opening the holes also. It is generally, used in the garden instead of fields.

5.4.5 Tractor/power tiller

'*Tractors*' and '*power tillers*' are used for tilling the unbroken soils or for final preparation of the soil before sowing crops, where mechanization is in vogue (All information would be given by Engineering section).

5.4.6 Seed-drill

'*Seed-drill*' is a device for sowing seeds of crops (keeping seeds in drum) uniformly in times with particular depth. This device may be attached to the tractor or bullock drawn ploughs. Starting from 2 to 8 rows, seed-drills are available, 'seeds-cum-fertilizer drill' is also available, where both fertilizer application and seed-sowing operations are done simultaneously. Of course, this is costly, hence rich-resource farmers may operate it, for sowing of wheat direct-seeded rice, *jowar*, barley, mustard, green gram etc. Specific, one-row seed drill is also available for sowing jute.

Practical Class : Drawing the mould-board plough and seed drill operation

Lesson 45 : Draw the part and label them of mouldboard plough and give the full description of the parts.

Drawing: As shown for demonstration.

Description of parts : (See at 2.5.2)

Lesson 46 : Draw the parts of jute-seed-drill with labelling and describe the parts therein [As displayed in the shed. See description of the parts as in 2.5.4]

Lesson 47 : Operate the jute-seed drill for sowing jute

Introduction

Jute is a bast fibre crop, which requires high temperature, high rainfall and high humidity for its proper growth. Hence it is grown in eastern parts of the country. It is generally, sown in the month of April. It is sown quickly because of frequent rainfall. It has been observed that huge amount of weed infestations occur in jute cultivation. Hence for line sowing with short time, specially seed drill is used for jute.

Objective

- O to sow jute quickly and efficiently,
- O to irradicate/control weed in the season, in jute
- O to facilitate the growth of jute (because of proper top-dressing)
- O to maintain proper number of jute plants in unit area.

Material required

✥ A ploughed and well prepared jute field, ✥ A jute seed drill fitted well with all parts, ✥ Fertilizer @ 40 kg N from urea, 40 kg P_2O_5 (SSP) and 60 kg K_2O (MOP) for 1000 m^2 area, ✥ Jute seeds @ 10 kg/ha (*Olitorius*) for 1000 m^2 area, ✥ Bowl – 4, ✥ Balance (spring balance 1, 2 kg capacity), ✥ Sack – 4, ✥ Note book, ✥ Pen/pencil.

Procedures

- Broadcast the required amount of fertilizer (for 1000 sqm)
- Plough and ladder the field,
- Check the jute-seed drill
- Put the seeds (small amount in the hopper) and check the seed rate,

- Start the sowing programme with seed drill and complete the area for sowing
- Note down the date of sowing, amount of fertilizers, amount of seed used etc.

5.5 MODERN CONCEPT OF TILLAGE

Modern concept of tillage i.e. special type of tillage, is a new develop-ment. Now-a-days, special type of tillages are done for maximum production with minimal tillage operations, under special situations, in order to mini-mize cost of production of crops. Among such special type of tillage, under different conditions, mainly are: (i) zero tillage, (ii) minimal tillage, (iii) stubble-mulch tillage (iv) no-tillage relay (i.e. *paira/utera* crops) and (v) skip-off tillage.

5.5.1 Zero tillage

Zero tillage/no-tillage is an extreme form of minimal (minimum) tillage. Primary tillage here is avoided completely and secondary tillage is also restricted in seedbed preparation, in the row/root zone only. It is also no-tillage (or no-till) and is resorted to places where soils are subjected to wind and water erosion, or other specific places, where time of tillage operation is too difficult and requirement of energy and labour for tillage is very high. Wheat, after transplanted *kharif* rice or cowpea, maize etc. could be cultivated by this method.

5.5.2 Minimal/minimum tillage

'Minimal/minimum tillage' is aimed at reducing tillage to a minimum necessary, ensuring good seed-beds, rapid seed germination, a satisfactory crop stand and favourable growing conditions of crops and ultimately good and economic yields of crops. Tillage could be reduced in two ways, such as: (i) by omitting tillage operations which do not give benefit compared to cost of cultivation or (ii) by combining agricultural and other operations like soil breaking, seeding, fertilizer application, weeding etc. wheat, after trans-planted *kharif* rice, could be grown with minimal or with two tillage or by making furrows in between the two rows of preceding transplanted *kharif* rice, without major tillage operations. Green gram, sesame, linseed, mustard, cowpea etc. can be grown economically with 1-2 tillage operations. Of course, concept of minimal tillage was started in USA (1974), due to high cost of oil (in mechanized cultivation). But now-a-days, this (minimal tillage) is in position everywhere, to reduce the cost of cultivation.

5.5.3 Stubble-mulch tillage

'Stubble-mulch tillage' is a system of minimal tillage, where the soil is covered with crop residues and minimal tillage is given in the soils, retaining the 'crop-residue cover' intact, keeping soil moisture higher and putting crop-seeds into the tilled operated areas.

It can be remembered that in advanced countries, crop-residues are left on the field itself, which latter act as 'soil cover' or 'crop-residue cover' mostly under rainfed conditions, under mechanized cultivation. In this way they grow maize, soybean etc. But it is not in vogue in India, as farmers will not leave any straw/stovers on the fields. In India, man consumes the grain and cattle (including buffalows, ass, goat) etc. consume straw/stover (stubbles). Hence, practically no crop-residue is left on the field for 'stubble-mulch tillage'. Hence it is not feasible in India.

5.5.4 *Paira/utera* (no-tillage relay crop)

The crop, which is sown by broadcasting seeds of some crops, over the preceding crops at maturity, under muddy/marshy conditions (sometimes optimum soil moisture conditions), without tillage operation and the said crop which is raised with residual soil moisture/rainfall, is called the rainfed *paira* or *utera* (no-tillage relays) crop, and the system with this crop, is known a `*paira / utera* cropping' system. Grasspea (*khesari*) lentil, linseed etc. could be grown in transplanted *kharif* rice as *paira/utera* crops. After *paira* crop, jute, green gram etc. may be grown.

5.5.5 Skip-off tillage

'Skip-off tillage' is a new system of tillage, where in one season, tillage is omitted, under rainfed multiple cropping system by growing *utrea / paira* crops, keeping other crops in the system with proper tillage.

Grasspea, (*khesari*), lentil or sunnhemp (for seed) may be grown as raifned *utera / paira* crops, after transplanted *kharif* rice, without any tillage. Other two crops in the system, such as *green gram* and rice, under '*paira / utera* crop (green gram and transplanted *kharif rice)*, are properly tilled. This 'no-tillage' for growing *paira / utera* crops, after transplanted *kharif* rice, is the 'skip-off tillage'. With the skip-of-tillage in *rabi*, 3-crop systems (such as *rabi / paira* crop, pre-*kharif* crops - *kharif* crops can be grown in eastern parts of the country.

Practical Class : Pactising minimal tillage in wheat and sowing of *paira/utera* crops

Lesson 48 : Practise `minimal tillage' in wheat after transplanted *kharif* rice

Introduction

Practising `minimal tillage' in wheat is very important, for food production with two cereals, particularly in eastern India, where rice is the major crop. Wheat requires winter season. More longer duration of winter is better for the yields of wheat. But in eastern part, winter is mild and short. After harvest of transplanted *kharif* rice, short and mild winter starts. Hence, no time is left actually for preparation of good tilth seed-bed in between the harvest of rice and sowing of wheat under this conditions, hence minimal tillage operation is advocated for raising wheat crop. In that case, wheat seeds @ 125 kg/ha are broadcast over the rice-stubbles in unploughed land, immediately after harvest of rice, followed by 2 criss-cross ploughings, followed by heavy ladderings twice, for covering the wheat seeds. Dwarf wheat possesses strong and stout crown roots that grow just below the soil surface. These roots are very essential for uptaking nutrients and water from the soils. Learners should go for practising minimal tillage in wheat.

Objective

- O to grow wheat in mild winter within a very short time,
- O to utilise the strong and stout crown roots which grow just below the soil surface,
- O to get more grain yield of wheat under mild and short winter,
- O to facilitate for incorporation of more crops in the cropping systems.

Material required

✤ Field after harvest of *kharif* transplanted rice – 1, ✤ Bullocks - 2 with ploughman–1, ✤ Plough (mouldboard) – 1, ✤ Ladder–1 or plank – 1, ✤ Wheat seeds (as calculated amount @ 120 kg/ha, for 1000 m^2 area), ✤ Basal NPK fertilizer (as calculated amount @ 60-60-60 kg N, P_2O_5 and K_2O/ha for 1000 m^2 area (rest 60 kg N should be given later at CRI stage). ✤ Bowls – 4 (big 3 + small 1), ✤ Balance – 1 (spring balance with 1 or 2 kg capacity), ✤ Sack – 3, ✤ Spade – 1, ✤ Note book ✤ Pen / pencil

Procedures

- Broadcast basal fertilizers uniformly (as per calculated amount) for 1000 m^2 area, over the unploughed land, immediately after harvest of transplanted *kharif* rice,
- Broadcast wheat seeds in similar manner (as per calculation),
- Attach and adjust the plough / *desi* plough with bullocks,
- Plough the land, length-wise and cross-wise (twice),
- Apply one laddering after each ploughing (ultimately two ladderings) after spading to loosen the soils at the corners, and
- Record the date of sowing, time (morning/noon/evening) and amount of seeds and fertilizer added.

Lesson 49: Practise sowing of rainfed *paira/utera* crops, after trans-planted *kharif* rice

Introduction

In the eastern parts of the country, rainfed transplanted *kharif* rice is grown in low lying areas, where a little amount of water is available in land at the time of harvest of rice crop. Under this condition, rainfed crops like, grasspea, oat, lentil, mustard, linseed, sunnhemp (for seeds) etc. could be grown, as *paira / utera* (no-tillage relay) crops without tillage operations. Learners should go for practising sowing of those *paira / utera* crops.

Objective

- o to grow some rainfed crops without tillage, on lands, having a little bit of water at the time of harvest of *kharif* transplanted rice.
- o to grow crops for increasing the cropping intensity under rainfed condition, in low lying areas.
- o to partially control weeds in the lands and to get more crop yield at the cost of weeds.

Material required

✤ A land with matured transplanted *kharif* rice, having a little bit of water above the soil surface, ✤ Seeds of *paira / utera* crop (say grass pea) @ 60 kg/ha, for 1000 sqm area, ✤ Small and big bowls – 2, ✤ Note book ✤ Pen/pencil.

[**N.B.** : Observe the water amount on the soil surface. If little bit water or soaked (with water) conditions, grasspea may be broadcast; seeds of other crops (say lentil/ mustard), should be broadcast in water-soaking condition].

Procedures

- Seeds of *paira / utera* crops should be broadcast uniformly, over the matured rice crop (transplanted *kharif*), in the evening, when the soils are in muddy or marshy conditions, without tillage, 1 day before the harvest of *kharif* rice,
- Rice crop (transplanted *kharif*) should be harvested at a height of 15 cm, above the soils, just 1 or 2 days after broadcasting seeds of *paira* crop, so that some seeds get inside the muddy soil by trampling of labourers during harvest of rice,
- Remove the harvested rice from the field immediately,
- Note down the date of broadcasting seeds for *paira* crop,
- Note down the amount of seeds required for 1000 m^2 area,
- Note down the date of harvest of rice crop,
- Note down the time of broadcasting seeds.

[Broadcasting of seeds should be done in the evening, as seeds will be soaked with the dew at night and will fall down by wind on soil, if they are adhered to the rice plant].

Lesson 50 : Practise `skip-off tillage' in "rainfed 3-crop system (Jute-transplanted *kharif* rice-*paira/utera* crops (no-tillage in rainfed *paira/utera* crop i.e. skip-off tillage)

Introduction

'Skiff-off' tillage is a new term for the rainfed *paira / utera* 3-crop system, where (out of 3-crop system) one season crop is raised without any tillage operations. This *skip-off* tillage is possible where transplanted *kharif* rice is grown in low lying conditions, with water obtained from rainfall. In eastern parts of the country, 'transplanted *kharif* rice-*paira*/utera (no-tillage) crops – green gram' is possible to be grown. In *rabi paira / utera* crops, tillage operation is skipped-off fully, hence, this system is known as *'Skip-off* tillage'. The learners, should know this and they can practise particularly in wet rice growing areas.

Objective

O to increase the cropping intensity,

O to produce higher yields annually,

O to utilise the wet rice-fallows under rainfed condition,

O to check the soil borne diseases of some crops,

O to improve the economic condition of the farming communities,

O to make up the productivity of crops, if other crop failure,

Material required and **Procedure :** as in lesson no. 49.

After rainfed *paira / utera* crops, grow green gram in pre-*kharif* and transplanted *kharif* rice in rainy season with proper tillage operations.

5.6 SPECIAL TILLAGE

Sometimes, special tillage operations are needed for some crops. It has been observed that these special tillage operations, gave better yields of the crops. Puddling, earthing-up, pitting, dibbling etc. are the special type of tillage that are done in transplanted rice, sugarcane maize/potato/groundnut, plant-propagules and wheat respectively.

5.6.1 Puddling (For rice)

'Puddling' is a special type of tillage operation, which is done under wet conditions, to destroy the soil structure and to check the percolation loss of water, for transplantation of rice/arum/paragrass etc.

Most of the rices (except direct-seeded rice), all over the country either in *rabi* or in *kharif* season, are grown in transplanted conditions with puddling the soil. In rice cultivation, water stagnation is there, over the puddled land; lower depth (5-6 cm) at earlier stage, higher depth (8-10 cm) at peak growing period and lower again at 5-6 cm depth, before harvest of the crop.

5.6.2 Earthing-up (Sugarcane/groundnut/poato/maize etc.)

'Earthing-up' is one type of tillage where soils in row crops in between rows, are lifted with the help of spade, and are taken and deposited to two sides of the crops, making furrows, both for drainage and irrigation.

(a) Sugarcane: In sugarcane deep furrows (about 10-12 cm) are prepared with deep earthing-up, in between the rows which are of about 60-90 cm width, with the help of big and long handled spades. This big furrows are used for either drainage for excessive rain or for irrigating in proper irrigation in sugarcane. Generally, light earthing-ups at earlier stage and deep earthing-ups in later stage, done by sides of the plants, with help of spades, depositing soil on two sides.

(b) Groundnut : Groundnut is planted at about 50-60 cm space in rows. In between rows, earthing-up is done twice at about 20 and 40 days after sowing (DAS), with the help of middle-handled-spade. These furrows, though not so deep like sugarcane, serve the purpose of drainage system for excess water and also for irrigation in due time, if necessary.

(c) Potato : In potato cultivation also light earthing-up is done, though not so deep like sugarcane. Earthing-up is done twice in potato, once at seedling emergence and the other at full-grown stage. In this earthing-up also, excess water is drained-off and irrigation is given on the same furrows at different times. In earlier stage, long-handled-spade is used; whereas at later stage, normally short-handled, bent type of spade is used.

(d) Maize : In maize, first earthing-up is done when the crop is 'knee high' and other earthing-up is, done at full growth of the crop. Earthing-up is done with spade. Generally, maize is planted at 50-60 cm space in rows where earthing-up is done, both for drainage and irrigations.

5.6.3 Pitting (*ol*)

'*Pitting*' is a term which comes from '*pit*'. Some crops are planted with the propagules, not with the seeds. Generally, (according to crops) a `pit' with dimension of 60-70 cm length x 60-70 cm breadth x 30-40 cm depth, is dug with the help of spade. Into that pit, some, organic manures are applied and there-after mixing with soils, is done and propagules, (say succur of banana, or *ol*) are planted and the root-portion is covered with soils. Pitting may be used for planting saplings also.

5.6.4 Dibbling (wheat)

'*Dibbling*' is a process of sowing some seeds in unploughed lands, where scooping of soil is done and in that scooped areas, seeds of wheat crop (or other crops like maize, cowpea, ricebean etc.) are put and the seeds are covered with the scooped soils. This dibbling is used in wheat sowing in unploughed lands after transplanted *kharif* rice in some places of eastern parts of the country.

Practical Class : Puddling operations for rice transplantation and earthing-up in some crops

Lesson 51 : Practise puddling operation for *kharif* transplanted rice

Introduction

Rice is a water-loving crop. The rice may be grown as (a) direct- seeded, and (b) transplanted conditions (both *aman* and *boro*) throughout the year. For transplanted rice, puddling is necessary, for retaining water in the fields by breaking down the soil structure. Generally, low/medium situated land is selected, having clay soils, for maintaining both good soil fertility and retention of water in the field, for throughout the rice-growing period. Therefore, the learner should know the '*puddling process*' for transplantation of rice, during *kharif* season.

Objective

- O to reduce the percolation loss of water,
- O to retain more water in the fields,
- O to keep the rice plant grow better,
- O to get more yield, because of retention of water throughout the growth period of rice with fertilizer [@ 50, 50 and 50 of N, P_2O_5 & K_2O kg/ha, for 1000 m^2 area].

Material required

✤ A low land, having clayey soils and stagnated water (irrigated/rain water), ✤ One pair of bullocks with one ploughman, a *desi* plough, yoke – 1, ✤ One plank/ladder-1, (with rope arrangement), or, a power tiller, having cage wheel), ✤ NPK-fertilizers @ 50, 50 and 50 kg of N, P_2O_5 and K_2O/ha, for 1000 m^2 area, ✤ Field spring balance (big) – 1 or 2 kg capacity, ✤ Plastic ropes (4 to 5 pieces, 50 ml long), ✤ Bowels – 2 (small and big), ✤ Sacks – 4, ✤ Pen/pencil ✤ Note book.

Procedure

- Set the bullocks, at yoke and *desi* plough for ploughing the land,
- Plough the land twice, length and cross-wise,
- Spread fixed amount of P_2O_5 and K_2O fertilizers together uniformly over the soils, by broadcasting, after weighing in 1000 m^2 area.
- Then broadcast the fixed amount of nitrogenous fertilizer (say urea), uniformly, separately on the soil surface, after weighing, for 1000 m^2 area.
- Plough the land again twice, length and cross-wise.
- Give the heavy plankings/ladderings twice, length and cross-wise. Now, the land is ready for transplantation of rice seedlings.
- Note down (i) the date of ploughing and transplanting, (ii) amount of fertilizer given, and (iii) number of ploughings and ladderings given, (iv) spacing and number of seedlings per hill.

Lesson 52 : Practise `earthing-up' in maize crop

Introduction

Maize is a C4-type crops, which means, it can be grown in different seasons of the year. This crop is generally irrigated, but in eastern part, where rainfall is high and its duration is long, it can be grown in pre-*kharif* and *kharif* seasons, under rainfed conditions, because of frequent and good amount of rainfall during that period. As it can not withstand water stagnation hence, earthing-up is necessary to drain out excess water. Therefore, the learners should be acquainted with the process of earthing-up of maize crop at 'knee-high stage'.

Objective

- O to drain out the excess water either from irrigation or from rainfall,
- O to help in growth and production of crop,
- O to facilitate the irrigation application,
- O to collect optimum amount of water in the soils for crop growth.

Material required

❖ An irrigated line sown field of maize, having knee-deep plants, with 'optimum moisture', ❖ Nitrogen from urea @ 60 kg/ha, for 1000 sqm area, ❖ Spade (long handed) – 1, ❖ Sack – 1, ❖ Bowels – 4 (three big and one small), ❖ Balance (Field type spring balance, big), ❖ Note book, ❖ Pen/ pencil.

Procedures

- Spread the nitrogenous fertilizers (urea) after weighing, over the soil-surface in between the maize rows uniformly.
- Hold the handle of the long medium handled spade and scrap the soils straight in lines in between the rows at a depth of about 4-5 cm and spread the soils (mixture of soils and fertilizers) just at both sides of the furrows (over the root zones of the plants).
- Note down the date of earthing-up, depth of furrow, amount of N-fertilizer spread for 1000 m^2 area, stage of the crop, etc.

[**N.B.** : It is assumed that N, P_2O_5 and K_2O @ 60, 60 and 60 kg/ha, have already been applied as basal at the time of sowing, only nitrogenous fertilizer @ 60 kg/ha, for 1000 m^2 area should be added now].

■■■■

CHAPTER 6

SEEDS, SOWING, STAND ESTABLISHMENT AND CROP-CLASSIFICATION

6.1 SEEDS (BOTANICAL AND AGRONOMIC VIEW POINTS)

Botanically a *'seed'* is a fertilized ripened ovary/ovule, while agronomi-cally a *'seed'* or seed material or propagule, is the living organ(s) of the crop in rudimentary form that is used for propagation; or in other words, a part of the crop from which a new crop will grow.

6.1.1 Seeds as botanical aspect

Botanically *'seed'* may be defined in many ways such as : (1) 'Seed' is a fertilizer ripened ovule; (2) it is the part of a flowering plant that contains the embryo and develops into a new plant, if sown; (3) a 'seed' is a young plant packed and ready for transport to wherever it may be desired to start growing; (4) anatomically a 'seed' is a an embryo plant, consisting of a rudi-mentary stem and root together with a supply foods sufficient to establish a plant in a new location, all encased in a protective coat; (5) physiologically, 'seed' is essentially a meristematic axis with storage organs covered with the membranes and the other one is the seed coat; (6) 'seed' is a mature ovule in a dormant state of plant, where the metabolic, synthetic and morphogenetic activities are suspended and (7) 'seed' is a unit of reproduction of flowering plants in general.

6.1.1.1 *Shape*

Different types of crops showed different seed shapes. The seeds of wheat, rice etc. are to some extent elongated; seeds of mustard prickly poppy etc. are round; whereas seed of maize is square shaped. Seeds of *olitorius* jute are triangle shaped, whereas those of *capsularis* jute are spherical. The seeds of sesame, linseed etc. are oval, whereas those of oat is very elongated. The seeds of greengram, blackgram etc. are spherical, while those of grass pea, *arahar* are round and bigger in size.

6.1.1.2 *Size*

The seed of *phalaris minor* (*phalaghash/gulidanda*), is very small, while that of maize is big, whereas wheat, rice, grasspea etc. are medium, and that of groundnut, *rajma* are very big in size.

6.1.1.3 *1000-seed weight*

1000-seed weight of the crops in g, are as follows : (1) Wheat – 40-50, (2) Rice – 20-30, (3) Barley –30-40, (4) Oat – 20-25, (5) maize – 80-120, (6) Groundnut 300-400, (7) Mustard – 5-6, (8) Jute 6-10, (9) Sesame linsed 5-10, (10) *Phala ghass* – 0.2 to 0.8 etc.

6.1.2 'Seed' in agronomic view points

Agronomically, a 'seed' or '*seed material* or *propagule* is the living organ (s) of the crop in rudimentary form that is used for propagation, or in other words, any part of the crop from which a new crop will grow.

6.1.2.1. *Setts (sugarcane), tubers (potato), stem cutting (sweet potato), popagules (banana and arum), corm (ol)*

Agronomically setts (sugarcane), tubers (potato), stem-cutting (sweet potato, paragrass etc.), propagules (banana), rhizomes (*Amorphophallus / ol*) etc. are the '*seed materials*' besides the actual 'seeds' of wheat, rice, jute, mustard, sesame, maize, groundnut etc.

Setts (sugarcane) : In sugarcane, seeds material '*sett*' is a top portion of sugarcane with both nodes and internodes, having 'eyes' on it. Setts may be placed in furrows or trench or seedlings and may be raised in puddled soils, for planting just before the harvest of potato or other crops, under multiple cropping systems.

Tubers (potato) : Smaller size of potato is the '*tuber*' which is the '*seed materials*' of potato. Of course, now-a-days *TPS* (true potato seeds) are used in potato cultivation, after raising seedlings of potato from seeds. Of course, this practice is good for the places, where long duration winter is available, but not suitable for the areas for short duration winter, like eastern part of the country.

Stem cutting (sweet potato) : '*Stem cutting*' is the seed material for planting sweet potato, paragrass etc. Generally 15-30 cm. long cutting of stems are placed in furrows made earlier, and the vine is then covered with the soils, keeping the leaves above the soil. The cuttings will survive, if moisture is available in soils. If the land is dry, irrigation should be given. Generally sweet potato is rainfed crop. Hence it should be planted in April/ May, after rainfall in eastern part of the country. Paragrass may be planted with stem cuttings, placing the cuttings slantingly in puddled lands.

Propagules (banana/arum) : *Propagules* are the '*seed materials*' for some crops, which do not bear fruits/seeds. Suckers/runners etc. are the propagules of some plants. Banana is propagated by *suckers*. Hence, *sucker* is the seed materials of banana crop and similarly in arum crops.

Corms (*amorphophallus/ol*) : The smaller size of *corms* of *amorpho-phallus* (*ol*) are the '*seed materials*'. In order to propagate *ol*, this '*seed material*' (corm) is used. The '*seed material*' is used to multiply or production of *amorphophallues* and put into the pits after using organic manures, in the months of April/May. This rainfed crop grows better if rainfall is high in pre-*kharif* /*kharif* season.

Practical Class : Collection and observations of sugarcane-setts, tuber, stem-cuttings, propagules and corm, from time-to-time in different seasons

Lesson 53 : Collection and observations of *setts* of sugarcane for planting

Introduction

Different crops are propagated either by seeds or by some vegetative means. The *setts* in sugarcane are the vegetative parts, which are used in propagation (i.e. cultivation of sugarcane). The sugarcane setts are of different types. Generally, the upper portion of the stem of sugarcane, is used for setts, with 2-4 internodes. Generally setts with 2 internodes are used for raising seedlings for planting after puddling the nursery-bed immediately, under transplanted condition. But setts of sugarcane with more internodes (long setts) are also planted, as such, in furrows or trenches during autumn or summer planting. The learners should know about this sett, after collection from the sugarcane fields and for that purpose, observations are needed.

Objectives

O to propagate sugarcane easily in cultivation,

O to identify the portion fort the selection of proper *sett*,

O to identify the 'eyes' of the sett of sugarcane,

O to identify old/medium/new portion the stem of sugarcane,

O to identify the length and strength of the cane,

O to identify the maturity condition of the stem.

Material required (for collection and observations)

✣ A well-grown sugarcane field, ✣ Sickle – 1, ✣ *Hasua* – 1, ✣ Scale (15 cm or 30 cm length) – 2, ✣ Lens –1, ✣ Field balance –1 (1-2 kg capacity), ✣ Note book, ✣ Pen/ pencil.

Procedures : (Collection and observation)

- From sugarcane field, select a sugarcane of medium size at maturity,
- Cut the sugarcane from the base,
- Remove all the leaves without the damage of the *'eyes'*,
- Remove the top-leave portion of the cane,
- Remove the lower portion where buds/eyes are damaged or rudimentary,
- Thus the top portion (except the extreme top), is selected as *sett* for cultivation,
- Note down date of collection and observation along with others like, length of sett, eyes, conditions etc.

Lesson 54: Collection and observation of `tubers' of potato for planting.

Introduction

The tubers of potato are the 'seeds' in agronomic sense, because by this tubers, potato production is raised, though tubers are the modified stems. Therefore, the learners should know how these tubers are collected what are the observations, what should be the criteria as *seeds* in potato production etc. Of course, true potato seeds (TPS) are used now-a-days.

Objectives

O to utilize the tubers for potato production,

O to minimise the 'seed rate',

O to look into the 'eye' of the tubers,

O to select disease-free medium sized tubers.

O to get more yield.

Material required

✣ A disease-free potato field under maturing conditions, ✣ Spade–1, ✣ Small plough–1, ✣ Balance –1, (Field type, spring), ✣ Lens –1, ✣ Scale (15 cm or 30 cm length), ✣ Note book, ✣ Pen/pencil.

Procedure (Collection and observation)

- Scoop-out the ridge portion of potato-bed carefully, with the help of spade,
- Pickup all types of potato available there,
- Pickup the middle-sized potato (about 75 to 100g), free from disease,
- Examine the *eyes* with the help of lens,
- Measure the length, breadth /diametre of the tuber,

(NB : Middle-sized disease-free potato, having about 100g weight, should be selected for cultivation),

- Note down all the characters like, length, width/diametre, more or less of special size and weight,
- Note down the date of collection and observation with characters of tubers for cultivation like number of *eyes* etc.

Lesson 55 : Collection and observations of 'stem-cutting' of sweet potato and paragrass for planting

Introduction

Some crops are grown with vegetative propagation, particularly with the stem *cuttings*, as in sweet potato or in paragrass (a fodder crop). Of course *stem-cuttings* are planted in different process under different conditions/ situations. Stem-cuttings segments of long vine in sweet potato, are planted in deep soil furrows, under rainfed condition, in summer/pre-*kharif* or *kharif* season; whereas stem cuttings of paragrass (small pieces of long stem), are planted under puddled or no-tilled murshy conditions at rainy season, in a slanting manner, for commercial cultivation.

Objectives

O to propagate either sweet potato or paragrass for commercial purposes,

O to get more yields of the crops,

O to economise the production of cost.

Material required

✣ A disease-free field of sweet potato/paragrass, ✣ Sickle –1, ✣ Spade – 1 ✣ Field spring balance –1, ✣ Furrow opener, ✣ Lens –1, ✣ Scale (15 cm or 30 cm length), ✣ Plough, with all arrangement for ploughing, ✣ Note book ✣ Pen/ pencil.

Procedure

(a) Sweet potato : Collection and observation

- Select the disease-free plants with vigorous growth,
- Cut the vine into long pieces 30 or 60 cm keeping, 3-4 leaves,
- Make the deep furrows on ploughed land,
- Put the segmented vines on the furrows continuously, with a spacing of about 50-60 cm. apart in lines,
- Cover the vines with side soils, keeping the leaves above the soils and pressing the soils lightly.

(b) Paragrass : Collection and observation and planting

- Select some healthy plants of paragrass from the field,
- Observe the eyes on the cuttings,
- Measure the cuttings,
- Puddle the land for transplantation or keep the low marshy lands under rainfed conditions,
- Under puddling in marshy conditions, put the *cuttings*, with 2-3 leaves, in muds by pressing the cuttings with the hands slantingly, so that one internodes must attached with the soils for rooting (as root will come from the internodes),

[**Note :** Roots come from the vines continuously in sweet potato, whereas roots come from the internodes only in paragrass].

Lesson 56 : Collection and observations of banana 'propagules' for planting

Introduction

From rhizome of banana, little 'banana-like-plant' (*teur*) comes out and this is *propagule* which is actually used for propagation for commercial cultivation of banana. The young learners should know the cultivation of banana and for this, collection and observations of *propagules* are essential.

Objective

- O to cultivate banana commercially,
- O to shorten the cultivation time/period,
- O to get more yields,
- O to economise the cost of cultivation,
- O to raise the ratoon crop year-after-year.

Material required

❖ A field of disease-free banana plant, ❖ Spade –1, ❖ *Khanta* –1 (a long wooden handle with sharp end points), ❖ Sickle –1, ❖ Bullocks (2), ❖ Ploughman (1), ❖ Necessary arrangement for ploughing, ❖ Big field spring balance – 1, ❖ Note book, ❖ Scale –1, ❖ Tape –1, ❖ Pen/ pencil.

Procedure

(a) Collection

- Select the disease-free small banana *propagule* from the well-grown plants in fields,
- Separate the *propagule* from the mother's body, with the help of spade/*khanta* or sickle with its rhizomes carefully so that rhizome is not damaged (damaged rhizome of the propagules will not serve the purpose later, hence carefully it should be done).
- Cut-off the old pale leaves, keeping only 2-3 leaves.

(b) Observations

- Propagule that has small but intact rhizome,
- Roots (big) are around the rhizomes,
- Some pale and green leaves,
- Stem (*Pseudo*),
- Leaf (big) with big ribs.

(c) Plantation

- Plough the land where planting will be done 2-3 times, followed by ladderings,
- Make a deep furrow at a space of 5 m apart in lines,
- Place the propagules on the line in furrows, at a space of 5m, after addition of organic manure in the furrows,
- The hole or furrow is covered with soils from sides (as was deposited earlier while furrow or pitt was prepared).

Lesson 57: Collection and observations of 'corm'/*ol* (*amorphophallus*) for planting

Introduction

The head-like-structure i.e. rhizome of the *amorphophallus* (*ol*) is the *corm*. This corm with carbohydrate and starch, is a good vegetable in India. In some African countries it is the major food like yam and others. This small *corm* is the seed material for commercial cultivation of *amorpho-phallus*. Therefore, the learners must know its cultivation fully. For this purpose, it is essential to know the observations and cultivation of *ol*.

Objective

O to multiply *amorphophallus* in commercial cultivation,

O to get higher yields,

O to retain as `seeds' for future use.

Material required

✣ A disease-free field of *amorphophallus*, ✣ Spade –2, ✣ Sickle –2, ✣ Baskets – 4, ✣ Bulky organic manure, ✣ Long coconut coir/string, ✣ Tine/ hoe, ✣ Big field-spring balance (20 kg capacity), ✣ Tape –10m-1, ✣ Scale – 30 cm - 1, ✣ Bullocks (2), ploughman 1, set for ploughing, ✣ Note book, ✣ Pen/ pencil.

Procedure

(a) Collection

- Select the disease-free small/medium sized *amorphophallus* plants,
- Remove the soils around the plants, and dig out the corms, with the help of spades carefully, so that the corms are not damaged by spading,
- Remove the pseudo-stem with the sickle,
- Remove the soils around corm by hand,

(b) Observation

- The corm, the head-like structure, is small, slightly depressed on the top where the pseudo-stem was attached,
- No roots around the corm,
- Weight of corm should be 1-2 kg,
- Colour of the corm is brownish to blackish,
- The entire body (corn), though hardy but turgid,
- Ready for vegetable-use.

(c) Plantation

- Plough the land 2-3 times, where the *amorphophallus* would be planted followed by ladderings, in April/May,
- Make straight furrows in lines at a space of 1m apart with tines/ hoe,
- Dig the points (ditch/pit) at a space of 1m on the furrow, for putting corm,
- Apply about 5-10 kg of bulky organic matter into each ditch/pit and mix it with the soils,
- Put the corms into the pit and cover them with soil from sides around the corm and press it gently,
- Cover the pits with some mulches (rice-straw).

6.2 SOWING

'Sowing' is the first step for any crop cultivation, just like the birthday of human beings. So it is very important. Duration of any crop is counted from sowing to harvest. Of course, sowing methodology differs from crop-to-crop, season-to-season and place-to-place. Sowing of cereals, pulses, oilseed, fibres, sugars and fodder crops are discussed.

6.2.1 Cereals (Rice/wheat/maize/*jowar*/*bajra*/barley/oat)

Rice, wheat, maize, barley, *jowar*, *bajra*, oat etc. are the main crops in cereal group. The sowing process of wheat, barley and oat are the same, whereas it varies in rice, maize, *jowar* and *bajra*, though *jowar* and *bajra* are under same sowing process.

Wheat, barley and oat : Wheat, barley and oat are sown more or less by the same method in *rabi* season. First seed-beds are prepared with 3 to 4 ploughings. Under irrigated condition 100 kg seed/ha is needed. Of course, in most of the areas wheat is grown under irrigated condition, whereas barley and oat are grown under rainfed condition. Oat and barley required 100 kg seed rate. Normally, even under mechanization, wheat is sown in lines at a space of line-to-line 20-25 cm and plant-to-plant 10-15 cm apart, whereas oat, barley rainfed wheat seeds are broadcast just before last ploughing and covered with ladderings/planking under rainfed conditions. All sowing programmes are completed by end of November as they are winter crops.

Rice : Rice is a cereal crop which is grown mostly in *kharif* season, under transplanted waterlogged conditions. Twenty-to-Thirty-day seedlings of rice are transplanted by the end of July, at 25 cm ×10 cm spacing. Before transplantation, puddling of soil with water, is done by giving 2-3 ploughings. In each hill (the planting point) 3-4 seedlings are transplanted in lines.

Maize : In well prepared seed-beds, having proper moisture, furrows are opened with tines/ploughs at a distance of 50-60 m apart and viable dry (high germination percentage) seeds are dropped manually on points at a space of 10-15 cm in the furrows apart, during May/June and cover the seeds with the soils manually.

***Jowar* and *bajra* :** Seeds of *jowar* and *bajra* are broadcast over the well prepared seed-beds just before the last ploughing and the seeds are covered by laddering or planking, in the month of June/July.

6.2.2 Pulses (Lentil/pea/blackgram/gram/greengram and grasspea)

Most of the major pulses like, pea, lentil, gram etc., are grown in lines, 25 cm apart with a seed rate of 30 kg/ha, in well-laid-out fields in earlier part of November but under similar conditions greengram/black gram in March/ April, (in June/July for blackgram) are grown under rainfed conditions. Of course, the tall plant *arahar* is grown under similar conditions in March/ April. Grasspea, lentil, gram etc. could be grown as *paira/utera* crop also in *rabi* season.

6.2.3 Oilseeds (Mustard/rape, linseed, sunflower, groundnut and sesame)

Oilseed crops are of different types, hence sowing methods are also different. Mustard/rape, linseed are grown in *rabi* under rainfed condition, by broadcasting seeds over well prepared seedbeds, just before last ploughing with seed rate of 8-10 kg/ha by middle of October. Sunflower is grown in March/April with a seed rate of 50 kg/ha, by broadcasting, in well moistured seed-beds. Groundnut is grown in June/July with or without shell, with a seed rate of 50-60 kg/ha in lines, 50-60 cm apart, in well prepared seed-beds, having adequate moisture.

6.2.4 Fibres (Jute, cotton, flax, mesta and sunnhemp)

Out of all fibre crops (jute, cotton, flax, mesta and sunnhemp), flax is grown in *rabi* but others (jute, mesta and sunnhemp), are grown in April/May or May/June. Jute, mesta and sunnhemp are the crops of eastern parts where rainfall is high, whereas cotton is the crop for drier areas, grown with irrigation. Jute, mesta and sunnhemp, of course, are grown in rainfed condition by broadcasting with a seed rate of jute - 10-12 kg; mesta and sunnhemp - 14-16 kg/ha; flask is grown by the end of November with a seed rate of 10-15 kg/ha by broadcasting.

6.2.5 Sugars/ and tubers (Sugarcane, sugarbeet and potato)

Sugarcane and sugarbeet are the sugar crops and potato is the tuber crop. Sugarbeet and potato are grown in *rabi* whereas sugarcane is grown during the whole year, though planted in pre-*kharif* /*kharif* season. Sugar-cane is grown either in March/April or September/October in trenches with high moisture.

Potatoes are sown in November with a seed rate (tubers) of about 20-25 q/ha, in furrows, made at a space about 50-60 cm apart in irrigated well prepared seed-beds.

Sugarbeet of course, is sown by the end of November in furrows in lines 40-50 cm. apart, with a seed rate of 10-15 kg/ha, in well moistured irrigated fields.

6.2.6 Fodder/forage (Napier, thinnapier, berseem, teosinte, lucern, blackgram and grasspea)

Among the fodder/forage crops, napier, thinnapier, berseem, lucern, teosinte, blackgram and grasspea are the main.

Berseem, lucern and grasspea are grown in *rabi* season, while, napier, thin napier, teosinte and blackgram are grown in pre-*kharif* /*kharif* seasons. Berseem and lucern are sown with a seed rate of 10-15 kg/ha in irrigated fine tilth, well moistured soils in the middle of November, while napier, thinnapier and teosinte are grown by the middle of April/May, in irrigated areas with seed rate of napier and thinnapier 10-15 kg and of teosinte 20-30 kg/ha. Grasspea is grown in rainfed areas with a seed rate of 30 kg/ha, by broadcasting.

Practical Class : Sowing of wheat, *aus* rice, jute, potato planting, planting of sugarcane and transplanting of rice

Lesson 58 : Sowing of wheat in *rabi* season.

Introduction

Sowing wheat, the winter-loving crop, is very important, because it is the second cereal crop in the country and in some places it is the first cereal crop. Its higher production is dependent on so many factors like sowing date, NPK-fertilizers, irrigation, soil management, weed control etc. Its higher production is very helpful to more or less to all concerned. The learners, should therefore, must know the sowing process of wheat. It is normally sown in *rabi* by the middle of November with the seed rate of 100 kg/ha in irrigated high lands with NPK fertilizers @ 100-50-50 kg N, P_2O_5 and K_2O/ha. It should be sown by the middle of November in major wheat growing zones, though in eastern part, wheat may be sown at the end of November. In the northern hilly regions this may be sown in October also.

Objective

- O to get higher yields of wheat per unit area,
- O to obtain higher number of effective tillers, higher number of grains/ plant and higher seed weight, and
- O to adjust the other crops, in the 3-crop systems,

Material required

❖ Irrigated high land after 2 ploughings with adequate moisture, ❖ Seeds of particular variety for 1000 m^2 area @ 100 kg/ha seed rate, ❖ NPK-fertilizers (basal) @ 50 (urea) kg N, 50 kg P_2O_5 (SSP) and 50 kg K_2O (MOP)/ ha, for 1000 m^2 area, ❖ Sacks-4, ❖

Bowls-4, ✤ Balance (Spring, field type, big) –1, ✤ Bullock (2), ploughman-1, with accessories, ✤ Ladder/plank, ✤ Note book and ✤ Pen/pencil.

Procedure

- Give the third plough with *desi* plough under adequate moisture,
- Broadcast the calculated amount of nitrogenous fertilizer (urea) uniformly on soil,
- Mix the required amount of P_2O_5 (SSP) and K_2O (MOP) fertilizer together, and broadcast again over ploughed land uniformly.
- Sow the wheat seed (as per amount) with *kera/pora* method where mouldboard plough should be used so that seeds are covered by next row sowing.
- Note down the date of sowing, method of sowing, tillage and laddering number, seeds amount, fertilizer amount, types of fertilizers etc. in the note book.

[**Note** : The rest 50 kg N/ha, should be applied as `top-dressing', at the first irrigation at CRI stage i.e. 21 days after sowing wheat].

Lesson 59 : Sowing of (a) direct-seeded rice (*aus*) and (b) jute in pre-*kharif* season, under rainfed conditions

(a) Direct-seeded rice (*aus* rice)

Introduction

Direct-seeded rice i.e. *aus* rice, is grown in rainfed conditions in pre-*kharif* (summer) season, particularly in eastern parts of India, where rainfall is available. The learners must know the sowing methodology of direct-seed rice in pre-*kharif* season particularly in eastern areas of India.

Objective

O to establish rice crop under rainfed conditions,

O to get more yield of cereal crops annually,

O to fit in the cropping system,

O to economise the cost of production,

O to utilize the rain water for maximum production of cereals.

Material required

✤ A medium situated-land, having medium soil fertility, ✤ Bullocks (2), ploughman –1, *desi* plough with accessories, ✤ Ladder/plank–1, ✤ NPK-fertil- izers (as calculated) @ 80 kg N, 40 kg P_2O_5 and 40 kg K_2O/ha for 1000 m^2 area, ✤ Seeds of rice of any variety, as calculated @ 100 kg seed rate/ha, for 1000 m^2 area, ✤ Wheel-hoe –1, ✤ Balance –1, ✤ Sacks – 4, ✤ Bowel– 4, ✤ Note book, ✤ Pen/pencil.

Procedure

- Plough the land two times, followed by 2 ladderings,
- Apply NPK-fertilizers (as calculated) immediately after the third ploughing, @ ½ of N, full of P_2O_5 and full of K_2O/ha uniformly, by broadcasting,
- Broadcast the calculated amount of seeds of any rice variety, uniformly after fourth plough No ladderings should be done after third and fourth ploughing.

(**Note :** Stagnation of water from rain is essential for good growth and good yield of rice. In pockets of seed beds, water will remain standing if ladderings are avoided; this water is very helpful).

- Note down in the note book the date of sowing of *aus* rice, the amount of N, P_2O_5 and K_2O fertilizers required, amount of seeds required. Record the amount of rainfall and date of rainfall in pre-*kharif* season.

(b) Jute

Introduction

Jute, the fibre crop, is generally grown in the eastern parts of the country in pre-*kharif* season, as this crop requires high rainfall, high humidity and high temperature, which are available in this region, during that period. This crop improves soil fertility, though not leguminous crop, in respect of nitrogen, organic carbon and available phosphate. This crop fits well in multiple cropping system in the region. Therefore, the learners, (particularly of eastern zone) should know the technique of jute cultivation.

Objective

O to get more fibre production for the industries,
O to utilize the climatic condition of the region in pre-*kharif* season,
O to improve the soil fertility,
O to be fit the crop in multiple cropping systems,
O to help the paper industry by providing jute-stick for manufacturing of paper/board,

Material required

✥ A land (low/medium/high), having medium soil fertility, ✥ Bullocks (2), ploughman –1, *desi* plough, ✥ Ladder/plank, ✥ NPK-fertilizers (as calcul-ated) @ 80 kg N, 40 kg P_2O_5 and 40 kg K_2O/ha, for 1000 m^2 area, ✥ Seeds (as calculated) @ 10 kg seed/ha, of any variety of *corchorus capularis/C. olitorius*, for 1000 m^2 area, ✥ Wheel-hoe/tine, ✥ Sacks – 4, bowel – 4, ✥ Spring balance (1, 2 kg capacity), ✥ Note book, ✥ Pen/pencil.

Procedure

- Plough the land two times, followed by 2 ladderings, in the later part of April,

- Apply NPK-fertilizer @ 40 kg N (urea), 40 kg P_2O_5 (SSP) and 40 kg K_2O (MOP)/ha (for N-fertilizer separate and then P_2O_5 & K_2O jointly) immediately after the third ploughing, as basal uniformly, by broadcasting, then ladder the soils,
- Broadcast the calculated seeds of *olitorius* jute @ 10 kg, for 1000 m^2 area, immediately after the fourth ploughing, uniformly,
- Apply the laddering after sowing,
- Note down the date of sowing, amount of seeds, amount of NPK-fertilizers, number of tillages, variety used, condition of moisture in the soils etc in the note book.

Lesson 60 : Planting of sugarcane in autumn (i.e. before winter)

Introduction

Sugarcane, the sugar crop, is cultivated twice a year, according to *eksali* (one year) or *adhsali* (one-and-a-half year) crop. From this sugarcane both sugar or *gur* is obtained. Bagasse is the by-product after crushing it. This bagasse is also used for board preparation etc. *Eksali* crop is planted in autumn i.e. after *kharif* but before winter. The learners should know the cultivation technique of sugarcane in autumn. Bagasse is also used for increasing soil fertility and mulching.

Objective

O to get sugar/*gur*, for the country,

O to get more yield of sugarcane,

O to improve soil fertility,

O to use the bagasse and leaves, as mulch.

[**Note :** Sugarcane in autumn is cultivated either in trench system or flat system. Here trench system will be followed].

Material required

✤ A medium or high land with medium soil fertility, ✤ Spade –2, ✤ Sickle – 3, ✤ Setts – agronomic seeds, (as per need for 1000 m^2 area), ✤ NPK-fertilizers (basal, calculated) @ 120 kg N, 60 kg P_2O_5 and 60 kg K_2O/ha, (Basal 60 kg N, 60 kg P_2O_5 and 60 kg K_2O/ha), ✤ Spacing continuous setts × 1.5 to 2 m in rows, ✤ Scale - 1, ✤ Tape –1, ✤ Balance – 1, (spring balance, 1, 2 kg capacity), ✤ Sacks – 10, ✤ Note book, ✤ Pen/pencil.

Procedure

- Cut the trench at a depth of 25 cm and width of 15 cm. with a spade, in lines at a space of 1-2m, depositing the soils on two sides of the trench,
- Put the setts (with 3-4 eyes) of sugarcane with about 30 cm. length, into the trench, keeping the eyes of the setts in the sides of the trench (so that the eyes can come out easily),

- Half of the soils, deposited in sides, should be used for covering the *setts* for aeration and moisture, for 'germination' / i.e. spreading the *eyes* for forming seedlings,
- After emergence of seedlings, nearly after one month, remaining soil should be used for further covering the base of seedlings.

Lesson 61 : Planting of potato in winter

Introduction

Potato, the major vegetable crop in India, is grown in plains during winter and during summer in hilly regions. As because, this is the important crop in the fields, the learners should know the cultivation process of potato in the plains. Though, variations are observed in cultivation of potato in plains and hills, but mainly ridge and furrows system is followed every where, because of nature of crops. Potato, the modified stem, grows in the under ground, just by the sides of the mother plant, below the soil surface.

Objective

O to get more yield of tubers (potato),

O to get rid of waterlogging,

O to enhance growth in low temperature,

O to obtain bigger sized potato,

O to fit in multiple cropping systems,

O to look for clear cultivation.

Material required

❖ A higher or medium land with loamy soils having medium soil fertility, ❖ Potato seed-tubers for planting as calculated for 1000 m^2 area @ about 8 q/ha, ❖ Bullocks (2), ❖ Ploughman –1, with *desi* plough having all acces-sories for ploughing, ❖ Ladder –1, ❖ Spade –2, long handled, ❖ Wheel-hoe/ tine –2, ❖ Bowels – 4, ❖ NPK-fertilizers @ 120 N, 60 P_2O_5 and 60 kg K_2O/ha (but only 60 N, 60 P_2O_5 and 60 kg K_2O/ha as basal), ❖ Spring balance – 1, 2 kg capacity, ❖ Note book, ❖ Pen/pencil.

Procedure

- Plough the land three times, followed by 3 ladderings; soil should be highly pulverised for good tilth,
- Apply NPK-fertilizer @ 60 kg N, 60 kg P_2O_5 and 60 K_2O/ha, for 1000 m^2 area, after ploughing again, followed by ladderings,
- Make the furrows with hoe/tines in lines at a space of 60 cm. apart,
- Put the tubers (40-50 gm each) in the furrows 10 cm apart,
- Cover the tubers with soils with spade from the sides slantingly,

- Give flash irrigations, at 3-4 days intervals, till seedlings are emerged, from the main channels.

Lesson 62 : Transplanting of *kharif* rice

Introduction

'Rice' is a water-loving crop. It is grown mostly under transplanted conditions, after puddling. *Puddling* is done with the stagnation of rain/ irrigation water, standing on the fields, with *desi* plough, in order to breakdown the soil structure and to minimise water percolation loss from the soils. Under these conditions, 3-4 ploughings are given, followed by 3-4 ladderings, creating the muddy condition, with slight accumulation of water above the soils. Then rice seedling is transplanted in the puddle land. The learner should be acquainted with this process of rice cultivation thoroughly.

Objective

O to breakdown the structure of the soil,

O to check the percolation loss of water, from the field,

O to get more growth and grain yield of rice.

Material required

❖ Bullocks (2), ploughman –1, *desi* plough with accessories, ❖ Ladder/ plank, ❖ Weighing field spring balance (1,2 kg capacity), ❖ NPK-fertilizer (N, P_2O_5 and K_2O @ 100, 50 & 50 kg/ha from urea, SSP and MOP respectively), ❖ Sacks – 4, ❖ Bowels –3, ❖ Coir/string 500 gm, ❖ Scale –2, ❖ Tape –1, ❖ Labours (for transplantation) – 10, ❖ Seedlings of rice (21 day–old), (as per 1000 m^2 area), ❖ Note book, ❖ Pen/pencil.

Procedure

- Collect water (either from rain or irrigation) about 6-9″ above the soil surface and keep it for seven days to decompose the weeds and to soften the soil,
- After seven days plough the land in water 3 times, followed by 3 laddering/ planking, so that land is in muddy condition but having some amount of water (say 4-5 cm) on the soil surface,
- Plough again for the fourth time in the same way, after applying measured amount of fertilizers @ 50 kg N, 50 kg P_2O_5 and 50 kg K_2O/ha, for 1000 m^2 area, by broadcasting, followed by 2 ladderings. Now the land is ready for transplantation,
- Some bunches of 21-day-old seedlings should be spread over the puddled land,
- Seedlings should be transplanted in lines 25 cm apart and plant-to-plant space should be 10 cm. 3-4 seedlings should be put in between three fingers and the fingers with seedlings should be put inside the muddy soil and leave that with attaching them with mud. It requires a good practice.

- Note down date of puddling, transplantation, seedling number/hill, amount of NPK-fertilizer, spacing (25 cm × 10 cm) etc.

6.3 STAND ESTABLISHMENT

Stand establishment is very important step for higher production of crops. Lower stand establishment per unit area gives lower yield, while higher stand establishment per unit area, gives higher yields of crops.

The stand establishment of crops, depends on so many factors such as (i) soil moisture, (ii) quality of seed /seed materials and (iii) timely management of crops with (a) irrigation application, (b) weed control, (c) fertilizer application etc., (d) timely sowing, (e) proper spacing etc. All these reflect on germination and later on population per unit area and higher yields.

6.3.1 Germination

'*Germination*' of any crop, is resultant effect of soil moisture, quality of good seeds/seed materials/propagules, time of agronomic management, timely good tillage, sowing, proper spacing etc. Among all the factors, soil moisture and quality of seeds are the most important steps for germination/ emergence of crop.

6.3.1.1 *Monocot plant*

Monocot-crops like rice, wheat etc. give fibrous roots system, for which '*good tilth*' of soil is needed in general, though deep ploughing is not necessary all the times. For germination of most of the monocot-crop seeds, soil moisture is most important factor. Some crops germinate with low content of moisture (as barley etc.), some need more content of moisture in soils (as in wheat).

6.3.1.2 *Dicot plant*

Dicot-crops like cotton, jute, *arahar,* greengram, etc., having the main long single tap roots, penetrate into the soils, for which deep ploughing is needed, along with well pulverised soil for secondary root-ramification. In dicot also, some crops require more moisture in the soil (as in jute) to germinate the seeds and some require less moisture in soil (as in sunflower).

6.3.2 Plant population

Plant population per unit area, is the final form of stand establis-hment of any crop. It depends upon firstly on germination and secondly on agronomic managements, such as space, seed rate, fertilizer, irrigation, weed control etc. along with soil adequate moisture for seed germination.

Providing more space : Too much space decreases the plant population per unit area. Therefore, that should be avoided. For wheat, generally ideal spacing is 22-25 cm × 10 cm, beyond that the population will be less, hence yield will be less.

High seed rate : Sometimes population is increased by giving higher seed rates, particularly under late sowing condition (as in wheat). Normally, 100 kg seed rate, under normal sowing, is the best, whereas for late sowing it should be 125 kg in wheat.

Higher NPK-fertilizers/organic matters/ irrigation etc. : Plant population in per unit area, may be increased due to application of manures, fertilizer (NPK), irrigation, weed control and timely sowing conditions. Normal plant population in per square metre area is : wheat – 400-500; jute 40-45, mustard 15-30 etc.

Practical Class : Increasing the plant population by (i) giving high seed rates and (ii) using higher NPK-fertilizer levels

Lesson 63 & 64 : Sow wheat with 50 kg and 100 kg seed rate per hectare in two plots and observe the tiller numbers/m^2 at about 45 days after sowing and compare the population between two plots

Sowing with seed rate : Sowing of wheat with different seed rates for increasing plant population.

Introduction

Generally, single seedling comes out from single seed in wheat. Therefore, it is obvious that increasing of seed rates, increases the plant population. Besides this, at the active tillering stage, tillers/plant increase heavily. Hence, more seed rates mean more basic seedlings plus more tillers/ sq. m. and ultimately more plant population (i.e. tillers/sq. m) are available. The learners should visualise practically, these observations :

Objective

O to increase the plant number per unit area,

O to increase the grain yield of wheat per unit of wheat.

Material required

✥ Bullocks (2), ploughman –1, plough (mouldboard) with accessories, ✥ Wheat seeds (of any variety) for 500 × 2 areas @ 50 and 100 kg seed rate/ha, ✥ Medium irrigated land, with average soil fertility, ✥ NPK-fertilizer, for 1000 m^2 area (500 + 500 for each treatment), @ 100 kg N, 50 kg P_2O_5 and 50 kg K_2O/ha (i.e. 50 kg N, 50 kg P_2O_5 and 50 kg K_2O/ha as basal), ✥ Sacks 4, ✥ Bowels – 4, ✥ Ladder –1, ✥ Spade (long handed) – 1, ✥ Tine/wheel-hoe, ✥ Pegs – 8, ✥ Scale (1m) – 1, ✥ Balance (spring type, for field, 10 kg capacity), ✥ Note book, ✥ Pen/ pencil.

Procedure

- Plough the land 2 times, after optimum moisture in the field, followed by ladderings,
- Apply the third plough on the same day as above,

- Apply NPK-fertilizer for 1000 (500 + 500) m^2 area @ 50 kg N, 50 kg P_2O_5 and 50 kg K_2O, as basal just after 3rd ploughing,
- Separate the plot A (50 kg) and B (100 kg) seed rate/ha,
- In both the plots, land should be levelled with ladderings/planking, and wait for 45 days,
- Sowing should be done with mould board plough,
- After 45 days of sowing wheat, in both the plots demarcate 4 areas putting 8 pegs in 1 m meter-row only,
- Separate the land with 500 m^2 area each in two pieces by making a furrow with spade,
- In one plot (A) sow wheat seeds @ 50 kg/ha and in another (B) sow wheat seeds @ 100 kg seed/ha, after ploughing,
- Count the tiller numbers from length area in between two pegs in both plots and convert the figure x 4 and put the record (i.e. same figure) on the following table:

Assignment

Observation on plant population (i.e. tillers/m^2)

Plot No. and seed rate in wheat		Tiller number per m^2 area (per samples)				
		1	2	3	4	Mean
(A)	50 kg seed rate					
(B)	100 kg seed rate					

Lesson 65 and 66

Apply NPK fertilizers @ 100 kg nitrogen, 50 kg phosphate and 50 kg potash per hectare in wheat, (using ½ nitrogen and full of phosphate and potash at sowing, as basal and rest ½ nitrogen use as top dressing just before first irrigation at CRI stage (i.e. 21 days after sowing) and in another plot with no fertilizers. About 45 days after sowing wheat, take observations, on tillers/m^2 in both the plots and compare the population (i.e. tillers/m^2) to see whether there is increase or decrease of the population of wheat with comments.

Sowing with high-fertilizers (NPK-fertilizers) application in wheat

Introduction

Tiller numbers increase, with the increase of NPK-fertilizers doses, ultimately grain yields of wheat increase. Therefore, the learners should know the fertilizer application method to wheat and increase or decrease of tillers there from. The observations are under assignment.

Objective

O to get more number of tillers, due to higher doses of NPK-fertilizers,

O to increase the yields of wheat.

Material required

✤ A medium irrigated field, having medium soil fertility, ✤ Bullocks (2), ploughman – (1), mouldboard (1) with accessories for ploughings, ✤ Ladder – 1, ✤ Seeds of wheat for 1000 sq/m. area @ 100 kg seed rate/ha (for 2 plots), ✤ NPK-fertilizers for 500 m^2 area, @ 50 kg N, 50 kg P_2O_5 and 50 kg K_2O/ha (as basal) (for one plot only), ✤ Sacks – 4, ✤ Bowels – 4, ✤ Balance (Field, spring balance, with 1,2 kg capacity), ✤ One metre scale-1, ✤ Pegs/sticks–8, ✤ Note book, ✤ Pen/pencil.

Procedure

- Apply 2 ploughings, followed by ladderings,
- Give the third ploughing and separate the area plot into 2, measuring 500 m^2 each. 'A' plot will be without any fertilizer and in the second plot- B, NPK-fertilizers will be applied,
- Apply N-fertilizer (say urea), on the B plot first and then both P_2O_5 (super phosphate) and K_2O (muriate of potash) together, followed by ladderings in the same plot. A plot will be without fertilizers.
- Plough again for the 4th time and sow the wheat seeds, in both the plots with mouldboard plough, line 25 cm apart, then wait for 45 days.

Recording observation on, tillers of wheat (due to variation of fertilizer apply) 45 days after sowing

Introduction

Generally 'recording observation' is a research oriented programme, but here, it has been done for comparison only, between fertilizer and no-fertilizer. Tillers in wheat, generally increase with fertilizerised plot. Comparison statement has to be visualised on tiller numbers of fertilized or non-fertilized wheat plots. The learners should visualise this comparison in terms of tillers number of wheat. Wheat should be sown with the spacing of 25 cm × 10 cm (i.e. row-to-row 25 cm and plant-to-plant 10 cm).

Objective

O to see the increased difference in tiller numbers which ultimately will increase the grain yield of wheat (as tiller or effective tiller are the components of grain yield of wheat), and

O to get more grain yield.

Material required

✤ Plot A and B, ✤ Wheat seeds for 1000 sqm (500 + 500) area @ 100 kg seed rate/ha, ✤ 1m scale, ✤ NPK-fertilize, @ 50 kg N (urea), 50 kg P_2O_5 (SSP) and 50 kg K_2O

(MOP)/ha as basal for 1000 m^2 (500 + 500), ✤ Pegs – 8, ✤ 50 kg N per hectare (Urea) as top dressing at CRI stage of wheat crop, ✤ Note book, ✤ Pen/pencil.

Procedure

- The plots 'A' and 'B', to be made available (A = without fertilizers and B = with fertilizers),
- One metre row should be pegged at two ends, in 4 places, in each plot (A+B), randomly,
- Count the number of tillers in wheat in both the plots, after 45 days of sowing wheat,
- Multiply the figure obtained by 4, for getting the figure in m^2 area,
- Tabulate the figure, make them average and give the comments on tillers/m^2 area of wheat.

Assignment

Observation recording (tillers of wheat)

Plot		Samples for number of tiller/m^2 area				
		1	2	3	4	Mean
1.	A (without fertilizer)					
2.	B (with fertilizer)					

Comments :

Plot A :

Plot B :

6.4 CLASSIFICATION OF CROPS

Crops may be classified in many ways, by different people, for different purposes, from different angles. Agronomists deal with the field crops. Therefore, field crops may be classified in different ways, such as :

(i) Classification according to ontogeny (life cycle), (ii) Classification according to season, (iii) Agrarian classification (agronomy/economic/ commercial), (iv) Botanical classification, (v) Classification according to leaf morphology, (vi) Classification according to tillage operation and (vii) Classification according to *special use*. These are as follows :

6.4.1 Classification of crop according to life cycle (i.e. Ontogeny) of the plants/crops

Every plant/crop completes its life cycle after passing through some progressive stages of development. Crop plants are, therefore, grouped together to their life cycles such as :

(a) **Annual crops :** The crops that complete the life cycle within a season or year such as wheat, rice, maize, mustard, green gram, pea, sunflower, jute etc. They produce seeds and then die. Some of the crops tillers which may grow next;

(b) **Biennial crops :** The crops, having a life cycle of consecutive two years. First year's is purely for vegetative growth, followed by reproductive growth in the second year for seed formation. After flower, seed maturity, the crops die. Cauliflower, sugar beet, cabbage, reddish etc. are in this group;

(c) **Perennial crops :** The crops that live for more than two years are the perennial crops. They may be seed bearing (*spermophytes*) or non-seed-bearing. The perennial crops may be propagated by either seeds or by vegetative means. Mostly the plantation crops, are the perennial crops.

6.4.2 Seasonal classification

Different annual crops are grown in different seasons. There are three different seasons, such as (i) *rabi* (winter), (ii) pre-*kharif* (summer) and (iii) *kharif* (rainy) seasons. Such seasons are prominent in the plains, in contrast to coastal or mountainous ranges, and the area having monsoon climate.

In *rabi* season, wheat, potato, mustard, pea etc. are grown, whereas in pre-*kharif* season. jute, *aus* rice, mesta, sunnhemp etc. are grown, while in *kharif* season transplanted rice, *jowar*, *bajra*, cotton etc. are grown.

6.4.3 Agrarian classification (Agricultural, agronomic, commercial and economic)

Classification according to use of crops and their product and these are: (i) Cereals, (ii) Pulses, (iii) Oilseeds, (iv) Fibres, (v) Fodders, (vi) Sugars, (vii) Medicine, (viii) Tubers, (ix) Timber, etc.

The main crops are rice, wheat, maize, *jowar*, *bajra* etc. in cereal; gram, lentil, pea, grasspea etc. in pulse; mustard, sunflower, groundnut, sesame etc. in oilseeds; jute, mesta, cotton, sunnhemp etc. in fibre; lucern, berseem, napier, black gram etc. in fodder; sugarcane, sugar beet etc. in sugar; potato, in tubers, tobacco, *Phyllanthus niruri* in medicine and so on.

6.4.4. Botanical (Taxonomy or families) classification

Any plant is dealt with under natural order or the families, under which they come, in a systematic arrangement (Toxinomy). This type of classifica-tion has certain advantages in the understanding of the morphological characters of any particular family of crop with different agro-botanical or morpho-agronomical characteristics / peculiarities. Field crops are grouped into some families of *Graminae* (rice, wheat, maize, *jowar* etc.); *Papilionaceae* (legumes, pea, lentil, cowpea, gram, bengalgram, greengram etc.); *Cruciferae* (mustard, radish, cabbage, cauliflower etc.); *Cucurbitaceae* (gourd, cucumber, ash gourd, bitter gourd, *parwal*, pumkin etc.); *Malvaceae* (cotton, lady's finger, mesta etc.), *Solanaceae* (brinjal, potato, tomato, chillies, tobacco etc.);

Tiliaceae (Jute); *Compositae* (sunflower, nigar, safflower etc.); *Chenopodi-aceae* (sugarbeet, beet, spinach, *bathuasag* etc.); *Pedaliaceae* (sesame); *Euphorbiaceae* (castor, tapioca etc.), *Convolvulaceae* (sweet potato); *Umbelli-fereae* (coriander, anise, cumin, carrot etc.); *Liliaceae* (onion, garlic etc.); *Zingiberaceae* (ginger, turmeric etc.) and so on.

6.4.5 Classification according to the leaf-morphology

All the crops are divided by leaf morphology into two groups, such as (i) Broad-leaved and (ii) Narrow-leaved plants.

(i) Broad-leaved plants / crops are : mustard, pea, jute, potato, tobacco, cotton, berseem, cowpea etc. All of them are the dicot plants.

(ii) Narrow-leaved plants / crops are : rice, wheat, maize, sugarcane, napier, paragrass, onion, garlic, ginger, turmeric etc. Sedges are also in this group and leaves are narrow. All of them are in monocot group.

6.4.6 Classification on tillage operations

Crops may be agronomically classified as per tillage requirements. These are : (a) Arable crops, (b) Non-arable crops and (c) Wet crops.

(a) **Arable** (dry soil) crops that require preparatory tillage and the crops are : potato, radish, jute, *aus* (direct-seeded) rice, berseem, mustard, tobacco, mustard etc.

(b) **Non-arable crops** (no-dry soil) that may not be requiring preparatory tillage and the crops are : blackgram, paragrass, cowpea, ricebean etc.

(c) **Wet crops that require puddling** (special tillage) for transplan-tation and the crops are : rice (both *aman* and *boro*), paragrass, arum, coax etc).

6.4.7 Classification of crops on special purpose

Some crops are grown only for special uses, but not only for food purposes. Jute is cultivated for bast fibre, cotton is for lint fibre and both the fibres are used in textile. Similarly timber is used for furniture, bamboo is cultivated for construction purpose. Cinchona is cultivated for drug (quinine).

6.5 SPECIAL PURPOSE CROPS

The crops that are classified based on serving some special/specific purposes, are the special purpose crops. These crops may or may not give higher yield but will serve the specific purposes. These are :

6.5.1 Catch crops

The '*catch crop*' may also called as *emergency crop* or *contingent crop*. The short duration crops that are cultivated to 'catch' the forthcoming season either for damaged preceding crop or to fill up the gap between two main seasons, are called the catch crops. Blackgram (for fodder) may be grown in between *kharif* and *rabi* seasons. Of course, the catch crop should be of short duration in nature and quick growing crop. Other examples are greengram. cow pea, ricebean, spinach, *radish*, *amaranth*, onion, coriander etc.

6.5.2 Restorative crops

The crops that are grown for some different purposes other than yields, after some heavy uptake of nutrients by some crops / cropping systems, only to enrich the soils for the next crops, are the *restorative crops / soil enriching crops*. All the leguminous crops, such as cowpea, ricebean, blackgram, greengram, grasspea etc are in this group. These leguminous crops fix atmospheric nitrogen by *rhizobium*, bacteria, harvouring in the nodules of roots. It has been observed that different legumes fix nitrogen in different amounts, ranging from 50 to 200 kg/ha and the nodules are of different sizes and shapes.

6.5.3 Exhaustive crops

The crops which exhaust the soils in nutrition view points, are called the 'exhaustive crops'. After the these crops, other crops can not be grown normally. These exhaustive crops are linseed, brinjal, sesame, thinnapier etc. To bring back the normal conditions, restorative crops like legumes etc. are grown.

6.5.4 *Paira / utera* crops

The crop which is sown by broadcasting seeds, over the preceding crop at maturity, under muddy, marshy or optimum soil moisture conditions, without tillage, and the said crop which is raised with residual soil moisture/ rainfall, is called the *paira / utera* (no-tillage relay) crop and the system with this crop is known as *paira* cropping system. After transplanted *kharif* rice, grasspea, lentil, linseed, coriander spinach (for seed) etc. could be grown as rainfed *paira / utera* crops in *rabi* season. After all these crops, jute, green gram, cowpea, ricebean etc. could be grown in pre-*kharif* season, after which transplanted *kharif* rice can be grown, paving the rainfed 3-crop systems (Transplanted *kharif* rice-*paira / utera* crops–jute/mung bean/cowpea/ sesame, under rainfed conditions). This is *paira / utera* cropping system.

6.5.5 Smother crops

Crops which are able to smother or suppress the other crops/weeds, by providing suffocation, by their dense foliage, are called the *smother crops*. Cowpea, ricebean, sunnhemp, grasspea etc. are in this group. It has been observed that grasspea, lentil, pea, sunnhemp etc. smoother the weeds in *rabi paira* crops and ultimately the crops give higher yields, under no-tillage conditions. Under arable conditions, cowpea, ricebean, blackgram etc. suppress the weeds.

6.5.6 Cover crops

The arable crops that are able to protect the soil surface from erosion (wind, water or both), through their ground covering foliage and or root mass, are called the *cover crops*. Groundnut, blackgram, ricebean, tetrakalai, grasspea, kudzu, sweet potato, paragrass, marvel grass etc. act as cover crops. Some of the legumes act both as soil-enriching crops or cover crop. The grass-group crops, having mass fibrous root system, can bind the soils also, besides covering the soils. It is known as soil 'binding crop'.

6.5.7 Nurse crops

The crops that help other crops in the nourishments by providing some nutrients or shade or acting as climbing sticks, are called the *nurse crops*. If oat is mixed with berseem and sown in *rabi* season, both crops will be benefitted (symbiotic effect). At earlier stage, oat will give shade to the berseem which is essential for it, at later stage oat is benefitted with nitrogen fixed by berseem, the leguminous fodder crop. Sometimes mustard also can be grown in the space between rows of sugarcane in *rabi* and it is known as companion crop also.

6.5.8 Guard / barrier crops

The crops that help the other crops, by giving protection against some enemies or providing shelter against wind/storm (that may damage the second crop), are known as *guard/barrier crops*. *Sarkathi, (Saccharum munju)* is planted around the sugarcane field, to guard it against the attack of jackles, etc. Safflower (tall oilseed crop, having spines on the leaves) is planted around gram/pea to protect against thieves. *Jhaw* (*Casuarina equisetifolia*) is planted by the sides of the sea/ocean/*bill/jhills* etc. to act as wind breaker.

6.5.9 Brake crops

The crops that are grown to break the continuity of the agro-ecolgical situations of the field, under multiple cropping system, are known as the *brake crops*. The inclusion of such crops in the cropping system helps reduce the inoculum of soil-borne harmful biotic-agents or to increase the availability of nutrients, and improve the soil conditions for higher crop yields. Rice-rice or rice-wheat or rice-maize etc. systems may be changed with jute-rice, *dhaincha*-rice-wheat, or cowpea-maize etc. system, for getting higher productivity and maintaining soil fertility.

6.5.10 Trap crops

The crops that are grown to trap soil-borne harmful biotic-agents such as parasitic weeds, *orobanche* and *striga* that are trapped by *Solanaceous* and sorghum crops respectively are called the *trap crops*. The weed seeds germinate when they come in contact with roots of these crops. Destruction of these crops is essential to reduce the problem in crop production.

Rotation of crops (different crops in sequences) once in three years, reduce the problems in crop production. Rotation of crops (different crops in sequences) once in 3 years, reduce the problems.

6.5.11 Mulch crops

The crops that are grown to conserve soil moisture from bare ground by their thick and multilayered foliage, trailing habit and sometimes, self-seeding nature, are called the *mulch crop*. Cowpea, ricebean etc. are the example of mulch crop The crops may draw water from the lower zone and improve soil moisture in the upper zone also. Grasspea, a hardy legume crop, draws moisture from lower zone of the soil, after its harvest, improves soil moisture in the soil surface, helping the next crop to germinate in a better way.

6.5.12 Sod / turf crops

The crops, belonging to *Gramineae* family, having sod type tillers with matted foliage and fibrous roots, close to the soil surface, are the *sod crops* or turf crops. They are grown to conserve soil from erosion, particularly in non-arable areas. Marvelgrass (*Dichanthium annulatum*), *digitaria* (*Digitaria songuinalis), cynodon*/doop (*Cynodon dactylon*) etc. are the sod-type crops.

6.5.13 Cleaning crops

The crops that clean the fields or seed-beds, by agronomic practices from weeds and stubbles, are called the *cleaning* crops. Potato, groundnut, ginger etc. are the examples of cleaning crops, as after cultivating the crops, the soils are more or less weed-free, due to excessive tillage operations or earth works like earthning-up at different times, or preparation of irrigation channels, application/placement of top dressing, harvesting of crops by digging etc.

6.5.14 Fouling crops

The crops, by whose cultural practices, develop the infestation of weeds in the fields intensively, are called the *fouling crops*. Direct-seeded rice (*aus*), jute, maize, cotton, sugarcane etc. at their earlier stages of growth, with wide spaces and small crop growth, allow grow with of weeds vigorously in the fields; as a result it is very difficult to keep the field weed-free.

6.5.15 Cash crops

The crops that are grown for sale, to get or earn hard cash, are called *the cash* crops. These crops require some processing which can not be done by individuals farmers. Thus, the processings of such crops after harvest, are done by the factory owners. The example of these type crops are jute, tobacco, cotton, sugarcane, tea etc. Individual factory processes the products for individual purpose. Jute or cotton fibre is used in textile etc.

6.5.16 Ware crops

The crops that are grown for temporary storing, as such, in ware-houses for future use or sale, are called the *ware crops*. Potato, beet, carrot, veget- ables, fruits, root crops etc. of high-value nature, are stored, after harvest, in ware house for future sale, in order to get more money and also for making the off-season products, available throughout the year, to the consumers.

6.5.17 Truck crops

The crops that are grown and sent to the market fresh, quickly in trucks, are called the 'truck crops'. Tomato, lady's finger, cabbage, cauliflower, greenpea, brinjal, berseem, spinach, radish etc. are the truck crops. To get the good market price, these fresh crops are sent to the market with the help of trucks. Now-a-days, fresh new tomato or other vegetables are going to areas where they are not produced at all.

6.5.18 Cole crops

The crops that are grown in cool areas, belonging mostly to *Cauciferae* family capable of withstanding considerable frost, are known as *cole crops,* Cabbage, cauliflower, brussels sprouts etc. The name 'cole' has been derived from 'cole warts', the ancestor of wild cabbage (cliff cabbage).

6.5.19 Silage crops

`The fodder crops that are grown to preserve in pits (Silopits) in a succulent form by a process of natural fermentation or acidification, for feeding the livestock, during the lean months/seasons, are called the *silage crop*. For silage purpose, generally grass-fodder such as, oat, napier, thin- napier etc. are good. Of course, legumes like berseem and lucern may also be used but risk is there for decomposition, as protein content is high in legumes.

6.5.20 Soiling crops

The crops that are grown to harvest while they are still green and fed fresh to livestock in stalls, are called the *soiling crops*. The main soiling crops are : berseem, lucern, maize, grasspea, napier grass, oat, blackgram, cow-pea, ricebean, teosinte etc. Generally, most of the soiling crops are harvested at flowering stage, because at that stage, nutrient contents in green fodders, are the maximum.

6.5.21 Ley crops

The fodder crops that consists of particular crop or combination of crops (grass + legume) grown either for grazing or harvesting for immediate or future feeding to live-stocks, are called the *ley crops*. The combination of 'marvel grass + berseem', 'dinanath grass + cowpea', 'berseem + mustard' etc. are good. For single crop, berseem, lucern, blackgram, ricebean etc. are also good.

6.5.22 Green manuring crops

The crops that are grown to be incorporated into the soil fresh, to increase the fertility of soils, are called the *green manure crops*. Some green manure crops that are incorporated into soil during wet season for wet crops, are (i) *dhaincha*, (ii) cowpea, (iii) blackgram, (iv) ricebean, (v) sunnhemp etc. Some green manure crops for dry areas are *guar*, cowpea, sunnhemp etc. In some areas (Tamil Nadu), leaf manuring is done, which acts as green manure in paddy field in wet season. For this, *glyricidia*, subabul/ subabool etc. are used. Green manure in *situ*, is a major practice in India.

6.5.23 Mixed crops

Two or more than two crops of different types, grown together in the same field in the same time, in the same season, simultaneously, are known as the mixed crops. The seeds of these crops may either be mixed together before sowing and broadcasted or drilled in rows or may be sown at a time separately and grown with the same management practices. They may be harvested together or separately. Examples are peas + *rai* mustard or wheat +mustard or gram + safflower etc. in *rabi*; maize + cowpea or *jowar* + cowpea or *arahar* + black gram in pre-*kharif* / *kharif*.

6.5.24 Inter-crops

The two or more than two crops that are grown simultaneously in alternate, rows in the same field, in the same season, in the same time, are called the `inter-crops' and the process of growing such crops, is inter-cropping. Of course, the crops are not necessarily sown at exactly on the same day and their harvest times may be different, but they are usually simulta- neous for a significant part of their growing period. These crops are grown normally, with the same agronomic practices without any separate identities, except for the respective rows. In two or more crops, at least one main crop will be there, such as, sesame + black gram, pigeon pea + groundnut, sorghum + pigeon pea, wheat + mustard, rice (*aus*) + pigeon pea etc.

6.5.25 Companion crops

The short duration subsidiary crops, grown in between the main long duration wide rowed crops, in the same season, in the same fields, are called the companion crops, where they are companions to each other, for some times of their growing period. Examples are these : Mustard/wheat/linseed/ lentil etc. may be grown in wide rows of autumn-planted sugarcane; basella/ lady's finger/*amaranths* may be grown in spring-planted sugarcane. Here long duration sugarcane is a wide-rowed crop and it grows slowly in earlier part, hence it is possible to grow short duration crops at earlier stage of sugarcane. Sometimes they can be grown for mutual benefit, as berseem + oat; here oat gives shed to berseem at earlier stage, but it is benefited by N from berseem in latter stage (as berseem fixes N from atmosphere).

6.5.26 Parallel crops

Two or more crops that are grown simultaneously in the same field, but do not have any inter-competitive effects, are known as the `parallel crops.' They are parallel in respect to competition. Examples are : 'Maize + cowpea', 'sorghum + pigeon pea', 'jute (for seed) + black gram', 'maize + ricebean' etc. In the parallel crops, one crop should be legume.

6.5.27 Mono-crop (only crop in the seasons)

The crops that are grown as pure or solid stands in the season, are the '*mono-crops*'. 'Mono' stands for whole or sole or pure crops. Transplanted *kharif* rice, jute, tobacco, ricebean etc. are the example of 'mono-crops'; if they are grown during the season of year. Sometimes '*mono-culture*' may be confused with '*mono-crop*'. 'Mono-culture' is the single crop cultivation in a year; whereas, in '*mono-crop*' may be in one season but for the year it may be 2 or 3 or more crops.

6.5.28 Multiple crops (different crops in different seasons)

Different crops in different seasons in a sequence, in the same land, are called the 'multiple crops' and the system of putting the crops in the sequence, is known as 'multiple cropping system'. In that multiple cropping system, at least one crop should be the legume, for maintaining soil fertility. (i) Wheat + green gram + transplanted *kharif* rice, (ii) Transplanted *kharif* rice + *paira* crops + jute, (iii) Wheat + jute + transplanted *kharif* rice, (iv) Wheat + cowpea + maize etc. are the examples of multiple crops and cropping systems (Jute, though not a legume, yet improves soil fertility). [Though 'maize + wheat' or 'rice + rice' or 'rice + wheat' may be grown, but fertility of the soils would be deteriorated because of all cereal crops in the systems].

6.5.29 Stub crops / first harvest

The perennial or annual multi-cut crops (mostly in forage/fodder production), where first harvest in the cycle is taken, are called the *stub crops*. These may be termed as '*first harvest*' also. In sugarcane, first harvest is good, but not second crop or ratoon crop, but all cuttings are equally good in berseem, lucern, marvel grass, guinea grass, paragrass etc.

6.5.30 Ratoon crop/stubble second or third crops

The perennial or multi-cut crops that are harvested subsequently from the regrowth of the stubbles or root-stocks or stumps, after first harvest, are called the *ratoon crops*. This may be termed as *stubble crop, second or third crops* also. Examples are sugarcane, napiergrass, berseem, lucern, ricebean, guinea grass etc. Such type of ratoon crops, reduce the crop cycle period, cost of cultivation and do not need 'seeds' for further sowing, though in some cases, production is less (as in sugarcane). In berseem and lucern, of course, ratoon crop is better than the original crop, provided irrigation and fertilization are done in long duration winter.

6.5.31 Seed crops

The crops that are grown only for seed production, are called the *seed crops*. The crops are specially raised for seed, hence special care is taken. Of course, some crops may be grown in some areas as food/vegetable production, but the same crop may be grown in different needs and in different areas. Examples are : seed production crops of jute, cabbage, cauliflower, potato, tobacco, berseem, cowpea, maize, lucern etc. Such crops require special management practices, for maintaining purity and identity. In some crops seeds are formed in hilly place (as cauliflower).

6.5.32 Leafy crops

The crops that are grown for leaves only for further processing or for leafy vegetables, are called the *leaf or leafy crops*. Example are : tobacco, spi-nach, basella, *amaranthus,* sisal, tea etc. In tobacco, tea, sisal etc. leaves are processed for different purposes, whereas spinach, basell (*Pui*), *amaranthus* (*data*) are used for leafy vegetables. Some types of arum (*arai kachu*) are used for utilizing leaves as leafy vegetables. Some soft twigs, along with leaves (as in pumpkin, gourd etc.) are also used as leafy vegetables.

6.5.33 Fuel / bio-fuel crops

The crops that are grown for obtaining fire woods or solid fuels, as a by-product, along with economic yields, are called the *fuel* or *bio-fuel crops*. Jute, sugarcane, pigeon pea, cotton, mesta, sunnhemp, maize, mustard, sesame *sesbania* etc. are the fuel crops. Besides these, some perennial crops like *acacia*, silk cotton, jackfruit, mango, bamboo etc. are also used as fuel crops. Seeds and plants of *karamja*, jatropha, *banverend*a, castor etc. are also used as fuel/bio-fuel crops. The day is ahead, when jatropha will be used as bio-fuel for many purposes.

6.5.34 Energy crops

The crops that are grown to obtain liquid materials for energy, such as ethanol or alcohol etc., are called the `energy crops'. Sugarcane, potato, maize, tapioca etc. are the examples of energy crops, as from them alcohol or ethanol may be obtained after processing. Generally, from all the cereal crops, alcohol may be obtained, though production will be less and quality will be inferior in some crops.

6.5.35 Border / outer crops

The crops that are grown around the borders of other crop-fields, to take advantage of economic aspects, along with partial checking against grazing by cattle or other enemies of the crops, are called the *border or outer crops*. Examples are: Safflower (spiny long oil-seed crop) around potato; radish/ barley/onion etc., on the border of cabbage, cauliflower plots; *amaranthus*/ lady's finger, basella (*pui*) in outer ridges of cucumber/sweet potato/gourd/ bitter gourd/ridge gourd plots; mesta/*sunnhemp* in the outer ridge of sugarcane.

6.5.36 Fodder / forage crops

The crops that are grown for feeding the cattle, goat, buffaloes, horse etc., are the *fodder / forage crops*. These fodder/forage crops may be of grassy or non-grass crops. Among the grassy fodder crops, napier, teosinte, maize etc. are the main, whereas, among the non-grassy group, berseem, lucern, cowpea, grasspea, ricebean, blackgram etc. are the main. They are mostly the leguminous crops. Of course, among the non-grassy group mustard (broad leaved good fodder) is included.

6.5.37 Weed-controlling crops

The crops that partially control the weeds in the fields, either by suffocation or by allelopathic effects, or by overcoming weeds, are called the 'weed-controlling crops'. Most of the weeds are partially controlled by cow pea, grasspea, ricebean, blackgram etc. *Vicia* weed (*hetka*, like grasspea) spp, are partially controlled by linseed crop, due to allelopathic effect. Sunnhemp (for seeds) in *rabi* season controls partially the weeds beneath the crop.

6.5.38 Soil-enriching crops

This (soil enriching crop) is also `soil-restorative crop' but with some speciality. Beside nitrogen fixation, it gives other crop-residue also, adding huge amount of organic matter to the soil which ultimately enrich the soils both in terms of nutrients and organic matter. Cowpea, ricebean, blackgram, horsegram, *kuthi* etc. are the crops in this group.

6.5.39 Alley crops

The crops that are grown in a long narrow passage, formed by the trees or shrubs, for beautification of the area, are called the *alley crops*, and the system by which alley crops are grown, is the alley cropping. Alley crops are generally the 'hedge crop' or creeper which is clipped after some days, or some flower plants.

6.5.40 Relay crops

The crops that are grown before the harvest of the preceding crops, are called the *relay crops*. This is analogous to a relay race, where one crop hands over the batton (to the next crop in quick succession). Sowing *toria* mustard in standing crop of maize is an example. Grasspea, lentil, linseed, sunnhemp (for seed) may be broadcast over the matured transplanted/*kharif* rice crop, under muddy/marshy condition without tillage in rainfed areas (i.e. *paira* / *utera* - no-tillage relay crop). All these crops are the relay crops.

6.5.41 Moisture-enriching crops

The crops that enrich the soil moisture by their cultivation, helping other next crops to germinate, are called the `moisture-enriching crops'. Grasspea, ricebean, blackgram, cowpea etc. are examples. Grasspea in *rabi*, heavily increases moisture in the soils and the next jute in pre-*kharif* season is benefited, with this moisture.

6.5.42 Riparian crops/plants

The crops that are grown along the irrigation and drainage channels or along with the water-bodies, like canals/rivers/ponds/*bills*/*jhills* etc., are called the *riparian crops*/ plants. Mostly they are the weeds which grow auto-matically in those areas. The plants are: (i) Kanchari/*sechisag* (*Alternanthera sessilis*), (ii) Swamp cabbage/water spinach/ *kalmi* (*Ipomoea reptans*), (iii) Moon flower/*dholkalmi* (*Ipomoea alba*), (iv) Mikania spp, (v) Beggar weed (*Desmodium tortuosum*), (vi) *Congress topi* / *Parthenium* (*Parthenium hysterophorus*), (vii) Jharber (*Zizyphus rotandifolia*), (viii) *Kalkasunde* (*Cassia sophera* / *tora*), (ix) *Sugandhiful* (*Lantana camara*), (x) Elephant grass (*Typha elephantia*), (xi) Indian sorel/*susni* (*Oxalis corniculata*) etc. They also protect the soils against water erosion.

6.5.43 Avenue crops

The crops that are grown along the farm roads, road of township, fences etc., are called the avenue crops. These are *Gliricidia maculeata, teprosia, milingtonia* (*Millingtonia hortensis*, casuarina (*Casuarina equisetifolia*), kadamba (*Anthocephalus cadamba*) etc. Of course, in farm roads, generally pigeon pea, sisal, glyricidia, sesbania etc., the avenue crops are planted.

6.5.44 Contour crops

The crops that are grown on or along the contour lines to protect lands from erosion in hilly areas, are the `contour crops'. For this, marvel grass, dinanath grass, anjan grass, small type of bamboo (reeds), like *nal* (*Phragmites karka*), common reed (*Phragmites communis*) etc. are grown on contour lines in rainy season.

6.5.45 Strip crops

The crops that are used for restricting/resisting, soil erosion in the form of strips, are called the `strip crops'. Strip crops are of different nature, for checking soil erosion, the following crops are grown across the slopes in strips alternately: marvel grass, groundnut, kudzu vine, *shyamaghash* (*Echinochloa crusgali*) etc.

6.5.46 Augmenting crops

The crops (second crop/sub-crop) that are grown/sown to supplement the yields of main crops, are called the `augmenting crops'. In 'Oat + berseem' or 'fodder mustard + berseem', oat and fodder mustard are the augmenting crop, because yields of fodder is made up with these crops, as berseem yield in first cutting, is less.

Practical Class: Some of the `special purpose crops' mentioned above (6.5) should be grown in farms and crops should be shown to students from time-to-time, during practical classes and students should take records on them in different seasons

Lesson 67 : Take the records of `special-purpose crops', grown in *rabi* season

Introduction

Special purpose crops are special in functioning of special type of work. If that work is successfully done, then the crops are successful in their functions. The blackgram which is used as a *catch crop*, is successful when its growth or fodder yield is complete, within a short period, if grown in between *aus* and *aman* paddy. Of course, the learners must know the growth behavior of these crops in different seasons.

Objectives

- O to get the benefit of special purpose crops,
- O to study the growth behavious of the those crops.

Material required

❖ Fields with special purpose crops, ❖ Scale (long and short), ❖ Note book, ❖ Pen/pencil.

Procedure

For recording observations on leaf, stem, inflorescence etc. observe the crops in *rabi* season at about one, two and three months after sowing the crops. Select the crops like grasspea (as moisture retention crop), sunnhemp (for seeds and weed control), oat (as companion crop), linseed (as exhaustive crop) and lentil (*paira* crop). Measure the leaf (length and breadth), cut the stem (for noting hollow/solidness), count seeds or pods if any) and observe fully minutely on the plant and write in the format with comments, on special purposeness of the crops.

[Note : The Farm Managers of the Farms of the Universities, with the consents from the concerned, Head, Agronomy Department, should grow the `special purpose crops' in different seasons, by the sides of 'farm-main-road', with common, English and scientific names of the crops, along with the date of sowing].

Assignment

Recording observations, on some special-purpose crops in '*rabi*' season (one month after sowing)

Crops with purposes	Crop characteristics of				Remarks
	Leaf	Stem	Inflorescence	Plant type	
	Size & shape	Hollow / Solid	Grain / Pod	Erect / Vine	
1. Grasspea (Moisture retention)					
2. *Sunnhemp (for seed & weed control)*					
3. Oat (Companion crop)					
4. Linseed (Exhaustive crop)					
5. Lentil (*Paira* crop)					

Give comments on 1-5 from above

Lesson 68 : Recording the observations of `special purpose crops', grown in pre-*kharif* season

Introduction; **Objectives** and **Material required :** All as in lesson 67.

Procedure

Out of so many special types of crops, grown in pre-*kharif* season, select the crops like cowpea (as cover crop), black gram (catch crop), rice-bean (fodder crop), *dhaincha* (green manuring crop) and green gram (as minimal tillage crop) for observations, on leaf, stem, inflorescence and type of crops (vine/erect), at one, two and three months after sowing crops. The observat-ions should be recorded in the format given, along with the comments, below.

Assignment

Recording observation on some `special purpose crop', during pre-*kharif* season (one to three months after sowing)

Crops with purposes	Crop characteristics of				Remarks
	Leaf	Stem	Inflorescence	Plant type	
	Size & shape	Hollow / Solid	Grain/ Pod	Erect / Vine	
1. Cowpea (Cover crop)					
2. Blackgram (Catch crop)					
3. Ricebean (Forage crop)					
4. *Dhaincha* (Green manuring crop)					
5. *Green gram (Minimal tillage crop)*					

Comment on 1-5 from above

Lesson 69 : Recording observations of `special-purpose crops', grown in *kharif* season

Introduction; **Objective** and **Material required :** All as in lesson 67.

Procedure

There are many special types of crops in *kharif* season. Out of all of them select, jute (as seed crop), napier (fodder crop), ricebean (smother crop), rice (cereal crop) and cotton (fibre crop). From them record the observations on leaf, stem, inflorescence etc. at one, two and three months after sowing in *kharif* season. The learners therefore, should know the growth condition of those crops. The observations should be recorded in the format along with the comments.

Assignment

Recording observations on some `special-purpose crops', during *kharif* season (at one month after sowing)

Crops with purposes	Crop characteristics of				Remarks
	Leaf	Stem	Inflorescence	Plant type	
	Size & shape	Hollow / Solid	Gram / Pod	Erect / Vine	
1. *Jute (Seed crop)*					
2. Napier (Fodder crop)					
3. Ricebean (Smother crop)					
4. Rice (Cereal crop)					
5. Cotton (Fibre crop)					

Comment on 1-5 from the above:

1.
2.
3.
4.
5

CHAPTER 7

MANURES AND FERTILIZERS FOR CROP PRODUCTION

7.1 MINERAL NUTRITION FOR THE CROPS/PLANTS

Plants/crops require some foods for their growth and development; foods are produced from nutrients from soils, air and water, hence it is necessary to know the nutrient elements.

7.1.1 Essential elements

Plants need 16 essential nutrient elements, human being needs 19 essential nutrient elements and other animals need 18 essential nutrient elements for their growth and production.

For plant essential elements required are, 16 in number, such as :

(1) Carbon (C), (2) Hydrogen (H), (3) Oxygen (O_2), (4) Nitrogen (N), (5) Phosphorus (P), (6) Potassium (K), (7) Calcium (Ca), (8) Magnesium (Mg), (9) Sulphur (S), (10) Iron (Fe), (11) Manganese (Mn), (12) Boron (Bo), (13) Zinc (Zn), (14) Copper (Cu), (15) Molybdenum (Mo) and (16) Chlorine (Cl).

Out of all these essential elements, C and O_2 come from air, O_2 and H from water and rest of other elements come from soil-minerals.

7.1.2 Macro and micro-nutrients

Oat of all 16 essential elements, six (6) are (A) **macro-nutrients** and seven (7) are (B) **micro-nutrients coming from soils besides 3** (Carbon, Hydrogen and Oxygen) come from air and water.

(1) Macro-nutrients are : (i) Nitrogen, (ii) Phosphorus, (iii) Potassium, (iv) Calcium, (v) Magnesium and (vi) Sulphur, whereas

(2) Micro-nutrients are : (i) Iron, (ii) Manganese, (iii) Boron, (iv) Zinc, (v) Copper, (vi) Molybdenum and (vii) Chlorine. Nitrogen, phosphorus and potassium are also called `major nutrients' and calcium, magnesium and sulphur are the secondary nutrients.

7.1.3 Nutrients availability

Except C, O_2 and H, all other essential elements come from mineral and organic soils. By the process of mineralization of soils, both macro- and micro nutrients are made available to the plants. Other sources of nutrients from organic matters are obtained under different conditions/ situations. When all these sources fail to provide nutrients to the plants, other sources are the fertilizers of different types. Besides this, some *bacteria* fix nitrogen from the air. Soil pH is the factor by which soil nutrients are either made available to the plants or made unavailable to the plants i.e. fixed.

7.1.4 Function of NPK-nutrients in plants

The functions of nitrogen (N), phosphorus (P) and potassium (K) are listed below in brief :

(a) Nitrogen : It is an essential constituent of protein chlorophyll and other compounds of physiological importance in plant metabolisms. It imparts dark green colour to plants, promotes leaf, stem, and other vegetative growth, but retains small root systems. It also produces rapid early growth, improves quality, succulence of leafy vegetables; increases protein content of foods/ fodders. For these, requirement of nitrogen is high in all cereals and to some extent to other non-leguminous crops.

(b) Phosphorus : It is a constituent of nucleic acid, phytin and phospho-lipids, enzymes etc. It is essential for development of roots in earlier stage and also for reproductive parts of plants. It is essential for metabolism of carbohydrate and also for respiration. It stimulates early root development and growth; it gives vigorous starting of plants and decreases lodging and brings about early maturity in cereals, stimulates flowering and seed formation. Requirement of phosphorus is very high in all legumes, including pulse crops.

(c) Potassium : Unlike others, it does not enter into the composition of any plant constituent but imparts increased vigour, resists disease to attack, produces stiff straw, imparts winter hardiness to legumes, regulates water conditions in plant, transfers starch sugar. For these requirement of K is high in potato, banana, turnip, tapioca, *amorphophallus* etc.

Hence, for higher production of crops, fertilization should be done (a) with nitrogen in all cereal crops, (b) with phosphorus in all legumes (including pulse crops) and (c) with potassium in all tuber crops, rhizomes, banana, root crops etc.

7.2 IMPORTANCE OF ORGANIC MANURES

Organic matter, always plays a vital role in the productivity and conditioning of soils. It serves as a source of food for bacteria, fungi and other microbes, responsible for converting the complex organic materials into simple substance, readily utilized by the plants/crops. This organic matter, ultimately improves biological and chemical and physical condition of soil, increasing fertility status of soil, ultimately giving the higher productivity of soil.

7.2.1 Application of bulky organic manures

The bulky organic manures are : (a) Farmyard manure (F.Y.M), (b) farm manure, (c) farm compost, (d) town compost, (e) night-soil, (f) sludge, (g) green manures and (h) other bulky sources of organic matter. All these manures, though bulky in nature, supply plant nutrients and organic matter, besides acting as binding materials in soils. Among all the various bulky manures, F.Y.M compost and green manure are by far the best. These 3 types of organic matters are applied to the soils in different forms. Well decomposed F.Y.M. and compost (about 100 q/ha) are applied in the fields, before ploughing the lands. After ploughing the land several times, sowing of seeds of crop is done. Most of the organic matters, work well in high soil moisture conditions. Green manure is prepared in the field itself, decomposed and used, generally under wet condition before transplantation of rice/arum in puddled conditions.

7.2.2 Application of concentrated organic manures

Concentrated organic manures are those manures which are organic in nature but contain higher percentage of major plant nutrients like, nitrogen, phosphorus and potassium. These concentrated manures are made from raw materials of animal or plant origin. The concentrated common organic manures are : *oilcakes*, *blood meals*, *fish meals*, *meat meals*, *wood-wastes* etc. Among all, the oilcakes are widely used in agriculture. Oilcakes are the quick acting organic manure and are applied either as such, after powdering, or after mixing the powder with the soils, just before last ploughing for sowing crops.

7.2.3 Process of decomposition and preparation of bulky organic manures (F.Y.M./compost/GM)

Preparation of 3 major bulky organic manure are stated in brief as follows :

F.Y.M (Farm yard manure): This is the decomposed or undercomposed mixture of dung and urine, along with litters (i.e. bedding materials) and left over materials. The undercomposed materials are put into some pit of desired dimension and process of decomposition starts if moisture, temperature and aeration are adequate.

Compost : It is obtained from a biological process in which micro-organisms of both types, namely *aerobic* (requiring O_2) and *anaerobic* (requiring no -O_2) act for decomposition to bring down the C : N ratio, with the help of moisture, temperature and O_2 in adequate amount. This is done as per desired dimension of the pit or space. Compost may be prepared from farm litters or town-refuses, or other sources. Nutritive value is high in town-compost than rural (farm) compost.

G. M. (Green manure): Green manuring may be defined as a practice of ploughing or turning the materials into the soil (i.e. undecomposed green materials) for the purpose of improving the physical structure of soil as well as to maintain fertility of the soil, for higher crop production. Under green manure in *situ*, green manure crops are grown and buried in the same field which is to be green manured, either as a pure crop or an inter crop with the main crop. The most popular green manure crops are *dhaincha*, sunnhemp, pillipesara, *guar*, blackgram etc. With *Gliricidia maculata*, leaf manuring (as green manure) is done, bringing leaves from out side. Green manuring is done normally in *kharif* season in central/northern India, for *rabi* crops, while it is done in pre-*kharif* season in eastern India for *kharif* transplanted rice crop.

7.2.3.1 *Compost making*

In farm or in the village, cow dung, litter, urine, weeds, stubbles, *bhusa*, sugarcane trashes etc. are collected and put them into the pit/trench/fields of desired dimension for composting and utilised in fields for crops. In the town, town-waste materials like vegetable-refuses, sewage-sludge, street and dustbin refuse, night soil, factory waste, wool and cotton waste etc. are collected and put them into pits of desired dimensions (15 to 30 m length, 5 to 8 m breadth and 2.5 to 3.5 m depth) and kept for 3-4 months or so, for decomposition. After decomposition, compost is used for plants/ crops.

7.2.3.2 *Crop cultivation and preparation of G.M. (green manure in situ and leaf manuring)*

Generally *dhaincha* crop is raised in the field in the month of May/June, by broadcasting seeds @ 60 kg/ha, on unploughed land and ladderings are done after two ploughings. The plants, about 45-60 days after sowing, are ploughed-under into the soils and kept them in the field, as such, for 15 days for partial decomposition and again after 15 days, the land is prepared by puddling 2-3 times, for the crop (say *kharif* rice). This practice is in vogue in eastern India, but other parts, green manure crop is raised in *kharif* for the *rabi* crops, because of dearth of water. For green manure crop, (decomposition) requires more water, which is not available in all places.

In some areas (T.N.), leaves from *Gliricidia* (legume) plant, are collected and placed into spaces between two lines of rice crop and put them into mud by trampling. It is useful for growing rice to some extent, as leaves contain high amount of nitrogen.

Practical Class : Compost making in pit/trench and its application in crops. Preparation of G.M. in *situ* for rice

Lesson 70 : Prepare compost in pit/trench

Introduction :

Compost is the bulky organic matter, prepared from refuse materials either from the farm-refuse from rural areas, or from the town-refuse materials from the city/towns. For any decomposition of organic matter bacteria, fungi, actinomycetes etc. act with the help of adequate moisture, air and temperature. Of course, to straw/trash etc. sometimes urea/ammonium sulphate, superphosphates are added, to get high quality compost. The learners must *know* the technique, to prepare compost in the farm through either pit or trench method from farm refuse-materials.

Objectives

O to get good quality compost/bulky organic materials for the crops (to be used as organic manure),

O to clean-up the farm areas for composting,

O to use the compost for maintaining soil aggregate and soil fertility both for single crops or for multiple cropping programmes.

Material required

❖ Refuse-materials from the fields, like straw/stover, weeds, litters, cow dung, urine, plant-residues etc., ❖ High lands, by the sides of the farm roads or nearer to the water sources, ❖ Cowdung, urine litters etc., ❖ Water (adequate) source, ❖ Urea, ammonium sulphate and superphosphate (for quick decomposition and for getting good quality compost), ❖ Field balance (spring, 10 kg capacity), ❖ Spring balance (long) – 20 kg capacity-1, ❖ Sticks (Bamboo)-2, ❖ Tape/long scale, ❖ Buskets- 4, Spade-4, ❖ Note book, ❖ Pen/pencil.

Procedure

- Collect the refuse-materials like crop-residues, straw/stover, weeds, trash, leaves etc. from the farms/farm areas,
- Collect cowdung, cattle urine or slurry to add to the composting materials,
- Dig out a pit or trench (1-2), measuring 3-4 m in length, 1.5 to 2 m in width and 1 – 1½ m in deep, by the side of the main farm road and nearer to water source,
- Put all the refuse-materials into pit/trench at about 2.5-3.0 m height,
- Give cowdung and cattle urine or slurry amounting to about 20 L over the materials, again give 25 –30 cm layer of refuse materials and slurry until the height become to about 1½ - 2 m.
- Spread on the heap, 2 kg urea/ammonium sulphate and separately 3 kg super phosphate or the compost materials, followed by sprinkling of water. Regular watering is necessary,
- Remove compost material after 4 – 6 months and heap on high land,
- After one month again, the decomposed bulky organic manure is used for crop cultivation.

Lesson 71 : Apply bulky organic manure (obtained from lesson 70) in transplanted rice crop

Introduction

Bulky organic manures/compost (rural) is very good both for soil aggregation and for maintaining soil fertility, for which crop productivity is high. Of course, the nutrient content in compost is variable according to the types of materials used. But in general, it contains about 0.50% nitrogen, 0.15% phosphate and 0.50% potash besides slight amount of other macro and micro-nutrients. The learners should know the method of application of compost in transplanted *kharif* rice.

Objective

- O to improve the soil structure,
- O to get higher yields of crops (i.e. to maintain soil fertility),
- O to retain moisture in the soil.

Material required

✤ The field, where *kharif* transplanted rice will be transplanted, ✤ Bulky organic matter/compost @ 30 q/ha, for 1000 m^2 area, ✤ Baskets 4, ✤ Spade – 4, ✤ Labourers – 4 (for 1000 m^2 area), ✤ Spring balance (10 kg capacity), ✤ Bullock (2), Ploughman–1, *desi* or mouldboard plough with accessories, ✤ Ladder/plank (1), ✤ NPK-fertilizers @ 50 kg N, 50 kg P_2O_5 and 50 kg K_2O/ha, as basal (for 1000 m^2 area), ✤ 25-day-old rice seedling (any variety), ✤ Note book, ✤ Pen/pencil.

Procedure

- Measured amount of compost should be heaped in different places in the field places and from there that should be spread over the entire field uniformly,
- The land then should be ploughed with mouldboard plough under arable conditions and then puddling should be done with *desi* plough 2 times, followed by ladderings twice, after accumulation of water,
- Again the land should be puddled after spreading nitrogen(urea), phosphate (super phosphate) and potash (meriate of potash) in the puddled land uniformly, followed by laddering. Now the land is ready for transplantation of *kharif* rice,
- Transplant the 3-4 seedlings (of 25-days-old) per hole or hill, under the spacing 25 cm × 10 cm, by the end of July.

[**Notes** : Rice seedlings are raised elsewhere. About 21 to 25 days after sowing, seedlings are uprooted by pulling, with the addition of water in the nursery beds by the labourers. Seedlings, with bundles, are spread over the puddled area for transplantation. From there, some amount of seedlings are kept in the left hands of labourers, from there 3-4 seedlings are put into holes by right hand fingers in land, spaced at 25 cm × 10 cm. This rice transplantation in puddled land is a specialised job].

Lesson 72 : Cultivate *dhaincha* crop in *situ* for G.M. and plough-under it, for the *kharif* transplanted rice

Introduction

Dhaincha, the leguems and also the green manuring crop, is used for transplanted *kharif* rice under wet conditions, for increasing soil fertility and soil health, by growing *in situ*. It is grown in the same rice field, before rice growing in the pre-*kharif* season. When its age is about 45 to 60 days, it should be ploughed-under into the soils and kept for about 15 days for partial decomposition. After that puddling is done with it. Thereafter, after puddling, *kharif* rice is transplanted. The learners should participate in the programme to gain knowledge, about rice transplantation.

Objective

O to improve soil fertility.
O to increase crop/plant (rice) health,
O to get higher productivity in *kharif* rice,
O to minimise the general dose of fertilizer.

Material required

✤ Paddy (low/medium lying) land, having standing water (either from rainfall or irrigation), ✤ Seeds of *dhaincha* for 100 m^2 area @ 50-60 kg seed rate/ha, ✤ Bullock (2), ploughman (1), *desi* plough/with accessories, ✤ Ladder /plank, ✤ Sickle/*hasua* – 4, ✤ Spade–1, ✤ NPK-fertilizers @ 50 kg N, 50 kg P_2O_5 and 50 kg K_2O/ha (as basal), (for 1000 m^2 area), ✤ Balance (spring, 10 kg capacity), ✤ Bowls (4), ✤ Sacks (4), ✤ Tape 1, ✤ Scale 1, ✤ String/coir – 500 gm, ✤ Note book, ✤ Pen/pencil.

Procedure

(a) Growing *dhaincha* in *situ*

- Plough the land twice, lengthwise and crosswise,
- Broadcast *dhaincha* seed (as measured for 1000 sq.m areas),
- Apply ladder/plank for covering *dhaincha* seeds,
- Be ensured for soil moisture for germination of seeds
- Allow the crop to grow for 45 days,

(b) Plough-under the '*dhaincha*' for green manure

- After 45 days, the *dhaincha* plants should be lowered down with bullocks and ladder so that plants touch the ground with rain accumulated or irrigation water, keep it for 7-10 days for partial decomposition of weeds, in the field itself,
- After 7-10 days, days the partial decomposed *dhaincha* should be puddled with 2 ploughings and ladderings and keep it for seven days,
- After about 7 days, puddle the land with fully decomposed *dhaincha*, after addition of NPK-fertilizer (as calculated for 1000 sq.m area),
- Apply ladder/plank twice, for uniformly mixing fertilizer and for leve-lling soils. Now, the land is ready for transplantation of *kharif* rice.

(c) Transplantation of *kharif* rice

- Uproot the 25-day-old rice seedlings from the beds, raised elsewhere,
- Spread the bundles uniformly on the puddled land,
- Transplant 3-4 rice seedlings at each point or hill, by holding seedlin- gs in left land and by putting them in between three fingers by the right hand into the muds, at a space of 25 cm × 10 cm in straight lin-es, prepared by holding coir/nylon thread from two ends of the string.

7.3 FERTILIZERS

Industrially manufactured chemical plant-nutrients, are called the `fertilizers'. Nutrient content is higher in fertilizers than those of organic manures. In fertilizers, nutrients are released nearly 7-15 days after their application.

7.3.1 Classification of fertilizers

Generally, the fertilizers are classified into three categories such as (i) strain, (ii) complex and (iii) mixed fertilizers.

(i) Straight fertilizers are those fertilizers, which supply only one primary plant nutrients, namely nitrogen, phosphorus or potassium etc. Urea, ammonium sulphate, potassium chloride, potassium sulphate etc are some of the straight fertilizers.

(ii) Complex fertilizers, generally contain two or three primary plant nutrients of which primary nutrients are in chemical combination. In general, these fertilizers are produced in granular form. Diammonium phosp-hate, nitrophosphates, ammonium phosphates etc. are the complex fertilizers.

(iii) Mixed fertilizers: These fertilizers are physical mixture of straight fertilizers. They are generally, two or three primary plant nutrients. Mixed fertilizers are made by thoroughly mixing the ingredients either mechanically or manually.

7.3.1.1 *Some macro-nutrients*

Nitrogen : Nitrogenous fertilizers are those fertilizers which contain nitrogen. They may contain secondary nutrients like calcium or sulphur.

Following are the nitrogenous fertilizers :

(i) **Amide fertilizers :** Urea (46% N), calcium cyanide (21% N),

(ii) **Nitrate and ammonical fertilizer :** Ammonium nitrate (33% to 34% N), calcium and ammonium nitrate (20% N),

(iii) **Nitrate fertilizers :** Calcium nitrate and sodium nitrate,

(iv) **Ammonium fertilizers:** Ammonium sulphate (20% N), ammo-nium chloride (24-26% N), and anhydrous ammonia (82% N).

All these nitrogenous fertilizers are readily available to the plants with moisture in adequate amount. Ammonical and amide fertilizers may be utilised in wet conditions, whereas ammonical and nitrate fertilizer may be used in soils under arable conditions. All the crops require nitrogenous fertilizers, but some need them in higher amounts.

Generally, in crops under cereals, oilseeds, fodder (except legumes) etc. higher amount of N-fertilizer is needed. For all vegetative growth, nitrogen is needed. Higher amount of N-fertilizers is needed for leafy vegetables.

Phosphate : Generally, phosphorus content in fertilizer is expressed in oxide form i.e. P_2O_5 (phosphorus pentaoxide), while its content in plant is expressed in elemental form. In soils, it may be expressed both as elemental or P_2O_5 form. The phosphatic fertilizers may be classified into following three groups, based on their availability to crops and solubility. These are :

(i) Phosphatic fertilizer having/containing water soluble phosphoric acid

(a) Single super phosphate/ordinary super phosphate- P_2O_5 content (in bracket) of single super phosphate (16% P_2O_5), double (32 % P_2O_5) and triple super phosphate (48% P_2O_5),

(b) Ammonium phosphate (20% P_2O_5 & 20 N),

(c) Ammonium phosphate-sulphate (20% P_2O_5, 16% N).

(ii) Phosphatic fertilizers containing citric acid soluble phosphoric acid

- basic slag (14-18 % P_2O_5), dicalcium phosphate (34-39 % P_2O_5).

(iii) Phosphatic fertilizers containing phosphoric acid (not soluble in water) – rock phosphate (20-40 % P_2O_5), raw bone meal (20-25 % P_2O_5), steamed bone-meal (22 % P_2O_5).

Category (a) fertilizers are easily available to plants and these fertilizers are good for wheat, rice, maize, pulses (all), oil seeds, sugar crops, fodders etc. Phosphate fertilization is essential for any pulse crop like pea, green-gram, lentil etc. But under rainfed conditions, *rabi paira* crops did not need it.

Category (b) fertilizers are good for long duration crops like tapioca, tea, coffee, sugarcane etc.

Category (c) fertilizers are also suitable for long duration/perennial crops like, tea, coffee, timber, plantation crops (like coconut, rubber, cocoa etc.).

Potash : Potassium fertilizers are classified into two groups, such as :

(i) Chloride form of potassium and (ii) Non-chloride form of potassium.

Potassium in the fertilizers, is expressed as K_2O (Potassium oxide).

(i) Chloride form of potassium : Fertilizer like potassium chloride (58-60% K_2O) is in this group. (ii) **Non-chloride form of potassium** : Fertilizer like potassium sulphate, potassium, magnesium sulphate, potassium nitrate etc. are in this group.

In the first group potassium chloride (muriate of potash, other name) is the common fertilizer for most of the crops and it is cheaper also. This fertilizer is good for wheat, rice, maize, *jowar*, *bajra*, mustard etc. but this fertilizer is not good for sugarcane, tobacco, potato etc. In sugarcane, sugar, tobacco, leaves and in potato, tubers are affected due to presence of chlorine. Hence, fertilization of K_2O with this fertilizer should be avoided.

In the second group, there are many fertilizers like potasium sulphate (48-50% K_2O and 17.5% S), potassium-magnesium-sulphate (22% K_2O, 11% Mg and 22% S.), potassium nitrate (or salt petre or nitre). (13% N, 44% K_2O). All these fertilizers are used in fruit trees and plantation crop cultivation, besides in sugarcane, potato and tobacco. Potassium polyphosphate (56% P_2O_5 and 24% K_2O) is produced in solid and liquid forms. This is good for most of the crops.

7.3.1.2 *Micro-nutrient (some)*

Micro-nutrients, though very essential, require very small amount, as our diet with common salt. Recently, it is observed that in some places, parti-cularly under multiple cropping programmes, some of the micronutrients are needed very much in different places.

Iron (Fe): For normal plant growth, requirement of iron (Fe) is in betwe- en 0.5 to 5.0 ppm. which is present in the soils particularly in organic matter.

Zinc (Zn) : The requirement of zinc (Zn) is in between 0.02 and 0.2 ppm in plants. This amount of Zn is available in soil and organic matter. But where multiple cropping programme is in vogue continuously (as in Punjab), deficiency of Zn will be observed.

Boron (B): In plant body, generally 0.1 to 1.0 ppm of boron (B) is present. Its availability is decreased with the decrease of pH of soils, hence in acid soil, its deficiency is observed. Hence in some areas wheat production is hampered for want of B.

Molybdenum (Mo) : Requirement of molybdenum (Mo) is near about 0.01 to 0.05 ppm in plant sap and this amount is present in soils and organic matter. Generally some legumes (berseem) require Mo in good amount. In general, soils under higher acidity, the availability of micronutrients is decreased.

Practical Class : Application of 'basal' and 'top-dressing' of 100-50-50 kg/ha dose of NPK fertilizer, on wheat

Lesson 73 : Apply 50 kg N, 50 kg P_2O_5 and 50 kg K_2O/ha, as 'basal dose' at sowing

Introduction

It has been observed from the experiments at national levels that under normal conditions, wheat requires 100 kg N, 50 kg P_2O_5 and 50 kg K_2O/ha and that had been recommended. For this requirement, generally, 50 kg N, 50 kg P_2O_5 and 50 kg K_2O/ ha is applied as basal fertilizer at sowing and the rest (50 kg) N is applied at later stage. Nitrogen, in general, increases total number of tillers (and effective tillers), number of grains per ear head and grain weight, and phosphate helps in root formation and as a result helps in formation of tillers, total grain number and grain weight of wheat, whereas potash, though not in chemical composition of any matter in wheat, yet its presence improves over all development in growth and yield of wheat. Therefore, balanced nutrients are applied as basal. The learners, should know the application of basal dose of NPK-fertilizers in wheat.

Objective

- to improve the tiller or effective tiller number, grain number and grain weight of wheat,
- to increase total grain yield production of wheat.

Material required

❖ An irrigated field for sowing wheat with fertilizers of N, P_2O_5 and K_2O. ❖ Bullock (2), ploughman (1), mouldboard plough (1) with necessary arran-gement for sowing, ❖ A ladder/plank, ❖ Variety – any variety, ❖ NPK-fert-ilizers for 1000 m^2 area, [@ 50 kg N, 50 P_2O_5 and 50 kg K_2O/ha (from urea, SSP and MOP respectively) for 1000 m^2 area], ❖ Spade-1, ❖ Scale–1, ❖ Balance (spring) – 1 (5-10 kg capacity), ❖ Wheat seeds (as per requirement for 1000 m^2 area, @ 100 kg/ha), ❖ Bowls (4), ❖ Note book, ❖ Pen/pencil.

Procedure

- Plough the land twice, at optimum moisture of soil, after irrigation, followed by ladderings,
- Plough the land for third time and apply the fertilizers, urea first by broadcasting uniformly, followed by SSP and MOP by mixing together, separately on soils,
- Apply laddering to mix the fertilizers, cover up the fertilizers and level the land,
- Again plough the land for the fourth time for sowing wheat seeds, with mouldboard plough, across the furrows, at a space of about 25 cm, either by *kera* or *pora* method or by tractor.
- Note down the date of sowing, amount of NPK-fertilizers and wheat seeds, number of ploughings and ladderings.

[**Note**: Seeds in earlier furrow will be covered by soils through mouldboard plough].

Lesson 74 : Apply 50 kg N/ha from urea, as 'top-dressing' in wheat, just before first irrigation at CRI (crown root initiation) stage

Introduction :

For dwarf wheat, crown root initiation stage (CRI) is very important both for application of first irrigation and top-dressing with nitrogenous fertilizers. First top-dressing of nitrogenous fertilizer (say urea) is broad cast over the soils, just at CRI stage of wheat, (generally 21 days after sowing wheat), followed by mixing with soils before irrigation. Ultimately this top-dressing of nitrogenous fertilizer, improves the tiller number, grain number and grain weight of wheat. As a result yields of wheat increase.

Objective

O to improve the number of tillers per unit area, number of grain per ear- head and grain weight of wheat,

O to increase the yields (both grain and straw) of wheat.

Material required

❖ A field of wheat which was sown at about 18-20 days before this operation, (top-dressing), ❖ N fertilizer, for 1000 m^2 area @ 50 kg nitrogen per hectare (say from urea),

✣ Wheel-hoe or tine, ✣ Irrigation arrangements, ✣ Spade–1, ✣ Balance (spring) – 2 (10 kg and 1 kg capacity), ✣ Sacks (4), ✣ Bowls (4), ✣ Note book, ✣ Pen/pencil.

Procedure

- Broadcast required amount (as calculated) of urea, on soils,
- Mix the urea fertilizer with soils, with the help of wheel-hoe/tine (The mixing with the hoe, helps in making furrows in between wheat rows, for easy passage of water from irrigation),
- Irrigate the land, immediately after top-dressing with urea at CRI stage (CRI stage starts from 18 to 25 days after sowing wheat).

[**Note**: 'Urea' fertilizer is granular and hygroscopic in nature, hence you need not be worried about the spreading it over the plants. Avoid scorching sun].

7.3.1.3 *Secondary nutrients (Ca, Mg, S)*

Calcium, magnesium and sulphur - these three essential elements, are called as secondary nutrients, because to the manufacturers of commercial fertilizers, these nutrients are of secondary importance, in the process of manufacture. Though these elements are essential but these are added through other commercial fertilizer. Recently chlorine (Cl) has been declared as essential element. Calcium content is about 19.5% in calcium nitrate, 195% in super phosphate (single), 33.9% in basic slag, in gypsum 29.2%. Magnesium is available 4.5% in CAN, 1.5 in calcium nitrate, 3.5% in basic slag. Sulphur is available in ammonium sulphate. About 18.60% of S is present in gypsum.

7.3.2 Calculation of fertilizers (N, P_2O_5 and K_2O) requirement in one hectare of land

For any higher crop production, now-a-day nutrients from fertilizers are needed. Therefore, it is essential to calculate the required amount of NPK-fertilizers, as per requirement of nutrients for the crops. Generally, specific fertilizer possesses specific amount of nutrients and this is represented as percentage. For calculation of fertilizers, to be needed, for a particular crop and amount, we must know the percentage of nutrient present in that fertilizer. Here some ideas are given for calculation of nitrogenous, phosphatic and potassic fertilizers. Generally, cereal crops, require more amount of nitrogen (N), pulses require high amount of phosphate (P_2O_5) tuber crops/rhizomes require higher amount of potassic (K_2O) fertilizers. The calculation of requirement NPK-fertilizers is as follows :

7.3.2.1 *Nitrogen (N) fertilizers*

Wheat, being a cereal crop, requires 100 kg N/ha. Now, how much amount of urea fertilizer will be required for that crops and for one hectare of land? Here we must know the percentage of N in urea. We know that nitrogen percentage of urea is 46 %.

Therefore, the amount of fertilizer will be required

$$\frac{100 \times 1{,}000}{100 \times 46} \text{ kg / ha } = 217.4 \text{ kg/ha}$$

∴ 217.4 kg urea fertilizer will be necessary for that crop, for fertilizing one hectare of land (One hectare of land = 10,000 sq.m area).

7.3.2.2 *Phosphatic (P_2O_5) fertilizers*

Suppose, in pulse crop, 60kg phosphate (P_2O_5) is required for one hectare of land, then how much amount of single super phosphate (SSP) will be required for one hectare area? Phosphate content of single super phosphate fertilizer is 16%.

Therefore, requirement of single super phosphate is :

$$\frac{60 \times 10{,}000}{100 \times 16} \text{ kg / ha } = \frac{6 \times 1{,}000}{16} = 375 \text{ kg/ha}$$

∴ 375 kg of single super phosphate fertilizer will be required for fertilizing one hectare of land.

7.3.2.3 *Potash (K_2O) fertilizers*

Suppose, the dose of any tuber/rhizomes crop is 60 kg K_2O/ha. Now, how much amount of muriate of potash (MOP) will be required for that crop in one hectare area? MOP (Muriate of potash) contains 60% K_2O.

Therefore, requirement of MOP for the crop and area will be :

$$\frac{60 \times 10{,}000}{100 \times 60} \text{ kg / ha } = 100 \text{ kg/ha}$$

∴ 100 kg muriate of potash (MOP) will be required for fertilizing one hectare of land.

Practical Class : Calculation of the amount of nitrogenous, phosph-atic and potassic fertilizers for the specific crops for specific areas

Lesson 75 : Calculate the amount of fertilizer to be required for ferti- lizing one hectare of land, with urea @ 100 kg N/ha in wheat

Introduction

Any nitrogenous fertilizer including urea, increases the vegetative parts in plants, like, leaf, stem, inflorescence, roots etc. ultimately gives the higher production. 'Nitrogen nutrient' and 'nitrogenous fertilizer' are two different terms. If we say 100 kg N is required by wheat, meaning different amount, in different fertilizers needed. In this case wheat crop will be fertilized with 100 kg N/ha. Now we will have to calculate the amount of urea fertilizer to be required. For this calculation, of course, the learners must know the percentage of nitrogen in different fertilizers. Urea contains 46% nitrogen.

Objective

- O to find out the amount of urea fertilizer to be needed for fertilizing one hectare of wheat crop @ 100 kg nitrogen per hectare.

Material required

✥ Note book/Exercise book, ✥ Pen/pencil, ✥ Book containing percen-tage of nutrients (like nitrogen, phosphate, potash and others in fertilizers)

Procedure

Calculation

One hectare of land is 10,000 sq.m area,

Therefore, the amount of urea fertilizer will be required —

$$\frac{100 \times 10{,}000}{100 \times 46} \text{ kg / ha} = 217.4 \text{ kg/ha}$$

Therefore, for fertilizing one hectare of wheat crop with urea, @ 100 kg N/ha, 217.4 kg 'urea' will be required.

Lesson 76 : Calculate the amount of fertilizer to be needed for fertilizing pulse crop, with single super phosphate, for one hectare of land @60 kg P_2O_5/ha

Introduction

Requirement of phosphate is high in any leguminous crop, including pulse crop, because of requirement by the *rhizobium* bacteria, harbouring in the legumes for fixation of nitrogen from the air. Besides this, phosphate increases the activities of roots at earlier stage of crops. Like other fertilizers, the learners should know the percentage of phosphate in phosphatic fertilizers for calculation. Phosphate content of single super phosphate (SSP) fertilizer is 16%.

Objective

- O to calculate the amount of single super phosphate for fertilizing pulse crop (say, green gram) @ 60 kg P_2O_5 (phosphate) ha for getting higher yields of green gram,
- O to encourage the root growth of the crop,
- O to help the *rhizobium* bacteria by supplying phosphate, for enriching the soil in terms of nitrogen (through N-fixation).

Material required

✥ Note book -1/ Exercise book, ✥ Pen / pencil, ✥ Book containing percentage of nutrients of the fertilizers.

Procedure

Calculation

One hectare of land is equal to 10,000 sq.m. area. That means we will have to calculate the amount of single super phosphate, which contains 16 % phosphate (P_2O_5), to fertilize the pulse crop (say, green gram) of one hectare of land, where pulse crop require 60 kg P_2O_5/ha.

Therefore, the amount of fertilizer to be required, will be :

$$\frac{60 \times 10,000}{100 \times 16} \text{kg / ha} = \frac{6 \times 1000}{16} \text{kg / ha} = 375 \text{ kg/ha}$$

∴ For fertilizing green gram crop of one hectare of land, @ 60 kg P_2O_5 / ha, requirement of 'single super phosphate' (SSP) is 375 kg /ha.

Note : Generally, legumes (including pulse crops) require high amount of phosphate, whereas non-legumes require less of it. The non-legumes again require a little bit high phosphate in tuber crops, oilseed crops, rhizomes etc. Of course, for balanced fertilization, it (phosphate) is required by all the crops, though amounts vary from crop-to-crop or place-to-place. In India, amount of available phosphate varies from place-to-place.

Lesson 77 : Calculate the amount of muriate of potash needed for fertilizing one hectare of potato, with 60 kg K_2O/ha

Introduction

Requirements of potash in some crops like tuber/rhizomes, fibres etc. are high, because of high production, though this (potash) does not go into the chemical composition in the plant body. [In some crops, of course, K_2O (potash) is not needed at all (e.g. wheat) where adequate K_2O is available in the soils. It has been observed that in most of soils in India, contain adequate amount of potash, to keep up the normal requirement of most of the crops]. Of course, some crops like tubers, oilseeds, rhizome, fibre etc. require more amount of potash. Therefore, the learners should know the calculation for finding out the amount of potassic fertilizers (say, muriate of potash) needed from the potash nutrient for one hectare @ 60 kg K_2O/ha.

Objective

O to increase some yields of crops,

O to utilize this fertilizer (potash) as a balanced nutrients like nitrogen, phosphate etc.

Material required

✣ Note book / Exercise book, ✣ Pen / pencil, ✣ Book containing percentage of nutrients in fertilizers (like nitrogen, phosphatic, potash and others).

Procedure

Calculation

10,000 sq.m area is one hectare of land; one hectare of potato crop (tuber crop) is to be fertilized with muriate of potash @ 60 kg K_2O/ha. Now, the learners should calculate the amount of potassic fertilizer (i.e. muriate of potash) needed for one hectare of potato. Muriate of potash contains 60 % potash. Therefore, the requirement of muriate of potash will be :

$$= \frac{60 \times 10{,}000}{100 \times 60} \text{ kg / ha} = 100 \text{ kg/ha}$$

Therefore, to fertilize one hectare potato crop @ 60 kg K_2O/ha, 100 kg 'muriate of potash' will be required.

7.4 INTEGRATED NUTRIENT MANAGEMENT

Organic manures improve the soil structure, stability, nutrient content of the soils for the growth and development of the crops, but the amount of nutrients is very less. As a result productivity is less. On the other hand, chemical fertilizers improve the productivity of crops, though soil structure and stability do not improve, rather they are gradually deteriorated fast. In order to keep pace with soil health and productivity of crops, integrated nutrient management is urgently needed. It is needed particularly in crops in multiple cropping system and under rotations.

7.4.1 Bulky organic manures plus NPK-fertilizers

It has been observed that NPK-fertilizers along with bulky organic manures improve both productivity and soil health in multiple crop systems both under irrigated and rainfed areas. Generally 50 : 50 : 50 for N, P_2O_5 and K_2O from fertilizers, along with organic manures/crop-residues is better, though huge amount of organic manure is needed. Of course, the efficiency depends on types of crops and types of soils.

Higher amount of organic matter is good for potato, rhizomes (*ol*), tuber crops in alluvial soils; whereas chemical fertilizers improve productivity in wheat, maize, *bajra*, *jowar* etc. in alluvial soils. Of course, rice gives higher yield with NPK-fertilizers, along with organic manures.

7.4.2 Concentrated organic manures plus NPK-fertilizers

Though concentrated organic manures, alongwith NPK-fertilizers, improve crop productivity of field crops, but cost of production is very high. Therefore, concentrated organic manures, along with NPK-fertilizers should be applied to the high-value field crops and to the vegetable crops, in high/ medium land, having alluvial/medium loamy soils, for higher crop production and economic gains, maintaining the soil health.

7.4.3 Concentrated/bulky organic manures plus NPK-fertilizers plus micro-nutrients

In chemical fertilizers, only specified types of nutrients, with specified amounts, are available; whereas in concentrated/bulky organic manures, all types of nutrients required by the crops are available in different amounts. Under multiple cropping systems, huge amount of NPK-fertilizers are needed, for maintenance of soil health. In those systems deficiency of micro-nutrients are also visible. In organic matters, those micro-nutrients are also available. Therefore, application of concentrated/ bulky organic manures, along with chemical fertilizers (including micro-nutrients) are essential for higher production of crops, particularly in multiple cropping system in most of soils and seasons.

General : For crop cultivation several types of fertilizers are to use for getting both major and minor nutrients. For saving time, one is to mix the fertilizer for spreading in fields, but all fertilizer should not be mixed together because of loss of costly nutrients. The chart for "ability and disability of guiding fertilizer mixtures" is very important for this (see I, under Important Information, at the end of this book).

■■■■

CHAPTER 8

IRRIGATION FOR CROPS AND CROPPING SYSTEMS

8.1 WATER REQUIREMENT OF CROPS IN STAGES

For any biological activities, water is essentially needed. In case of plant and animal (including human beings) life, water is actually the "*life*". It is the part of all organisms and some of which contain more than 90% water. Water is an essential part of photoplasm. In plant, it is an important ingredient for *photosynthesis*. About 400 to 500 litres of water is necessary for the production of a kilo of plant-dry matter. It is required for all activities in plants.

Generally, water requirement of crops is (figure in mm) : Rice 900-1200, sorghum 500, wheat 300-350, maize-625, finger millet-350, pearlmillet –350, groundnut-510, sugarcane-2800, cotton-500, banana-2000 etc. At all stages, water is needed but some critical stages are there in crops. These are : CRI, tillering and booting stages in wheat; panicle initiation heading and flowering stages in rice; booting, blooming and milky dough stages in sorghum; silking and tasselings to soft dough stages in maize; heading and flowering stages in pearl millet; primodial initiation and flowering stages in finger millets; tillering and growth period in sugarcane; light flushing at early stage and irrigation in growth stage of potato etc. Rice is grown under water logged condition. Water depths (cm) for rice at early, medium and latter stages are 3-4, 7-8 and 2-3 cm, respectively.

8.1.1 Rice (pre-*kharif*/*kharif*/*rabi* i.e. *aus*/*aman*/*boro*)

According to growing conditions, rice is classified as : (i) *Aus* (direct-seded under rainfed conditions, (ii) *Aman* (transplanted in *kharif* season) and (iii) *Boro* (transplanted in winter season, but harvested in summer).

Rice, under normal conditions, is grown during *kharif* season, with the rain water in eastern parts of the country, actually under rainfed condition. If rainfall is less, then and then only *kharif* rice is irrigated there. In *boro* rice water requirement is high, as it is grown in dry season but under water logged conditions. For both *aman* and *boro* season rice, water stagnation is highly needed, though in *aus* season rice is grown with natural rainfall - water only. Of course, a good amount of water is needed at panicle initiation stage, heading and flowering, stages though it needs waterlogged soils.

8.1.2 Wheat

In wheat critical stages are: (i) crown root initiation (CRI), (ii) tillering, (iii) flowering and (iv) dough stages, though all through its growth, adequate moisture is essential. It is a dry crop, grown in winter season and water/ irrigation is essentially needed during these phases.

Irrigation numbers differ in different areas. Where growth period of wheat is less say, 120 days, 4 to 5 irrigations are given, while duration is long say, 185 days, then 5 to 6 time-irrigations should be given in loamy soils.

8.1.3 Sugarcane

Sugarcane is a long duration crop (1-1.5 years) and its water requirement is also high. It requires irrigations frequently upto maturity. Critical periods are (i) tillering and (ii) growth periods. As because the crop is tall (1.5 to 2.0 m), hence spacing is also large (1m × 50 cm). Immediately after establishment, light earthing-up is done and second earthing-up is heavy. Irrigations are given in the furrows, made by earthing-ups. Generally 12-20 time irrigations are given during the growth period of *adsali* (2 years) crop. Of course, for *eksali* (one year) crop irrigation number is less. Sugarcane is grown in loamy/clayey loam etc. soils, where water-holding capacity of soil is more or less high. In these soils, water stagnates for many days (about 15-20 days), if an irrigation is given.

8.1.4 Potato

Potato is a tuber crop. Tubers are in general, planted in small furrows in flat lands and thereafter furrows are made wider after emergence of seedlings. Before emergence of seedlings of tubers, the land is irrigated by flushing water with the help of bowls/cans/ dish etc. After emergence of seedlings the furrows are again made by earthing-ups of lands, in between furrows that was made earlier, for putting tubers. After earthing-up big furrows are automatically formed. Then 7-8 times irrigations are given in the furrows; 4-5 time flushing are needed for emergence of seedlings of tuber. Generally, two times earthing-ups are done, during plant growth in loamy soils.

8.1.5 Pulses (lentil/pea/greengram/ grasspea/gram and *arahar*)

Pulses, in general, require less amount of water, hence mostly they are grown under rainfed conditions. Of course, sometimes water/irrigation is needed under dry conditions. In that case, one or two irrigations should be given before and after flowering the crops.

Among the pulses, other main and important crops are : pea, lentil, grasspea, gram etc. are grown in *rabi*; blackgram, *arahar* etc are grown in pre-*kharif* / *kharif*. Of course, blackgram is grown in *kharif* but harvested before the start of *rabi* season. *Arahar* is a tall crop requires sometimes irrigation at flowering. Of course, *rabi arahar*, is grown now. This *arahar* is a short duration crop, with high production potentiality.

8.1.6 Oilseeds (mustard/sunflower/groundnut and sesame)

Oilseed crops, like pulse crops, require less amount of water/irrigation. In general, it is cultivated under rainfed conditions. But now-a-days, some newly varieties require some water/irrigations. In general, oilseed crops (high yielding varieties) should be irrigated two times, one before flowering and the second time after flowering. Mustard, rape, niger, safflower etc. are grown in *rabi*, whereas sunflower, groundnut, sesame, jatropha etc. are grown in pre-*kharif* / *kharif* season in loamy soils.

8.1.7 Fibre crops (jute/cotton/mesta/sunnhemp and flax)

Fibre crops are of two type, such as : (i) Lint fibre and (ii) Bast fibre. Cotton is the lint fibre crop, whereas, jute, mesta, ramie, sunnhemp and flax are the bast fibre crops. All the bast fibre crops (except flax) are grown in the eastern region of the country, where rainfall is high and their durations, are long. All these crops, therefore, are grown under rainfed condition. If rainfall is less, then irrigation is needed and that too at growing stage. Cotton is a dry area crop. It requires 4 to 5 time irrigations, if rainfall is less in *kharif* season, though it varies from species-to-species and variety-to-variety in black cotton soil/ loamy/clayey/loam soils.

8.2 SCHEDULING OF IRRIGATION IN CROPS

Scheduling of irrigations in crops are very important for higher production. This scheduling is done on the basis of some methods/conditions, under different approaches. These approaches differ from place-to-place and crop-to-crop and duration of crops also. Generally, 5 approaches are followed in India and these are :

(i) Soil moisture application approach, (ii) Climatological approach, (iii) Open-pan-evaporimetre approach, (iv) Critical stages or physiological stages of crop-growth approach and (v) Combination and leaf-rolling approach. Different crops are to grow with different approaches, under relevant situations.

8.2.1 Soil moisture depletion approach

For '*soil moisture depletion approach*', tensiometer reading is essential or some percentage of moisture in the soils is also essential. With tensiometer, depletion of soil moisture is measured and according to that data, irrigation schedule may be done. *Soil moisture tension* is the force with which the water is held by soil against gravity. It is generally expressed in atmospheres or in mm of mercury.

Soil moisture depletion is measured by the percentage of moisture in soils (Gravimetric method) determined by ordinary method like this :

$$\text{Soil moisture content (\%)} = \frac{(\text{Weight of moist soil} - \text{Weight of oven dry soil})}{\text{Weight of oven dry soil}} \times 100$$

Some crops like wheat, *jowar*, *bajra*, maize etc. are sometimes irrigated by, this approach.

8.2.2 Climatological approach

The '*climatilogical approach*' is sometimes adopted for irrigation in some crops like wheat, maize etc. In this approach, irrigation is given in some crops in some areas, if moisture in soil is less, due to less rainfall. In Punjab, rainfall is inevitable during wheat growth, but if it fails, then irrigation should be given, if the crop is in critical condition. Similarly rice, maize or jute are irrigated in *kharif*, if rainfall is less in eastern part of the country, but in other times in *kharif* season, irrigation is not essential.

8.2.3 Open-pan-evaporimeter approach

This '*open-pan-evaporimeter approach*' is done in the commercial mechanised cultivation areas, as in Punjab. In this system, 'open-pan' is set in the field, with the water measuring arrangement. Water loss is there, which is obtained from evaporimeter reading, is replenished after some days, through irrigations. In wheat, this system is adopted in our country or in some foreign countries also. This system will work in all the big cultivated areas, with the same crop under cultivation.

8.2.4 Critical stages or physiological stages (of crops) approach

This '*critical stage*' or '*physiological stages*' approach, is in vogue in most of the crops and in most of the countries, because it is easy to recognise the critical stages, once requirement of irrigation is fixed. In case of wheat in India, critical stages are: (i) Crown root initiation (CRI), (ii) Late tillering (LT), (iii) Flowering (Fl) and (iv) Dough stages. Hence, in eastern, central and southern parts of India, 4 to 5 irrigations are needed, whereas in Northern, North-Western and Hilly zones. 5 to 6 or 7 irrigations are given, due to long duration of wheat crop. Of course, in other crops like pulses, oilseed etc. irrigation is given with this approach.

8.2.5 Combination and leaf-rolling approach

Sometimes '*combination and leaf-rolling approach*' is followed in some crops, in some particular areas, where soil type is peculiar. Some times critical stages are considered, some times tensiometer approach is adhered to, along with the leaf-rolling approach. In some areas of sugarcane cultivation, *leaf-rolling* approach is considered for applying irrigations in the fields. The tip of the smallest leaf is rolled with deficiency of water in the soil and gradually intensity increases with more leaf-rolling. This leaf-rolling is sometime considered for irrigations in sugarcane, because of long duration.. Of course, in no other crops, this approach is considered now.

Practical Class : Irrigation of wheat and sugarcane crops, at different stages

Lesson 78 : Irrigate the wheat crop, specifying the time for irrigation by observing the physiological stages

Introduction

Wheat, both in India and abroad is very important crop. Therefore, importance has been given in India and abroad for irrigation. In advanced countries, wheat is irrigated either according to the availability of moisture in the soils or as per the evaporation rate in the wheat field. But in India, mostly `physiological stages' are used for irrigating the wheat crop in the field, because in some stages irrigation requirement is high, though some amount of moisture in soil is highly necessary for all the growth periods of the crops. It has been, experimentally verified that the following stages/ phases in wheat, are very important in respect of irrigation.

Critical stages for irrigation in wheat (with specific times)

(1) Tillering phase (CRI) : This phase is known as CRI (crown root initiation) stage or earlier tillering stage. This CRI stage is very important (for the tiller formation, in respect of application of nutrients and irrigation. Under normal conditions [i.e. normal sowing, (mid Nov.), with normal seed rate, (100 kg seed rate/ha), normal fertilization (NPK-@ 100, 50,50 of N, P_2O_5 and K_2O/ha), normal irrigations (4-6) times], the CRI stage in wheat comes at about 18-25 days after sowing. This period is very vital for tiller formation. Irrigation is needed very much during this period.

(2) Late tillering/jointing phase (LT) : This phase/stage (LT) is observed in between tillering and flowering period of wheat. The plants are a little bit pale in colour at this stage and elongate the plants very rapidly. This stage comes around 40 to 45 days after sowing but in long duration areas (i.e. long winter) the plants look green and the stage comes around 45 to 50 days after sowing.

(3) Flowering (Fl) : This flowering stage is easily recognised by the inflorescence of the wheat plants. This period comes at about 62 to 65 days after sowing wheat in short winter, but 70-80 days after sowing wheat in long winter areas. Irrigation is needed very much during this period.

(4) Milk stage (Milking) : After flowering, the grain formation starts in the form of milk. Hence this stage is known as milk/milking stage and it takes place generally during 85-90 days or 100-110 days after sowing, in short and long winter areas respectively. [In long winter-season areas, two more irrigations are required at (5) **dough** (after milk stage) and (b) **booting** (after late jointing but before flowering stages). These two phases come in between about 120 to 130 and 60 -70 days after sowing wheat, respectively].

Objective

- O to increase the number of tiller per unit area, number of grain per ear head and grain weight of wheat,
- O to produce higher yields in wheat.

Material required

✥ Wheat field for irrigations, ✥ Irrigation appliances, ✥ Source of water from high level, ✥ Urea fertilizer, for 1000 sq.m area [@ 100, 50 and 50 kg N, P_2O_5 and K_2O/ha respectively, where 50 kg N will be top-dressed at CRI stage)], ✥ Balance (spring balance, 10 kg capacity) – 1, ✥ Sacks (4), ✥ Bowls (4), ✥ Spade (4), ✥ Wheel-hoe (4), ✥ Note book, ✥ Pen/ pencil.

Procedure

(a) Irrigation for short/medium duration wheat crop

(1) First irrigation at CRI stage : First irrigation in wheat should be applied just after application of urea and mixing it with soils, during about 3 weeks at CRI stage after sowing, at a depth of 6-7 cm.

(2) Second irrigation at late jointing (Jt) : Second irrigation in wheat should be applied in the similar way as in (1), during about 6 weeks at late jointing stage, after sowing wheat (without fertilizer) at a depth of 6-7 cm.

(3) Third irrigation at flowering stage (Fl) : Third irrigation should be applied at flowering stage at about 9 weeks after sowing wheat (with out fertilization), at a depth of 6-7 cm.

(4) Fourth irrigation at milk stage (milk) : The fourth irrigation should be applied during milk stage, at about 12 weeks after sowing wheat, at a depth of 6-7 cm. In all cases water depth should be in between 6 to 7 cm. [If the weather is dry, then one more irrigation should be given at dough stage of wheat].

(b) Irrigation for long duration wheat crop

Irrigations should be applied at a depth of 6-7 cm in wheat at the following physiological stages of wheat :

(1) CRI, (2) at late tillering (LT) or jointing (Jt), (3) at booting, (4) at flowering (FL), (5) at milking and (6) at dough stages. Note down the date of irrigation, physiological stages, amount of water added etc.

Lesson 79 : Irrigate the sugarcane crop at different stages, by climatological and leaf-rolling approach

Introduction

Sugarcane, the sugarcane crop, can be grown either as *eksali* or *adhsali* crop. In both the systems, irrigation requirement is high. Generally it requires nearly 110 to 120 cm. depth of water in 12 to 14 time irrigations. Water requirement is high in dry season than the wet season. All arrangements are done, considering the *eksali* crop, which is planted in autumn season (i.e. after rainy season).

In some countries, sugarcane is irrigated either by sequence (i.e. climatological approach) or by leaf-rolling method. In the sequence, irrigation is given once in a month (according to weather), whereas in leaf-rolling method, irrigation is given by seeing the 'tip-leaf-rolling', indicating deficiency of water in the soil. Now-a-days, in most of the areas, these two methods are combined and irrigations are given accordingly. Sugarcane, though requires high amount of irrigation water, but it is not the water-loving crop, like transplanted *kharif* rice which is grown in water-logged condition. Of course, some sugarcane can be grown in waterlogged conditions in Bangladesh or other low lying areas with different varieties. In the trench system, irrigation is given in the trench itself for germination of 'eyes' in the formation of seedlings. Then trench is filled up with the side-soils. After this, earthing-up is done (i.e. furrow is created by adding the soils and two sides of the clumps. Then irrigation water is passed through the furrows. Again second time earthing-up is done afterwards, but only irrigation water is passed through the furrows at a depth of 8-10 cm. depth.

Objective

- O to increase the number of cane per unit area, for sugarcane,
- O to increase the '*cane*' yield of sugarcane.

Material required

✤ Sugarcane field, where sugarcane has been planted in (autumn later part of *kharif) kharif* in *trench* method, ✤ Irrigation arrangement (from the higher levels), ✤ Source of water from higher areas, ✤ Spade – (4), ✤ Labourers 10 (for spading), ✤ NPK-fertilizers (urea, SSP and MOP) for 1000 sq.m. area @ 60 kg N, 60 kg P_2O_5 and 60 kg K_2O/ha should be applied as basal, but rest N (60 kg) at first and the second earthing-up, followed by irrigations), ✤ Balance (spring balance 10 kg capacity), ✤ Sack (4), ✤ Bowls (4), ✤ Note book, ✤ Pen/pencil.

Procedure

Irrigation schedule as under, should be followed methodically with sequences, along with leaf-rolling method, if needed :

First irrigation after seedlings emergence

- After first earthing-up in the month of November, first irrigation water should be applied through the furrows, after applying NPK-fertilizers,

Second irrigation

- After second earthing-up irrigation water should be passed through the furrows, after top-dressing with urea only, in the month of December,

Third irrigation

- In the third irrigation water should be applied through the furrows (as was done earlier), in the month of January.

Fourth to Twelfth irrigations

- All these irrigations should be applied in the month of February (4th), March (5th), April (6th), May (7th), June (8th), July (9th), August (10th), September (11th) and October (12th), as usual, with 8-10 cm of water. In the months of March, April and May, leaf-rolling may be observed and for which more irrigation water may be required. On the other hand, in the rainy season months (i.e. June, July, August etc.) requirement of irrigation water will be less.
- Sometime excessive water may be stagnated in rainy season.

- This should be removed or drained out of the furrows by cutting the end portion of furrows,
- Write down the dates of operations, with the amount of water applied and with the method adopted and other things in detail.

8.3 METHOD OF IRRIGATION

'*Method of irrigation*' is very importation for proper use of irrigation water and for higher production of crops. Again methods of irrigations depend upon so many factors like, need of the crops, place, source of water, slope of land, availability and amount of water etc. Major irrigations are done, generally through (i) Surface or wild irrigation, (ii) Sub-surface irrigation, (iii) Flash irrigation, (iv) Strip irrigation, (v) Sprinkler irrigation, (vi) Drip irrigation and (vii) Spot irrigation.

8.3.1 Surface or wild irrigation (as in wheat)

In flat lands, '*surface irrigation*' or '*wild irrigation*' is given in most of the crops like wheat, rice, mustard, pulses etc. For this type of irrigation, the crop-plots are divided in smaller plots like 10m × 50m or more. In between two big plots, an irrigation channel is passed through. Water sources should be on the upper side, then individual plots of 10m × 10m size should be irrigated. Dimension may be different according to crops and water source. In this method application of water is easier, provided the water sources are in slight higher areas. Of course, wild irrigation is applied in transplanted rice (*kharif* and *boro*) where some small plots are prepared (say, 10m × 10m or 10m × 20m and so on).

8.3.2 Sub-surface irrigation (as in sugarcane)

The '*sub-surface irrigation*' is given in the crops which are earthed-up once or twice. Generally, sugarcane, potato, maize, groundnut etc. are earthed-up. In those furrows (after earthing-up), irrigation water is applied and these amounts of water are soaked by the two sides of the furrows and ultimately plant utilize this water. Of course, in 'sub-surface irrigation', water-loss is high because of deep furrows made by earthing-up of the crops.

8.3.3 Flash irrigation (as in potato at earlier stage, later sub-surface irrigation)

Some times flashing of water is done with the help of some bowls, dishes/plates etc. to some crops, which do not germinate, due to lack of moisture in the soils or in some surface soil crops like potato. In potato two types of irrigations are given, one type i.e '*flash irrigation*' is given for germination of the potato which is planted flat first, then first earthing-up is done and there after '*surface irrigation*' is given through the furrows.

8.3.4 Strip irrigation (as in wheat)

If the land is a little bit sloppy (1 to 2%), then '*strip irrigation*' is done, particularly in wheat. From the upper side of slope, irrigation channels are prepared at a distance of 10-15 m width, at a length of about 50 to 100 m. Across the irrigation channels, some bunds are given in between two channels at a space of about 10 to 15 m (or so according to slope). Then water is passed through the channels at the end. From there wheat field is irrigated and one should come towards up gradually from down portion in irrigating the land.

8.3.5 Sprinkler irrigation (as in betelvine or in some vegetables)

In some crops like *betelvine*, *'sprinkler irrigation'* system is applied, to sprinkle water on the leaves or stems of the plants/crops. Here, of course, water requirement is less, but it is costly because of high cost of machine and pipes etc. For irrigation on lawn also this system of irrigation is used. No irrigation is given like this, in any field crops, other than some vegetable crops.

8.3.6 Drip-irrigation (as in grapes)

'Drip irrigation' system is followed in irrigating the grape vine or other crops. Here, entire area is not irrigated; with small amount of water near about stem, soil is soaked, so that moisture is taken by the roots of the high valued crops, particularly in dry areas. In this system also water required, is very small but continuous only on root zone of crops. Of course 'drip irrigation' is not given to any field crop.

8.3.7 Spot irrigation/micro-irrigation (as in seedlings of some crops)

This type of *'spot irrigation'* is given to newly planted seedlings like, brinjals, chillies, cauliflower, cabbage etc. but not in any general field crop. In this system of irrigation, water is given in a small amount to the base points of seedlings, for some days, for establishment of the crops properly. After establishment, of course, the furrows are prepared by light earthing-up of the crops and thereafter surface/sub-surface irrigation is given. This 'spot irrigation' is normally given by pitcher, tubs, mug or buckets containing water.

8.3.8 Drainage system for removal of excess water

Irrigation and *drainage* go hand in hand in any irrigation programme/ system. Excess water in irrigation, sometimes cause some damage to the crops, hence that excess water is drained out. Besides, in some areas (like in eastern parts of the country) heavy rainfall accumulates water in the root zones of some crops and due to water stagnation, some of the crops die. Hence drainage system is essential for proper growth of the crop.

Practical Class : Practising flash-irrigation in potato and spot irrigation in chillies

Lesson 80.Practise the flash irrigation in potato, before emergence of seedlings

Introduction

'Potato', a tuber and winter crop is grown under ground, in ridge and furrow method. But in earlier stage the 'seeds' (seed tuber) are placed in flat land with highly tilled conditions. For germination, these 'seeds' require flashing water several times, for moisturing the soils.

Of course, under normal conditions, when adequate soil moisture is available, 3-4 flashings are necessary; but in dry soils, 5-6 flashings are needed during 15 days from sowing time. Of course, after emergence of seedling, the line-sown potato crop is earthed-up and then irrigation water is passed through furrows. But before that flash irrigation is given by spreading water from the main channels with bowels. Actually potato is grown normally with potato-tubers, the vegetative stems ('seeds'), though potato can be grown with tubers, but now-a-days, that can be grown with `true-potato-seeds' (TPS) also.

Objective

O to germinate the 'seeds' into seedlings,

O to enhance the growth of the stems,

O to get more tuber yields from the plants.

Material required

✣ A potato field which had been just sown, ✣ Spade (4), ✣ Bowels (4), ✣ Water sources for irrigation, from higher position, ✣ Note book, ✣ Pen/ pencil.

Procedure

- *First flash-irrigation* water should be spread over the flat land care-fully, where *tubers* have been planted, with the help of bowels, from the main channels which is to be prepared immediately after sowing potato,
- Wait for two or three days,
- On the 4th or 5th day of planting potato, apply the second irrigation water, again over the flat land where *tuber* has been planted (like previous flashing),
- Again wait for further two 2 or 3 days,
- On the 7th or 8th day of planting potato, apply the third irrigation as the earlier one. Observe visually the germination of seed-lings. If it is enough, then stop irrigation further, if not, then continue the process as usual upto ten days, after sowing, and
- Note down the number, time of irrigation and number of emergence of seedlings per unit of area.

Lesson 81. Practise for spot-irrigation in chilli crop in the field (after transplantation of seedlings)

Introduction

Chilli crop (spice) is generally cultivated in the field, with the seedlings, raised in the nursery beds, mostly in *rabi* season, when its water requirement is high. After the transplantation of seedlings of chilli, small amount of irrigation water is necessary/

needed for their establishment. Therefore, the learners should know the process of spot-irrigation in chilli in the field, after transplantation for its establishment. Spot-irrigation is generally done, with small amount of water, applied with the help of some pitchers or other pots, just around the plants, planted on the ploughed land. The plants are kept under sheds for some days, till they are established. Generally, with 7-10 days after transplantation, water is required for their establishment and during that period, water is urgently needed. [For sheds, different materials are used, according to availability. In the eastern part of the country, generally, banana pseudo stem is used].

Objective

- O to establish the roots,
- O to develop the plant growth,
- O to flower and bear the fruits,
- O to get more yields of chilli.

Material required

✤ A field with chilli, just planted, ✤ Water source, ✤ Small piece (about 16 cm) of banana-pseudo-stem (after harvest of banana), ✤ Spade (1), ✤ Small sticks (40,000/-), ✤ Pitchers/tubs (4), ✤ Mugs (4), ✤ *Nirani* (1), ✤ Note book, ✤ Pen/pencil.

Procedure

- Lift water from the water source, and apply it by the sides of the chilli plants, through the tubs/pitchers,
- Apply a mug of water just on the root area of the chilli plant,
- Thus apply few mugs of water around all the chilli plants, planted in 1000 sq.m area,
- Prepare the 15-16 cm long, pseudo-stem brackets of banana, amounting to 40,000 pieces (for 1000 sq.m area, brackets about 10 cm x 15 cm),
- Make ready 40,000 pieces of stick (10 cm length x 10 mm dia),
- Put one stick by the side of each chilli plant (at about 10 cm away from the stem of the chilli plant),
- Put the banana stem pieces just over the chilli seedlings planted, keeping one end of it on the end of the stick (The stem pieces will cover the new seedlings and help against scorching rays of the sun in day-time and at night this should be removed),
- Water the plant in this way, in the evening, after removing the bracket and keep the bracket in the morning every day,
- Watering should be done every alternate days, for 7-10 days,
- After 7-10 days, loose the soils by the sides of the plant without damaging the roots of the crop with the help of *nirani/kharif* etc.,

- Apply more water (about double) this time (4th time) around the plant from the pitcher/bucket with mug, and
- Remove both the brackets and stick, and keep the seedlings as such.

[**Observations :** Generally, the seedlings become pale in earlier 7 days but at later dates after 8th day onwards, the seedlings would be slightly green, indicating the striking roots of the seedlings].

8.4 IRRIGATION SCHEDULE FOR CROPS

Under irrigation, most of the crops require some irrigation schedules; i.e. in particular time particular number of irrigation is needed, for higher production of crops. Sometimes light irrigations or sometimes heavy irrigation for a particular crop, are needed.

8.4.1 Wheat

Wheat is a winter-loving crop, on which huge amount of works, on national basis, have been done and schedule of irrigation has been establ-ished. These schedules are as follows. Generally, wheat requires 4-6 irrigations according to duration of the crop. In the eastern parts of the country 4 irrigations, whereas in the northern parts 6 irrigations are needed, such as :

(1) *First irrigation* at crown root initiation stage i.e. CRI stage, 18-21 days after sowing, (2) *second irrigation* at late tillering (LT), 42 days after sowing, (3) *third irrigation* at flowering i.e. 63 days after sowing, (4) *fourth irrigation* at milk/dough stage at 84 days after sowing. These 4 irrigations are needed in eastern parts of the country, where duration of wheat crop is about 120 days. But it can be extended upto (5) *fifth* and (6) *sixth* irrigation at dough and late-dough stages where duration of wheat crop is about 140 to 160 DAS, particularly in northern parts of the country.

8.4.2 Potato

In potato, the winter crop, two types of irrigations are given. *First type* is flash irrigation, at earlier stages of growth upto emergence of seedlings and *second type* is deep irrigation in furrows at later stage of growth. In the first type, 2 to 3 flashes are given with 3-4 cm of water depth; in the second type i.e. deep irrigation – 4-5 times in the furrows of about 6-8 cm depth of water. For first type of irrigations, generally dishes or bowls are used for water flashing but in second type of irrigation, water from upper side is passed through channels in furrows, without any container.

8.4.3 Maize

Maize is a C_4-type of crop; hence it can be grown at every time of the year but winter crop gives the maximum yields. During winter period like wheat, it requires 4-6 irrigations, about 20 days intervals. In eastern part, it requires 4 irrigations, whereas it requires 6 irrigations in other places where long duration winter season prevails. Water is channalised into furrows. Maize can also be grown without irrigation in eastern parts, during pre-*kharif* and *kharif* seasons. During these seasons, in other parts of the country, 4-6 irrigations at 20-25 days interval, should be given.

8.4.4 *Boro* paddy

Transplanted *boro* rice is transplanted in winter season, by November and for that purpose water stagnation is needed upto 5-6 cm depth of water. After transplantation in November upto harvest in March/April nearly 10-12 times of irrigations about 12-15 days interval, at a depth of 7-8 cm, should be given. During earlier stages of growth, water requirement is less, whereas in later stage water requirement is high. Of course, nearly 20-30 cm of water is needed for puddling the soil. In *boro* paddy water requirement is very high; wild irrigation is given.

8.4.5 Sugarcane

For both *eksali* and *adsali* sugarcane crop, water requirement is high. Generally irrigation is given in furrows through the channels. Generally 12-15 times irrigations are applied at an interval of 30 days, at a depth of 8-10 cm. During rainy seasons, irrigation should be less, whereas in dry season it should be high.

8.4.6 Cotton

Cotton is dry-area crop. It requires about 5-6 irrigations or more at an interval of 20 days. Irrigation is given in furrows, through channels. At each interval 8-10 cm depth of water is given. First irrigation at 20 days after sowing is essential. Irrigation at flowering the plant is also essential.

Practical Class : Scheduling for the irrigation in wheat and potato and practising them at stages of growth in wheat and surface irrigation in potato

Lesson 82. Prepare the irrigation schedule for wheat and practise on them at all stages

Introduction

Schedule of irrigation in wheat is very important for maximum grain yield, because in critical stages or phases of wheat plants improve the growth and yield of plants. All these critical stages are as follows :

(i) Crown root initiation stage (CRI), (ii) Late jointing (LT), (iii) Flowering (Fl), (iv) Milking stage, [From (i) to (iv) irrigations are required for wheat, grown in shorter/medium duration areas] but for long duration crop particularly in long duration winter, irrigation at (v) Booting and (vi) Dough stage are also essential.

[From (i to iv) and (v to vi) irrigations are required for wheat, grown in shorter and longer duration areas respectively] (see as in lesson 78).

Objective; **Materials and methods** and **Procedure :** All as in lesson 78.

Lesson 83. Prepare the irrigation schedule for potato and practise them at later stage of the growth (i.e. channel irrigation)

Introduction

Potato is a crop where two types of irrigations are needed. These are :

(i) Flash irrigation (at earlier stage i.e. for germination of 'seed tubers' for the formation of seedlings).

and (ii) Channel irrigation (4-5 times) (i.e at later stages after preparation of channels).

'Tuber potato' is harvested 3 to 3.5 months after sowing 'potato tuber' as seeds, under normal condition in winter. If the plant growth is more, yield is more.

Objective; **Material required,** as in lesson 80; plus ✣ Channels ✣ Furrow and ridges ✣ Irrigation arrangements.

Procedure

(i) Flash irrigation : as in lesson 80.

(ii) Channel/furrow irrigation :

(a) **First channel/furrow irrigation** : should be applied after channel formation, after first earthing-up of the soils in lines and it is done/ made after optimum moisture conditions, through the furrows (about 15-20 days after sowing), giving about 8-10 cm depth of water (i.e. at the ridge height),

(b) After first channel irrigation, when optimum moisture comes, second earthing-up should be done,

(c) **Second channel/furrow irrigation** should be applied, in the same way (as in the above), through the same furrows, giving about 10-12 cm depth of water, at about 30-40 days after sowing,

(d) **Third channel irrigation**, should be applied as in (c), at about 10-12 cm depth, about 45-55 days after sowing,

(e) **Fourth channel irrigation**, should be applied, as in (d), at about 65-75 days after sowing, with a depth of 8-10 cm,

(f) **Fifth channel irrigation**, should be applied, with a depth of 6-8 cm, at about 80-90 days, after which, before harvest of the crop, no irrigation should be given. [Before the harvest of the crop, land should be dry so that the quality of the potato is good].

■■■■

CHAPTER 9

WEEDS AND WEED MANAGEMENT FOR CROP PRODUCTION

9.1 WEEDS

The man who first used the word *'weed'* in the present-day-meaning, was Jethro Tull. Therefore, subsequently, several definitions of weed, of course, were proposed by different scientists at different times and these are :

(i) A weed is a plant growing, where it is not wanted, (ii) Weed is an unwanted plant, (iii) A plant that is extremely noxious, useless, unwanted or poisonous, is the weed, (iv) Weed is a plant out of place, (iv) Any plant or vegetation, excluding fungi or bacteria, interfering with the objectives or requirements of people, is the weed.

The weeds rob the crops of the plant nutrients, water and light, as a result, yield is reduced. It is menace in most of the cases in crop production, but of course, in some cases it is beneficial too.

9.1.1 Different dominant weeds

There are some dominant weeds both in broad and narrow-leaved weeds (including sedges), in different seasons in different crops in the fields. As these weeds are dominant in nature, crops are affected in respect of growth and yield, due to depletion of nutrients, space, light and moisture/water, allotted for the crops. Different dominant weeds under broad, narrow-leaved weeds, in different seasons, are stated below :

9.1.1.1 *Dominant broad-levelled weeds in the seasons*

***Rabi* season :** Among the dominant broad-leaved weeds in *rabi* season, following are very important.

(i) Pig weed (*Chenopodium album*), (ii) bengal vetch (*Vicia* spp.), (iii) white-sweet clover/yellow-sweet clover/ sweet clover (*Melilotus alba / indical parviflorus*), (iv) anagallis (*Anagallis arvensis*) etc.

Pre-*kharif* season : In pre-*kharif* season, the dominant broad-leaved weeds are :

(i) Physalis (*Physalis minima*), (ii) black-night-shade (*Solanum nigrum*), (iii) bhangra (*Eclipta alba*), (iv) kanteri (*Commelina benghalengis*), (v) kanchari/sechisag (*Alternanthera sessilis*) etc.

***Kharif* season :** Following are the dominant broad-leaved weeds:

(i) Wild brinjal (*Solanum khasianum*), (ii) swamp cabbage (*Ipomoea raptans*), (iii) *mikania* (*Mikania scandens*), (iv) beggar weed (*Desmodium tortuosum*), (v) parthenium (*Parthenium hysterophorus*) etc.

9.1.1.2 *Dominant narrow-leaved weeds in seasons*

***Rabi* season :** Main dominant narrow-leaved weeds in *rabi* are :

(i) Bermuda grass (*Cynodon dactylon*), (ii) *chiria kabajra* (*Pharis minor*), (iii) wild oat (*Avena fatua*), (iv) darnel grass (*Lolium temulentum*) etc.

Pre-*kharif* season : Following are the dominant narrow-leaved weeds in pre-*kharif* season :

(i) Barnyard grass (*Echinochloa crusgalli*), (ii) digitaria/*Digitaria sangui-nalis*), (iii) gamba grass (*Andropogon gayanus*), (iv) goose grass (*Eleusine indica*), (v) hilo grass (*Paspalum conjugatum*), (vi) cogon grass (*Imperata cylindrica*) etc.

***Kharif* season :** Main dominant narrow-leaved weeds in pre-*kharif* / *kharif* season are as follows :

(i) Sporobolus/*chiria kadana* (*Sporobolus diander*), (ii) rice grass (*Leersia hexandra*), (iii) brachiaria (*Brachiaria deflexa*), (iv) bermuda grass (*Cynodon dactylon*) etc.

9.1.1.3 *Dominant sedges in seasons*

***Rabi* season :** The following dominant sedges in *rabi* season are :

(i) Asphodelus (*Asphodelus tenuifolius*), (ii) mootha (*Cyperus rotundus*) etc.

Pre-*kharif* season : The main dominant sedges in pre-*kharif* season are as follows :

(i) *Mutha*, (ii) *Jalmutha* (*Cyperus iria*) etc.

***Kharif* season :** Following are the main dominant *kharif* season sedges:

(i) *Mutha*, (ii) *jalmutha*, (iii) *rosonia* / *scirpus* (*Scirpus erectus*) etc.

9.1.1.4 *Major weeds in wheat, rice, jute and rainfed paira / utera crops in rabi seasons*

There are some major weeds (both broad/narrow-leaved) in some crops like rice, wheat, jute and rainfed *paira* / *utera* crops, which reduce production of crop substantially. *Aus* rice (direct-seeded), wheat, jute and *paira* crops are heavily affected by some weeds, and these are as follows :

Wheat : Wheat crop in *rabi* season, is affected by the following broad and narrow-leaved weeds :

(a) Broad leaved weeds : (i) *Bathua*, (ii) *Vicia* spp., (iii) *Melilotus* spp, (iv) *Anagalis* etc.

(b) Narrow leaved weeds : (i) *Chiria-ka-bajra*, (ii) *Phalaris*, (iii) Darnel grass, (iv) Bermuda grass etc.

Rice : The following broad and narrow-leaved weeds affect the pre-*kharif* / *kharif* season rice crop :

(a) Broad-leaved weeds : (i) *Physalis*, (ii) *Eclipta*, (iii) *Croton*, (iv) *Solanum*, (v) *Tridex* etc.

(b) Narrow-leaved weeds : (i) Bermuda grass, (ii) *Sporobolus*, (iii) *Shyma* grass, (iv) *Paspalum* etc.

Jute : (a) Broad-leaved weeds : (i) *Physalis*, (ii) *Eclipta*, (iii) *Solanum* spp., (iv) Croton, (v) Tridex etc. infest the jute crop in pre-*kharif* season heavily.

(b) Narrow-leaved weeds : (i) Doop grass, (ii) *Cyperus*, (iii) *Sporobolus*, (iv) *Shyama* grass, (v) Paspalum etc. attack the jute crop in pre-*kharif* season heavily.

Rainfed *rabi paira / utera* crops

(a) Broad-leaved weeds : (i) *Chenopodium* spp., (ii) *Vicia* spp., (iii) *Melilotus* spp., (iv) Sun berry, (v) *Jangli palak* (*Rumex maritimus*), (vii) Wild tobacco (*Nicotiana* plum*baginfolia*) etc. are the dominant weeds.

(b) Narrow-leaved weeds : (i) *Doop* grass, (ii) *Sporobolus*, (iii) *Cyperus* spp. etc.

Notes : Instead of menace, some of the weeds are beneficial to the crops under rainfed *rabi paira/utera* conditions. The crops benefited are grasspea, lentil, sunnhemp (for seeds) linseed etc. The weeds that give support to the crops are, *Cyperus*, *Melilotus*, *Chenopodium* etc.

Practical Class : Identification of some weeds in crops (like wheat, rice, jute and rainfed *rabi paira / utera* crops)

Lesson 84 : Identify with scientific names, at least 3 broad-leaved and 3 narrow-leaved weeds in wheat fields

Wheat

(a) Broad-leaved weeds

(i) ***Bathua*/pig weed** - *Chenopodium album*; Family Chenopodi-aceae; annual; flowers white/yellow as per varieties; height about 10-70 cm; leaves soft, used as vegetable; stem solid; damage the wheat crop heavily; tap root with branches.

(ii) ***Senjimethi*** - *Melilotus alba / indica / perviflorus*; family *Legu-minoceae*; annual; vine type; leaves are small; stem solid, small tendrils; flowers red/pink/purple, small in size in cluster forms, vines grasp the wheat; can be used as fodder.

(iii) ***Solanum nigrum*** – *Solanum nigrum*, family *Solanaceae*, annual; straight plant with cluster of fruits in sacks; height about 30-50 cm; tap root with other strong roots; leaves oval type; more branchings; stem solid; flower white but in cluster; damage the crop at later stage of wheat.

(b) Narrow-leaved weeds

(i) ***Mutha***– *Cyperus rotundus*, family *Cyperaceae*; sedge, perennial, with nuts hence called *nut grass*; root fibrous, nuts in the roots; height about 6-10 cm; grows in all the season; a long stick comes as inflorescence with star-like brownish flowers; stem is solid with 3 ridges.

(ii) ***Doop / durba*** – *Cynodon dactylon*, family Gramineae; grass of small size grows every where; perennial, root fibrous, height about 4-6 cm, grasps the soils heavily, grows in all seasons, leaves small; stem had; inflorescence also small with star-like flowers; used as soil binding material.

(iii) ***Goolidanda/Phalaghash*** – *Phalaris* minor; family Gramin-eae; plant is like wheat; fibrous roots, stem hollow; inflorescence – like cat-tail with white flowers (clusters), diameter about 1-2 cm, length 4-5 cm, flower earlier than that of wheat; an obnoxious weed in wheat.

Lesson 85 : Identify at least 3 broad-leaved and 3 narrow leaved weeds, of *aus* rice (direct-seeded) with scientific names from the fields

(a) Broad-leaved weeds

(i) **Sunberry** – *Physalis minima*, family Solanaceae; annual, height about 50-60 cm. fruits in winter but growth in summer /rainy season; bigger leaves in winter, smaller leaves in rainy season; stem solid, leaves big, inflorescence in clusters, fruit in sack, yellow in colour when ripe.

(ii) ***Eclipta alba* (*Kesut*)** – family *Compositae*, annual, height 40-50, grows in warm climate; leaves are small; stem solid; flower – star-like (like aester), weed of all crops in summer and rainy season.

(iii) ***Tridex***– Scientific name *Tridex procumbens*; family Composi-teae; annual, grows in warm climate; leaf broad; inflorescence stock, bending; stem solid.

(b) Narrow-leaved weeds

(i) **Nut grass/*mutha*** – *Cyperus rotundus*, family Cyperaceae, parrenial, leaves thin long, pseudostem, hardy, nuts in the ground, inflorescence, star – like, damages the crop heavily.

(ii) **Bermuda grass / *doop* / *durba*** – *Cynodon dactylon*; family Gramineae; perennial, leaves small, stem fine, hardy, inflore-scence small, star-like, grows in all types of soils in all seasons. Grows both in warm and cool climate; may be used for soil conservation.

(iii) **Barnyard grass / *'shyamaghash'*** – *Echinochloa crusgalli*; family Gramineae, annual, grows in warm climate, height about 30-80 cm, obnoxious weeds, but may be used as fodder, plant is like rice plant. It may also be used as organic manure in pit/trench.

Lesson 86 : Identify, with scientific names, at least 3 broad leaved and 3 narrow-leaved weeds of jute

(a) Broad-leaved weeds

(i) *Physalis minima* : family *Solanceae*, annual, big plant in pre-*kharif* and *kharif* seasons but small in *rabi* season. Inflorescence yellowish, fruit is in sacks, stem solid, damages the crops heavily.

(ii) *Solanum nigrum* : (as in lesson 84 a iii)

(iii) Croton / *lankeswar*

Scientific name *Croton sperciflora*, family *Euphorbiaceae*, annual, grows in warm climate, smaller plant, leaf-broad, stem solid, small fruits are seen.

(b) Narrow-leaved weeds

(i) **Nut grass / *mutha*** – *Cyperus rotandus*, family Cyperaceae, height – 30-40 cm, nuts in underground attached to roots, grows in all the season, leaf small, stem-speudo, long stock for inflorescence, flower very small.

(ii) ***Sporobolus*** – Scientific name *Sporobolus diander*, family- Gramineae, height 50-60 cm, grows in warm climate, leaf small, stem solid but then, inflorescence on long stock.

(iii) ***Doop* grass** – *Cynodon dactylon;* (as in lesson 85 b ii)

Lesson 87: Identify weeds with scientific names at least 3 broad leaved and 3 narrow-leaved weeds of rainfed *rabi paira / utera* crops (mainly in grasspea field) Describe the beneficial aspects of the weeds in rainfed *paira/utera* crops

Rabi paira / utera crops

(a) Broad-leaved weeds

(i) *Bantamak* / wild tobacco – *Nicotiana*; Scientific name – *Nicotiana plumbaginifolia*, family- Solanaceae, height- 50-80 cm, annual, grows through-out-the-year, but big leaves in *rabi*, season, flowers in summer; flowers like tobacco plants, leaf spherical, stem solid, flower cluster.

(ii) *Physalis minima*: (as in lesson 85 a i)

(iii) *Hetka* / *Vicia* – Scientific name *Vicia benghalensis*; family- *Papilionaceae*, annual; height 30-50 cm, though weed in *rabi paira* crop, but may be used as fodder;

broad leaves but small; stem solid; inflorescence cluster; seed coat black/greyish, vine type, grows in *rabi*, require low/medium alluvial/clayey soil.

(b) Narrow-leaved weeds

(i) *Mutha* – *Cyperus rotundus* : (as in lesson 86 b i).

(ii) *Cynodon* : (as in lesson 85 b iii).

(iii) *Sporobulus diander* : (as in lesson 86 b ii).

9.2. WEED CONTROL IN CROPS / CROPPING SYSTEM BY HERBICIDES

9.2.1 Wheat

Weed control in wheat by the following chemicals, was found effective in nation wide experiments. Details are as follows :

(i) 2, 4-D @ 0.5 kg ai-/ha, at 30-35 DAS (days after sowing), (ii) Isoproturon (arelon) post emergence @ 0.75 kg ai/ha, 30-35 DAS, (iii) Isoproturon + 2,4-D (0.75 + 0.18 kg ai @ 1875 q/ha formulation, (iv) Fluroxy pyr (strane) @ 0-20 kg ai/ha 30-35 DAS, (v) Tribunil @ 1.5 kg ai/ha, pre-emergence, (vi) Terbutryn @ 0.75 kg ai/ha, pre-emergence etc.

[Note : Terbutryn should not be used in wheat in eastern parts of the country because of a little bit higher temperature].

9.2.2 Rice

Recently new herbicides have arrived for transplanted *kharif* rice cultivation, where weeds are controlled with very small amount of herbicides. The following two new herbicides are very effective in controlling weeds of *kharif* transplanted rice :

(i) Butachlor @ 1.5 kg ai/ha, at 6 days after transplantation (DAT) and

(ii) Cinmethylin @ 0.08 kg ai/ha at 6 DAT. For this herbicide, the amount needs, is very less.

9.2.3 Jute

Fluchloralin @ 0.6 kg ai/ha. at 3 days after sowing, is effective in controlling major weeds in jute in pre-*kharif* season. This herbicide also requires very small amount.

9.2.4 Weed control in cropping system

In the '*rice-wheat-jute*' multiple cropping system, weeds of rice fields in *kharif* season, are controlled by (i) Butachlor @ 1.5 kg ai/ha, at 6 days after transplantation and (ii) Cinmethylin @ 0.08 kg ai/ha, at 6 DAT in transplanted rice. In wheat, weeds are controlled by (i) 2, 4-D @ 0.5 kg ai/ha, at 28 days after sowing (DAS) and (ii) Tribunil @ 0.75 kg ai/ha at 28 DAS. Thereafter, in jute, the weeds are controlled by Fluchloralin @ 0.6 kg ai/ha at 3 DAS. Thus, the weed control is complete in '*rice-wheat-jute*' cropping system.

Practical Class : Herbicidal formulation/calculation and spraying herbicides, on wheat and on cropping system with wheat

Lesson 88 : Calculate the amount of 2, 4-D salt required for spraying on one hectare of land under wheat, @ 0.5 kg ai./ha, 28 days after sowing

Calculation of herbicides

For calculation of wettable powder or water soluble powder/granular herbicides (to be used in one hectare of land)

the formula is –

$$= \frac{100 \text{ x Rate (kg/ha) x Area (in ha)}}{\text{Concentration (\% a.i.)}} \quad \text{(say, concentration is 80\%)}$$

Therefore, amount of 2,4-D salt is

$$= \frac{100 \text{ x } 0.5 \text{ x } 10{,}000 \text{ x } 100 \text{ gm/ha}}{1000 \text{ x } 80} = 625 \text{ g/ha}$$

For spraying 2,4-D herbicides in wheat, 625 g sodium salt will be required for one hectare of land.

Lesson 89 : Practise spraying cautiously, the same amount of herbi-cides (as in lesson 88), with 600 L of water in wheat field

(as in lesson 88) and for spraying 625 gm of 2,4-D salt cautiously, the following are required).

Introduction

Objective; **Material and methods** and **Procedure** : as in lesson 91.

Lesson 90: Calculate the amount of herbicides for one hectare of land

(a) **Butachlor** @ 1.5 kg a.i. /ha in rice,

(b) **2, 4-D** @ 0.5 kg a.i./ha in wheat and

(c) **Fluchloralin @ 0.6 kg a.i./ha in jute, under '*wheat-jute-rice*' cropping** system, along with the required amount of water [@ 500 L of water/ha]

Calculation of herbicides (to be required for the crops) in the multiple cropping programmes)

(a) Butachlor @ 1.5 kg a.i./ha in rice

Amount of herbicides to be required -

$$\frac{100 \text{ x } 1.5 \text{ x } 10{,}000 \text{ x } 100}{1000 \text{ x } 50} \text{ g/ha (when concentration is 50\%)}$$

$$= \frac{150000000}{1000 \times 50} = \frac{150000}{50} = 3000 \text{ g/ha}$$

i.e. 3 kg/ha

Therefore, for rice, 3 kg/ha of butachlor will be required.

(b) 2,4-D @ 0.5 kg a.i./ha in wheat

Amount of herbicides to be required

$$\frac{100 \times 0.5 \times 10{,}000 \times 100}{1000 \times 80} \text{ g/ha (when concentration is 80\%)}$$

$$= \frac{50000000}{1000 \times 80} = \frac{50000}{80} = 625 \text{ g/ha}$$

Therefore, for wheat, 625 gm/ha of 2,4-D will be required.

(c) Fluchloralin @ 0.6 kg a.i./ha in jute

Amount of herbicides to be required -

$$\frac{100 \times 0.6 \times 10{,}000 \times 100}{1000 \times 20} \text{ g/ha (when concentration is 20\%)}$$

$$= \frac{600000000}{1000 \times 20} = \frac{6000}{20} = 3000 \text{ mL/ha}$$

i.e. 3 L/ha

Therefore, for jute 3 L/ha of fluchloralin will be required.

For 'rice-wheat-jute' cropping system (a) 3 kg/ha butachlor in rice, (b) 625 g/ha of 2,4-D in wheat and (c) 3 L/ha of fluchloralin in jute will be required.

Lesson 91 : Spray cautiously, the above mentioned amount (as in lesson 90) in the same cropping system, applying 600 L of water/ha, at 6 DAT (days after transplantation) of rice, 28 DAS (days after sowing) in wheat and 3 DAS in jute

Introduction :

In the multiple cropping programmes, different crops are grown in different seasons. The different herbicides are applied in different crops, in relevant seasons, under different situations. Here in "rice-wheat-jute" cropping system, **butachlor** in rice, **2,4-D** in wheat and **fluchloralin** in jute, are applied in *kharif*/rainy, *rabi* and pre-*kharif* seasons respectively under different situations.

Objective :

Herbicides are used –

O to kill the weeds in crop,

O to increase the number of yield-components and

O to get more yield from the crops

Material required :

✣ The weed infested fields will be required in *kharif, rabi* and pre-*kharif* season for rice, wheat and jute, ✣ Sprayer–1 (with nozzle and other fittings), ✣ Balance (1 kg) (or fine balance – 1), ✣ Volume measuring cylinder – 1, ✣ Labour – 1, ✣ Cloth about 1 m x 0.5 m (for applying on mouth), ✣ Good water source, ✣ Tub – 1, ✣ Note book, ✣ Pen/pencil

Procedure

- Measure the calculated amount of herbicides (use measuring cylinder),
- Put these herbicides into the sprayer (Through mouth of cylinder),
- Add water into the sprayer (through the mouth of cylinder),
- Pump the water to increase pressure (by giving pressure with handle),
- Place the cloth piece over the mouth to avoid drifting
- Grip the handle of the sprayer (after placing the spray or on back)
- Spray herbicides over the crops and weeds uniformly by holding the handle of the nozzle.

9.3 CULTURAL MANAGEMENT FOR WEED CONTROL/CHECKING

9.3.1 For checking the obnoxious weeds, by sowing crops

Recently some new obnoxious weeds, have been observed with the arrival of new dwarf wheats and these weeds are damaging the wheat crop, resulting in lower yields. They are damaging the wheat crop, as they are almost the same to look at. For this, either chemical or cultural practices are needed for controlling those obnoxious and other weeds. In this regard, rotation or change of crops gives better performance. In the next year, in the field other crops like mustard or gram or, lentil, or potato etc. but other than wheat crop, should be sown, to check the obnoxious weeds like *Phalaris* etc.

9.3.2 Weed-checking by higher seed rate of crops

It has been observed that higher seed rate (1.5 time more than the normal) checked the population of obnoxious and other weeds, as a result, the production of wheat crop was higher. Therefore, for wheat crop, give 150 kg/ha seed rate in place of 100 kg/ha, where obnoxious/other weeds, a big problem. Generally, *Phalaris minor* (*gullidandal sitti/ phalaghas),* wild oat and *temulentum*, the obnoxious weeds, are big problems in Punjab, Haryana, U.P. etc.

[**Note :** *Phalaghash* (*Phalaris minor*), wild oat (*Avena fatua*) and *karatghash/lolium* (*Lolium temulentum*) are in Gramineae family and resembles with wheat plant; hence very difficult to identify and to eradicate them. It should be chemically controlled].

9.3.3 Weed-checking by agronomic manipulations

Agronomic manipulations like (i) closer row-spacing, (ii) hoeing, (iii) increasing of seed rate, (iv) cross sowing etc. improve grain yields of wheat, as these cultural methods decrease the weed-population in wheat, though weeds are not controlled fully. (i) Closer wheat-spacing with 15 cm × 10 cm, (ii) hoeing time, (before irrigation but after application of fertilizers at CRI in wheat), (iii) increase of seed rate @ 125-150 kg/ha, (iv) Cross-sowing (Criss-cross method) etc. improve grain yield of wheat. These principles will also hold good in other crops to some extent.

9.3.4 Weed-checking by growing cover/tall crops

Crop yield of wheat may be increased by growing some cover crops or tall crops to minimise weeds population and to get more yields. Lentil, gram, pea etc. may be grown in between wide rows of wheat, after sowing of wheat, to check the population of weeds and to get more yield of wheat. Cowpea decreases weed population in other pre-*kharif/kharif* seasons. It has also been observed that tall mustard crop (B 85), decreases weeds in wheat field and increase yields. In these cases both wheat and mustard crops are in better conditions with high yields. Any cover crop (ricebean) is better for checking weeds. Sunnhemp as *paira* crop for seed, checks weeds in *rabi* season.

Practical Class : Practising for running wheel-hoe, before first irriga-tion in wheat and growing cover crops for checking weeds

Lesson 92 : Run the hoe or wheel-hoe or tine, before first irrigation of wheat but after applying N for top-dressing in wheat and observe the weed condition at 30 days after irrigation

Introduction

Running wheel-hoe or hoe or tine is a good practice in wheat, just after application of top-dressing with urea [@ 50 kg N/ha], but before the first irrigation at crown root initiation (CRI) stage, because this practice is very helpful for (i) mixing urea with soils and for (ii) decrease weed numbers in wheat fields. Therefore, the learner should practise this for getting higher yield of wheat.

Objectives

Hoeing should be done immediately after fertilization but before first irrigation (CRI) in wheat

- O to mix the fertilizer with soils,
- O to increase efficiency in application of fertilizer,
- O to enhance the thorough passing of irrigation water,
- O to minimise the weed population.

Material required

❖ A wheat field which has been sown in lines (25 cm apart rows) just around 20 days, before this operation, ❖ Wheel hoe/hoe/tine (fitted properly), ❖ N (urea) fertilizer @ 50 kg/ha, for 1000 sq.m area, ❖ Spring balance (1,2 kg capacity), ❖ Sack (2), ❖ Bowl (4), ❖ Note book, ❖ Pen/ pencil.

Procedure

- Spread over the required urea (granular) fertilizer on soils.
- Run the hoe, in between the wheat rows immediately after applying urea, by holding the handle, where soils will go to two sides, nearer to wheat root zones, making small furrows,
- Irrigate the land at CRI stage, by introducing water into the furrows.

Assignment

Observations on weeds (to be recorded by the learners)

Hoeing with wheel-hoe/hoe	Wheat tiller and weed numbers	
	Wheat tillers sq. m.	Weed numbers/sq.m.
(i) Before hoeing		
(ii) After hoeing 30 DAS)		

Note : DAS = days after sowing

Conclusion

Note: (Generally, weed numbers will be less, 30 days after irrigation; while wheat tiller numbers will increase).

Lesson 93 : Practise sowing of rainfed *paira / utera* crop of sunnhemp (for seed) in *rabi* and observe the weed population conditions at 30, 45 and 60 days after sowing

Introduction

Rainfed *paira / utera* crops, are grown just before harvest of the preceding crops (say transplanted *kharif* rice), under marshy condition, without any tillage operation. In this case, sunnhemp (seed) will be sown as *paira* crop for seed purpose. Normally sunnhemp is grown in pre-*kharif* season for fibre). At sowing sunnhemp seeds in murshy/muddy land, practically no weed is available, but gradually weeds grow. As because the sunnhemp is the bigger crop, in competition weeds can not dominate the crop. Hence, weed problem in sunnhemp as *paira* crop, is less. The learners, therefore, should practise the sowing methods and observe the weed conditions there.

Objectives

Sowing of rainfed *paira/utera* sunnhemp (for seed) is done

O to increase the cropping intensity,

O to utilize the marshy/muddy land under rained condition,

O to control, partially the weeds in the field,

O to get over all higher crop productivity.

Material required

✣ Matured rice-field in low-lying areas having slight water in it, ✣ Seed of sunnhemp, as calculated for 1000 sq. m. area, @ 60 kg/ha, ✣ Bowls (4), ✣ Spring balance –(1,2 kg capacity), ✣ Sickles (4), ✣ Labourers (4), ✣ Scale (1), ✣ Note book, ✣ Pen/pencil.

Procedure

- Broadcast the required seeds of sunnhemp (in bowls) over the matured *kharif* transplanted rice, uniformly, after weighing the amount of seeds for 1000 sq.m. area, @ 60 kg/ha, in the evening,
- Harvest the rice crop, next day, at a straw height of 15 cm (above the ground), with the help of sickles,
- Take the count of weed-population (sq.m) at sowing,
- Record the weed-population/sq.m. area at 30, 45 and 60 days after sowing,

Assignment :

Observation on weeds (in sunnhemp, grown as rainfed *paira / utera* crop)

Paira / utera crop	Weed number / sq.m at		
	30 DAS	45 DAS	60 DAS
Sunhemp			

Notes : DAS = days after sowing.

Conclusion :

[**Note:** After 30 DAS, weeds will be there, heavily again at 45 and 60 DAS but by this time, sunnhemp will be bigger in size, hence no damage by weeds].

9.4 UTILIZATION OF WEEDS IN DIFFERENT FORMS

Weeds that are checked by some cultural operations or by chemical control, we can not get the weeding materials i.e. weeds but if hand-weeding is done, a good amount of weeds are available. Hand-weeding, of course, is the best among the cultural practices, though this process is very costly. From hand-weeding operations, a huge amount of weeds are available for utilization in different forms. The weed-materials may be utilized as fodder, vegetable, medicine, organic manure etc. This weed may also be utilized for garden beautification.

9.4.1 Fodder / hay purpose

Some types of weeds are utilized as fodders after uprooting them at different stages, if hand-weeding is done. Barnyard grass/*sanwak*/ *shyamaghash* [*Echinochloa crusgalli* (L) Beauv.], an obnoxious weed in rice fields is exactly similar to rice plant, grows well with rice crop and it is very difficult to control by any means. In that case, hand-weeding is the best for controlling it. If hand-weeding is done at different stages, huge amount of weeds are available. This weed may be used as a good fodder for the cattle. Similarly in wheat, *gullidanda*/*sitti*/*phalaghash* (*Phalaris minor* Retz.) is also used as fodder in some areas, where it is a big problem.

9.4.2 Medicine or perfumery purpose

Some weeds, if they are uprooted, may be utilized as medicine or as perfumery. *Jangli amli*/*hazardana* (*Phyllanthus niruri*), a small broad-leaved weed, grows in summer, may be utilized as medicine for jaundice or stomach ache or diarrhoea or urinogenital disorders. Nut grass/*motha*/earth-almond (*Cyperus rotundus*/*esculentus* L.), the perennial sedge, grows all-the-year round, in most of the areas in India, may be utilized for perfumes, as the nuts give essential oils. The drug '*punarnava*' is produced from hog weed/ *Punarnava* (*Boerhaavia diffusa*).

9.4.3 Vegetable purpose

Some weeds may be utilized as leafy-vegetables, where leaves and twigs are eaten. Pig weed/lambs quarters/*bathua* (*Chenopodium album* L.), a big broad-leaved weed in winter season, is used as leafy-vegetable. *Kanchari*/ *sechisag* (*Alternanthera sessilis*/*amoena* Voss) a weed in summer/ rainy season, is also used as vegetable. Swamp cabbage/water spinach/*kalmi sag* [*Ipomoea reptans* (L) Poir], a weed running over the surface of water, is used as vegetable. Indian sorrel/*Amrul*/*susni*/ chuka-tripati (*Oxalis corniculata* L.), a broad-leaved weed in summer, is used as leafy-vegetable. It has got medicinal property (for piles/dyspepsia). Hog weed/ *Sant*/*santhi*/*punarnava* (*Boerhaavia diffusa*/ *purpurea* L. is used as both vegetable and medicine).

9.4.4 Organic manure (compost) purpose

Organic manures are prepared from either plant or animal materials. As the weeds are the plant materials, hence with weeds, organic manures are produced. The weed materials obtained after hand-weeding, may be utilized for farm composting. Composting may be done with the available weed-materials, obtained from hand-weeding, particularly with *shyamaghash* (*Ech-inochloa*) or *phalaghash* (*Phalaris*) etc. Composting may be done by the side of the field within a pit, in dry season and on high flat land in rainy season.

9.4.5 Beautification purpose in gardens

Some of the weeds give good flowers or decency. The flowers of prickly poppy [*Argemone mexicana* L.), *kamat* (*Launaea pinnatifida* Cass) are very beautiful to look at. These plant may be planted in the gardens also. The *lantana* (*Lantana camara* L.) an aromatic shrub-weed, having good flowers, is planted in some gardens, lawn etc. for beautification.

Practical Class : Uprooting the weed `*shyamaghash*' by hand weeding from the *aus kharif* rice fields and use them as fodder for cattle and some of them should be kept for pit/trench method of composting and later use compost for using as organic manure. Use *Lantana* or *launaea* (weed) for flower purpose in the garden

Lesson 94 : After hand weeding in *aus* rice, use '*shyamaghash*' (weed) as fodder. Give it to the cattle and observe their consumption processes and palatability by them (cattle)

Introduction

'*Shyamaghash*' (*Echinochloa crusgalli*), the obnoxious grassy-weed, like rice-plant, damage the *aus* rice crop (direct-seeded) in pre-*kharif* season. If *aus* rice is weeded manually, then the weeds can be kept for the consumption process by the cattle, as this is a good fodder crop also. Therefore, the learners should weedout the crop manually (by hand-weeding) and should be fed to the cattle as fodder. To observe the consumption process of this weed by the cattle, the learners should be involved and they should record the palatability test from this studies (Just observation).

Objectives

Hand-weeding should be done in *aus* rice

O to keep the weed as fodder (this weed is like rice-plant),

O to feed the '*shyamaghash*' to the cattle as fodder,

O to conduct the palatability test (by involving the cattle),

O to minimise the weed problem in the field.

O to keep rice-crop weed-free

Material required

✥ An *aus* rice field, where age of rice is about 30-45 days after sowing, ✥ Labourers (4), ✥ *Nirani* (4), ✥ Cows (but not the calf) (4), ✥ Rope – 500 gm, ✥ Spring balance (with 10 or 20 kg capacity), ✥ Stop watch (for recording time for consumption, ✥ Note book, ✥ Pen/pencil.

Procedure

- Uproot the weeds including '*shyamaghash*, with the help of *nirani*,

- Keep the weeds by the side of the field,
- Tie all these weeds, but separate *shyamaghash* from other weeds,
- Give this *shyamaghash* weighing to 20 kg, to the cattle for feeding,
- Record the time for consumption of *shyamaghash,*
- Observe how relishly it (fodder) is consumed by the cattle,
- Weigh the refuse materials of fodder given to cattle after some time,
- Calculate the consumed fodder material by substracting the amount of refused materials from the total amount, given to the cattle (20 kg here).
- Note down the time for consumption and the amount of fodder, consumed by cattle.

Assignments

Observation

Weed (*Shyamaghash*) as fodder for cow	Total amount (kg) given	Refuse matter (kg)	Consumption	Remarks on palatability (Good/bad)
			Amount (kg) with time (mts)	
Sample 1 (cow)	20			
Sample 2	20			
Sample 3	20			
Sample 4	20			
Mean	20			

Conclusion

[Generally, 80-85% of total amount of fodder is consumed within 10-15 minutes; the rest is refuse materials. Write on the remark column : 'Good' or 'bad' in terms of palatability, if consumed within short/long period].

Lesson 95: After hand-weeding in direct-seeded rice (*aus*), get *shym- aghash* (weed) and use the weeds ('*shyamaghash*') for compos-ting either in pit or in trench and use that organic manure for multiple cropping system (wheat-jute-rice) after 6 months

Introduction

'*Shyamaghash*' the big grassy-weed of *aus* rice, can also be used as good organic matter for using in crops, in the multiple cropping programmes. The decomposition may be done different methods. Here it should be decomposed by pit method by the side of the field and the organic matter/ manure should be applied in crops in the multiple cropping programmes. The learners, therefore, should be present at the time (a) *process of weeding* in *aus* rice where 'shyamaghash' is a weed, (b) would be decomposed in pit and

(c) organic manure obtained there, should be applied in crops, for higher crop-production and maintaining soil fertility.

Objectives

O to keep the weeds as composting materials,

O to prepare compost in the pit with weeds,

O to use the organic manure obtained from compost, to the crops.

Material required : (as in lesson 94) plus

✣ Weed materials including '*Shyamaghash*', ✣ Land for pit, ✣ Baskets - 4, ✣ Spade - 1, ✣ Tape - 1, ✣ Tub - 1, ✣ Coir (500 gm) ✣ Slurry of cow dung and urine (all composting materials needed), ✣ Strong and long bamboo stick, ✣ Note book ✣ Pen/pencil.

Procedure

(as in lesson 94 upto accumulation of weeds) plus

- Dig a pit, measuring 10m-15m in length, 3-4 m in width and 1-2m depth, keeping the soils by the sides of pit, near the road sides,
- Collect all the weeds, along with *syamaghash* and put all these into pit,
- Give slurry over the weeds, and again put the weed materials over the stack and add slurry over the weeds and continue till the height is about 1½m,
- Pour water from the top of the stack time-to-time for moistening,
- Keep the stack for 6 months and remove the well-decomposed manure in another place,
- Apply this organic manure (compost) in crops/cropping system, particularly under wet conditions in rice under 'wheat-jute-rice' cropping system.

[**Notes :** The utilization of weeds as organic manure, requires three distinct divisions such as: **(i) Weeding and weed processing** for composting, **(ii) Preparation of compost** and **(iii) Application of organic manure (compost)** to crops and cropping system like 'wheat-jute-rice'. Any organic manure preparation is better when high moisture is available in soils, hence it should be used in crop under wet condition, particularly in pre-*kharif* or *kharif* season.]

Lesson 96 : Sow seeds or plant the clump of weeds (*launia* or *lantana*), in rainy season, in the garden, for beautification (for having good coloured scented flowers)

Introduction

There are some plants which are weeds in crops in the fields, but may be good plants for gardens, for decoration purposes. *Kamat* (*Launaea pinnatifida*) gives good star-like yellow flowers in latter part of winter. Prickly poppy (*Argemone mexicana*) gives good yellow flowers (though spines on the body). The lantana (*Lantana camara*) is a bushy shrub which gives small mixed coloured flowers, having a good scent. All these plants (weeds) may be planted, for beautification of the gardens. The learners should be acquainted with some of the weeds which can be planted in the gardens for beautification in different seasons in different places. *Lantana* will be grown in the garden for the practical purpose.

Objectives

Sowing / planting of weeds in the garden

O to beautify the gardens,

O to increase the aesthetic value of the persons in the garden,

O to fresh the minds of the visitors with scents, available in the plants.

Material required

✥ A garden, having space in it for plantation, ✥ Selection of plants in the fields / places which acts as weeds, (in our case *Lantana carara*), ✥ Spade (1), ✥ Sickle (1), ✥ Organic manure – 20 kg, ✥ Basket (2), ✥ Balance (10 kg capacity), ✥ Watering can, ✥ Water source, ✥ Note book, ✥ Pen/pencil.

Procedure

- Uproot the plant (*Lantana camara*) with the help of spade for stumps/ clumps, very carefully, so that roots are not fully damaged, in rainy season, from the field,
- Dig out a pit measuring about 15 cm × 15 cm × 15 cm, keeping the soils on the sides,
- Weight 10-15 kg of organic manure and put this in the pit,
- Mix manure with soils (from the sides),
- Make a big hole in the pit, where mixed materials (organic manure and soils) are present,
- Put the clumps of the plant into that hole and fill up the vacant portion with soil, slightly press it,
- Put the water over the plant and soil, till the soil is saturated, with can,
- Give a shed over the plant for some days (say 7 days), to protect against the sun,
- Irrigate the plant every alternate day, upto 7 days, if no rainfall.

[**Notes :** This lantana plant, perennial in nature, flowers throughout the year / seasons. The small flowers having mixed coloures, emit scent which is very exciting. In the nature, it is the shruby-weed which is very difficult to control, but when planted in the garden, it acts as 'beautification'.

■■■■

CHAPTER 10

SOIL MOISTURE MANAGEMENT UNDER RAINFED CONDITIONS

'*Soil moisture management*' under rainfed condition, is very important for higher yields of crops. 'Soil moisture management' may be done with (i) mulch, (ii) crop-residues, (iii) soils, (iv) crops/cropping manipulation, and (iv) other agronomic manipulations.

10.1 MULCHING

'*Mulching*' is one of the best methods for conserving soil moisture, for higher production of crops. This may be done with (i) straw mulch, (ii) other plant materials, (iii) organic matters, (iv) plastic sheet etc.

10.1.1 Straw-mulch (wheat/rice)

Straw mulch improves the soil moisture conditions, as a result the crop yield is higher. It has been observed that the rainfed wheat yields increased heavily, with the addition of mulch with rice-straw at the rate of 3-4 t/ha. Similar observation was noted with wheat straw also.

Under rainfed '*wheat-jute-rice*' cropping system, application of rice-straw as mulch @ 4 t/ha in wheat, improved wheat yield because of improvement of soil moisture. With this soil and moisture, germination of next crop 'jute' also improved was and as a result, production of jute very much also higher, in the rainfed "wheat-jute-rice" cropping system.

10.1.2 Other plants / plant materials

Mulching with other plants/plant materials, may be done, for improve-ment of soil moisture and ultimately for higher production of crops, under rainfed conditions. In sugarcane cultivation, refuse materials like leaves, bagasses etc. are utilised for mulching of the same crop or other crops. For cultivation of jute, leaves are shed on the fields itself and the leaves are spread over the soils either for mulching or for addition of organic matters to the soils.

10.1.3 Organic matter

Organic matters, as such, is very helpful for both conservation of soil moisture and plant nutrients. If under composed plant - materials are spread over the soils, they act as soil-mulch in the dry seasons and decompose later in rainy season and then act as organic manures again. Any other plant materials may also be utilized in the same fashion, under rainfed conditions. Clods over the soils act as mulch, known as `soil-mulch'.

10.1.4 Plastic sheets

Though costly, plastic sheet is also very helpful for conserving soil moisture. Generally, black polyetheline sheet/plastic sheets are used in conserving soil moisture in high-value-crops. In some advanced countries, black sheets of polyetheline, are used both for weed-control and conserving moisture in the soil. Cucumber/*khira*/*sasa* (*Cucumis sativus* L.) is cultivated with this black sheet. Some portion in lines of the sheet are pierced and seeds are sown there and spot irrigation is given, through those holes.

Practical Class : Arrangement of rice/wheat (straw)/mulch materials and spreading over the materials on soils, in rainfed wheat

Lesson 97 : Calculate the amount of rice-straw, to be required for mulching the rainfed wheat in 1000 sq.m area @ 3 t/ha and procure the same before sowing of rainfed wheat in November.

Introduction

Under rainfed conditions, mulching improves the crop productivity. In the eastern parts of the country, rice-straw is easily available before the sowing of wheat. Hence, for the learners, calculation for mulching materials in particular area, is necessary, provided requirement is given for one hectare area. Procurement is necessary for using that before sowing of rainfed wheat. Before procurement of rice-straw, calculation is necessary. Hence 3 types of works to be done : (i) calculation, (ii) procurement and (iii) spreading over soil with the mulch, in rainfed wheat.

Objectives

Mulching is necessary -

O to improve moisture conditions of rainfed crops,

O to enhance growth of plants,

O to increase yield of rainfed wheat,

O to increase the cropping intensity.

Material required

(i) For calculation

✣ Note in exercise book with pen/pencil, ✣ Records for *area* to be covered, at a rate (tonne/quintal per hectare).

(ii) Procurement

✤ Hanging spring balance 50/100 kg capacity-1, ✤ Rice-straw (measured amount), ✤ Rope or other tieing materials, ✤ Carrying arrangement (like trolley/truck/van).

(iii) Spreading over

✤ Land for sowing rainfed wheat, ✤ Labourer – 5

(iv) Miscellaneous

✤ Long rake, ✤ Note book with pen/pencil.

Procedure

(i) Calculation of rice-straw for 1000 sq.m area

For 10,000 sq.m (i.e. one hectare), requirement of rice-straw is 3 tonnes (i.e. 30 quintal). Therefore, for 1000 sq.m. area, rice-straw requirement will be –

$$\frac{30\ \text{q} \times 1000}{10{,}000}\ \text{q/ha} = 3\ \text{quintal / h}$$

Therefore, 3 quintal of rice-straw will be required for 1000 sq.m. area for growing rainfed wheat.

(ii) Procurement

- Procure the calculated amount of rice-straw from other nearby places,
- Tie-up and carry it (rice-straw) to side of the land where rainfed wheat will be sown; carrying should be done by truck/van in the farm,

(iii) Spreading the mulch-materials

✤ Spread the rice-straw materials (as calculated) over the soils, immediately after sowing rainfed wheat, uniformly.

[**Note :** It has been observed experimentally that 3 t/ha of rice-straw gave the maximum grain yield of rainfed wheat at BCKV Farm, West Bengal].

Lesson 98 : Spread over the same calculated amount of rice –straw (as in lesson 97) over the soils, immediately after sowing rainfed mustard

Introduction

Mustard, the oilseed crop, consumed very much by the people, is grown mostly under rainfed conditions in the eastern and other parts of the country, during *rabi*, season in high land. Under this condition, moisture condition is depleted mostly in high land. Hence, mulching is necessary for having good yield of mustard, particularly with the long duration and tall varieties. Hence, the learners should calculate the amount needed for 1000 sq.m area, procure and spread over the same, for getting more seed-yield of mustard, under rainfed conditions.

Objectives; Material required and **Procedure** same as lesson 97.

(**Note :** Only in case of wheat, it should be mustard).

Lesson 99 : Calculate the production economics (in Rs/ha) of rainfed wheat cultivation from the following data :

(a) Cost of general rainfed wheat cultivation (without mulch) – Rs. 6000/ha,

(b) Grain yield of wheat is @ 10 quintal/ha,

(c) Straw yield of wheat @ 18 q/ha,

(d) Price of grain of wheat - Rs.1400/q,

(e) Price of straw of wheat - Rs.100/q,

(f) Cost of mulch materials @ Rs.100/q,

(g) Application cost of mulch @ Rs.100/ha.

Calculation

For calculation of production economics (in Rs/ha) of rainfed wheat, we must know: (i) total crop production (i.e. grain and straw), (ii) their price and (iii) cost of cultivation. In this case, additional cost of mulching (with mulch application cost) should be added to production cost.

Total production

(i) **Grain yield (q/ha)** – 10 q and

(ii) **Straw yield (q/ha)** – 18 q.

(b) Price (total)

(i) **Grain (Rs/ha)** – 1400/q

(ii) **Straw (Rs/ha)** –100/q.

(c) Cost of general cultivation of wheat

(i) **Cost of cultivation** – Rs. 6000

(ii) **Mulch + application cost** – Rs. 100/q + 100/ha (for 3 t/ha).

Production economic (Rs/ha) (i.e. Net returns Rs/ha)

= Gross returns – Cost of cultivation = Net returns

= Total productions and their prices – cost of cultivation

= Grain yield × price + straw yield × price

– cost of cultivation + cost of mulch and its application

= 10 q. × 1400 q. + 18 × 100 q – 6000 + 3000 + 100

= Rs.14,000 + 1800 - 9100

= Rs. 15,800 – 9100 = Rs. 6,700 per hectare.

∴ Net returns from cultivation of rainfed wheat is Rs. 6,700 per hectare.

Benefit cost ratio (B/C) for cultivation of wheat

= 15,800 / 9100 = 1.76

∴ B/C = 1.76, indicating production economics.

Total production economics is obtained from

(i) Net returns = Rs. 6,700 per ha and

(ii) B/C of wheat cultivation = 1.76

10.2 CROP-RESIDUE MANAGEMENT

Crop-residues in fields, act as soil mulch as such in dry season, preserving soil moisture and the same materials in rainy season, act as manure for the next crop. Some portions of some crops are left on the ground and act as moisture conservation for yield improvement of crops. The crops like rice, wheat, mustard, greengram etc. are harvested with the sickles and some lower portions are left on the soils. Hence management of these crop-residues from different types of crops, improve soil moisture. These are as follows :

10.2.1 With cereal crops

Cereal crops such as rice, wheat, barley, oat etc. are harvested with sickles and some of the lower portions of stems (stubbles) are left behind in the fields. The stubbles/straws improve the soil moisture, to some extent. Other cereal like *jowar*, *bajra* or maize are cut at the base of the stem and entire materials are carried for threshing but some leaves are left and these act as mulch which improve soil moisture in the soils. Some rice plants which are very tall like **'deep water paddy'** which leave good amount of straws in the field, which act as mulch in preserving soil moisture.

10.2.2 With pulse crops

Most of the pulse crops, are harvested only on the top portions and the rest portion is a good crop-residue. This crop-residue improves the soil moisture conditions maintaining soil fertility also. Most of the pulse crops act as a cover crops also, hence the crops check evaporation and increase soil moisture. In this regard, crops like grasspea, lentil, greengram blackgran etc. work in a better way. The *arahar* (pigeonpea) is very important for leaf-fall which conserves moisture in the soil maintaining soil fertility, further it gives, stick for fuel.

10.2.3 With tuber crops

After the harvest of 'potato' – the tuber crops, the entire *haulm portion* is left in the field. This *haulm*, when dry, act as mulch in the summer and improve soil moisture which is essential to grow the next rainfed crops in summer and the same haulm, when decomposed in rainy season, acts as organic manure.

10.2.4 With fibre crops

Most of the bast fibre crops like jute, mesta, sunnhemp, etc. leaf shedding is a must for fibre extraction. In that case entire leaves are left on the land. These leaves act as soil mulch and at later stage, after decomposition, act as organic manure. Lint fibre crop like cotton also leave the leaves on soil, which ultimately improve soil moisture conditions in the soils.

10.2.5. With sugar crop (sugarcane/sugarbeet)

Sugarcane, the tall crop, under *Gramineae* group, leaves the leaves on the furrows, when cleaned the canes. These leaves act as 'mulching material' for conserving moisture in the soils. This conserved soil moisture ultimately improves the next ratoon crop.

Again after crushing the canes, 'bagasses' are obtained and these bagasses not only improve soil moisture, but also improve soil fertility. Some bacteria increases soil nitrogen by fixation of N from the atmosphere, after getting energy from *bagasse* and sunrays.

10.2.6 With oilseed crops

Most of the oil-seed crops, are cut in the top portion and rest lower portion is left on the field. Rape and mustard crops are harvested the top portion, keeping the lower portion on the field. This lower portion may be utilized as crop-residues, for improving soil moisture and nutrient for the next crops. The lower portion of the mustard crop may be ploughed-under for crop-residue management. Of course, these stick portions of mustard may be used as fuel also.

Practical Class : Utilization of crop-residues for crop production

Lesson 100 : Harvest the top portion of any pulse crop (specifically green gram) and plough-under the crop-residues.

Introduction

Utilization of crop-residues is very important, in respect of maintenance of soil fertility, soil health and crop-productivity. For this leguminous crops give more beneficial effects because they initially improve (i) **soil health** by fixing nitrogen from atmosphere and at later stage by (ii) **adding more nitrogen** to the soils, because crop-residues contain more nitrogen. Therefore, the learners should be acquainted with harvesting technique of crops and the incorporation of crop-residues into soils, for maintaining soil fertility, soil health and crop-productivity.

Objectives

Crop residues should be utilized -

- to maintain soil fertility,
- to improve soil health,
- to obtain higher crop-production in the next crops.

Material required

(a) For harvesting and processing of crop

✥ Land with matured green gram crop, ✥ Sickles (4), ✥ Rope/string/coir (for tieing harvested materials), ✥ Truck/van, ✥ Branches of trees (for beating), ✥ Buckets/ containers – 2,

(b) For incorporation of crop-residues

✥ Ploughing arrangement (Bullocks, plough) along with accessories, ✥ Ladder (1), ✥ Note book, ✥ Pen/pencil.

Procedure :

(a) For harvesting of crops

- Cut the top portion containing the pods of green gram, with the help of sickles, and keep that by the sides of the field,
- Remove the top portion of the crop by tieing them up, by the truck/van, to the threshing floor of the godown for processing,
- Spread the top portion of the green gram on the threshing floor for sun drying for 3-4 days.
- Remove the grains from the top portion of the plants, by beating with stick,
- Keep the grains in the sun for proper drying for two days.

(b) For incorporation of crop-residues

- Give the laddering over the lower portion of the green gram that was left in the field, after harvest of the top portion, for partially breaking down the green gram plants,
- Plough the field along with the first laddering, followed by second ploughing accross the earlier ploughing,
- Apply the ladder again, for covering the gap due to ploughing,
- After seven days, again plough the land. Now the land is ready for sowing the next crop. [say, jute (for seeds) or maize],
- Sow the next crop jute (for seed) or maize, by broadcasting seeds under prevailing conditions, and
- Ladderings should be given twice length- and cross-wise after sowing.

Lesson 101 : (a) Harvest the jute crop, (b) Stack them on the field after bundling them, (c) Shed the leaves, for utilizing as manures and remove the plants (without leaves) and (d) Plough-under leaves before transplantation of *kharif* rice.

Introduction

Jute, the bast fibre crop, is grown in eastern parts of the country abundantly, during pre-*kharif* season. After the harvest of this crop, transplanted *kharif* rice is grown. In order to make the soil fertile, the jute-leaves are incorporated into the soils, after shedding leaves on soil-surface of the field and after spreading over the leaves on entire soil surface, then ploughing/puddling is done for transplantation of *kharif* rice. This is a common practice in eastern parts. Therefore, the learners should be acquainted with the processes of (i) *harvesting*, (ii) *stacking*, (iii) *leaf-shedding*, (iv) *spreading leaves*, and (v) *ploughing* and puddling before transplantation of *kharif* rice.

Objective

- O to harvest crop, for getting fibre yield,
- O to bundle-up the stems and stacking them in one place, keeping them standing slantingly,
- O to shed the leaves from the jute plants (jute leaf is a good manure),
- O Removing the bundles for 'jak' (accumulation of bundles in water),
- O Spread over the leaves over the soil surface uniformly,
- O Plough the land for incorporation of leaves,
- O Plough/puddle the land, after seven days, for transplantation of *kharif* rice with basic NPK-fertilizers @ 50-50-50 kg of N, P_2O_5 & K_2O kg/ha,
- O Transplant the rice seedlings.

Material required

✣ One jute land with crop of about 120 days after sowing, ✣ Sickles/ *hasua* (4), ✣ Rope/string/coir/jute thread (2 kg), ✣ Busket of bamboo/cane (4), ✣ Bullock (2), Ploughman (1), plough (1), ✣ Ladder/plank, ✣ Balance (Spring balance of 10 or 20 kg capacity), ✣ Sacks (4), ✣ NPK-fertilizers for 1000 sq.m. area, @ 50-50-50 kg N, P_2O_5 & K_2O/ha, ✣ Note book, ✣ Pen/pencils.

Procedure

- Harvest the individual jute plants by cutting them at the base, with the help of sharp sickles (or *hasua* if strong and big plants),
- Keep the individual plants lying on the soil surface, clusterly and go on harvesting the entire field,
- Bundle them (the plants) with a diameter of 4 to 5 cm [with coirst or by jute plants],

- Collect all the bundles in one place,
- Stack them, by keeping the bundles standing with a slight slanting keeping the heads above,
- Wait for 4-5 days for shedding leaves,
- Remove the bundles (without leaves), by keeping the shedded leaves on soil surface, to the place where water is available, for steeping in '*Jak*',
- Put the accumulated leaves into the baskets, for spreading them on soil-surface of entire area uniformly,
- Plough the land for incorporation of leaves into soils,
- Wait for some days for accumulation of water,
- Apply calculated amount of fertilizer [@ 50 kg N, 50 kg P_2O_5 and 50 kg K_2O/ha], as basal, on the soil surface,
- Puddle the land (after some time for accumulation of rain water) for 2-3 times, for transplanting *kharif* rice,
- Transplant the rice seedlings in puddled land,
- Note down the dates of harvest, stacking, removal of bundles, spreading of leaves of jute and ploughing date and date for transplantation along with methods in short.

10.3 SOIL AND CROP/CROPPING MANIPULATION

Soil moisture may be improved, to some extent, by manipulating soils and with the help of multiple cropping programmes. Soil manipulation and multiple cropping programmes have been discussed in brief.

10.3.1 Soil manipulation (for moisture preservation)

Sometimes soil manipulations improve soil moisture; this soil moisture is very helpful to some next crops. It has been observed that soils after harvest of transplanted *kharif* rice, under rainfed conditions, shrinks heavily and cracks are formed, indicating the loss of moisture from the soil. In that context, if the land is scratched with the help of tine or by light ploughing and ladderings heavily, the soil cracking and loss of moisture is checked, ultimately, after sometimes soil moisture is improved. Even if ploughed land is heavily laddered/planked, with high pressure, soil moisture will be higher inside, though dry soils on the surface.

10.3.2 Cropping manipulation for moisture preservation

'*Cropping manipulation*' is very important for preserving moisture in the soil, under rainfed conditions.

Rabi leguminous crops, like pea, grasspea, lentil, sunnhemp (for seeds) etc. improve soil moisture in the rainfed cropping system like (i) Rice-grasspea-green gram/jute, (ii) Rice-sunnhemp (for seed)-green gram, (iii) Rice-lentil-green gram etc. if grasspea, sunnhemp and lentil are grown as rainfed *rabi paira/utera* crops. After the cultivation of *tetrakalai*, cowpea, ricebean etc., improvement of soil moisture was observed.

Practical Class : Soil moisture conservation by scratching the rainfed 'rice-fallow' soils with tine, for multiple cropping programmes

Lesson 102 : Scratch the soils in 'rice–fallows' with the help of tine/ hoe/wheel-hoe and observe the moisture condition, 30 days after scratching the soils. Note down whether moisture is lost or retained, by visual observation, roughly.

Introduction

In the eastern parts of the country, there is abundant water from rain fall, during pre-*kharif* and *kharif* seasons, but there is a dearth of water in *rabi* season, in 'rice-fallows' (fallow land after transplanted *kharif* rice) under rainfed conditions. *Kharif* transplanted rice (about 80%) is grown mostly in eastern parts in India, under rainfed conditions, as a result practically no crop is grown in those lands, for want of water (irrigation), during winter season. Here, 3-crop systems may be continued in rainfed areas, provided *paira / utera* crops are grown and some moisture is retained in soils by any means. It had been observed that moisture may be retained with *scratching the soils*. Hence, the learners must be acquainted with the method of moisture conservation in rainfed 'rice-fallows' by scratching the soils.

Objective

Scratching the soils in *rabi* season to sow crops in pre-*kharif* season.

O to conserve/retain soil moisture,

O to increase cropping intensity in rainfed rice-fallows,

O to increase crop production ultimately.

Material required

✣ A 'rice-fallow' land after harvest of *kharif* transplanted rice, under rainfed conditions, ✣ Wheel-hoe or tine (4), ✣ Bullocks (2), a ploughman, and plough, ✣ Ladder (1) / plank – 1, ✣ Note book, ✣ Pen/pencil.

Procedure :

- Scratch the soils (1000 sq.m area during the month of November/ December), with the help of wheel-hoe/tine,
- Immediately after scratching, laddering two times length-and cross-wise is necessary,
- Keep the land 'as such' for 30 days after scratching,

- Visually observe (after opening the soils with *khurpi*) the soil moisture conditions,
- Note down date of scratching the soil, date of laddering with numbers,
- Observe soil moisture visually,
- Conclude about the soil moisture, whether next crop-seeds can germinate with that amount of moisture or not (Green gram or blackgram may be sown for the next crop).

Lesson 103 : Sow rainfed '*tetrakalai*' (giant blackgram/blackgram in *kharif* season) in rainfed condition, after its harvest (as fodder) immediately sow rainfed wheat in November, for proper germination of wheat seeds. Observe visually whether good soil moisture is there, enough for germination of wheat seeds and count the number of wheat plants/m^2 – 15 days after sowing.

Introduction

In has been observed that in the eastern India, mid-November, is the best time for proper sowing time of rainfed wheat in upland, but moisture may be a limiting factor for germination of seeds of the wheat crop. To retain some moisture, and utilize it for germination of seeds of wheat in *rabi*, a short duration crop like blackgram/ or giant blackgram (*tetrakalai*) is grown for fodder, then after its harvest in mid-November, wheat is sown in *rabi* immediately for germination of seeds and higher production of wheat. Therefore, the learners should know the technique of soil moisture conservation for sowing and getting higher yields of rainfed wheat.

Objective

In *kharif* season, particularly in high lands, blackgram or giant black- gram is grown for fodder and is harvested in Mid-November.

- O to get more soil moisture for germination of seeds of rainfed wheat,
- O to increase the cropping intensity,
- O to get more grain yields of wheat under rainfed conditions,
- O to improve soil health.

Material required

✤ The field with blackgram/giant blackgram (*tetrakalai*) sown in August, ✤ Sickle (4), ✤ Coir/string, ✤ Truck/van, ✤ Bullocks (2), ploughman (1) with other materials for ploughing, ✤ Ladder – 1, ✤ Wheat seeds as required @ 125 kg seed rate, for 1000 sq.m area, ✤ Balance (1 or 2 kg. capacity), ✤ NPK-fertilizer as required for 1000 sq.m area, @ 40-20-20 kg of N, P_2O_5 and K_2O/ha respectively in the form of urea, superphosphate and muriate of potash.

Procedure

(a) For harvesting of blackgram in *kharif*

- Harvest the entire crop of giant blackgram for fooder, during November, with the help of sickle,
- Remove the harvested materials immediately, for feeding to the cattle

(b) For sowing rainfed wheat

- Plough the vacant land, immediately after harvest of blackgram,
- Add NPK-fertilizer by broadcasting on soils, as calculated for 1000 sq.m area,
- Laddering is done after fertilizer (urea, super phosphate and muriate of potash) application,
- Again, plough the land (for broadcasting wheat seeds),
- Broadcast wheat seeds of required amount, on soils (as per calculat- ion) uniformly,
- Ladder two times (Length - and cross-wise) heavily,
- Observe the soil moisture in the field visually, for seed germination,
- Count the number of wheat seeds germinated/sq.m. area, from 4 samples 15 DAS, add together and make them average and give the comments on it.

(**Note :** Generally, 40-50 or more seed germination/sq.m means 150-300 ear head (effective)/sqm area at the end, from which 10 to 12 or more q/ha may be obtained).

10.4 AGRONOMIC MANIPULATION

Some agronomic management practices may also improve soil moisture, to some extent, under rainfed conditions. These are : (i) adjustment of sowing date, (ii) high seed rate etc. and these are discussed in brief below :

10.4.1 Adjustment of sowing date

Some crops may be adjusted for sowing, provided adequate moisture in soil is available. Jute crop is to be sown in April, in general, but due to lack of adequate moisture in soils, sowing may be delayed for some rainfall and adequate moisture upto earlier part of May. Similarly particular variety of mustard (*rabi*) may be sown in October, when soil moisture is there in soil, instead of November where moisture is less.

10.4.2 Use of higher seed rate

Higher dense canopy of crops can check to some extent, the evaporation of soil-water from the soil. For this dense canopy, high seed rate of crops is needed. Loss of water, due to evaporation is higher than the need of moisture required by crops/plants. Therefore, checking of loss of moisture, invariably improves further moisture condition of the soils.

10.4.3 Retention of weeds 'as such' in *rabi* rainfed *paira/utera* crops for moisture conservation

Rainfed *rabi, paira / utera* crops get more moisture from the soil, where some weeds are retained, as such. Therefore, retention of some weeds like *vicia* spp. and *melilotus* spp. etc. are very helpful, for rainfed *paira* crops such as grasspea, lentil, linseed, sunnhemp etc. in terms of soil moisture retention in the soil. Hence weed though menace, is very important for moisture retention in soils and also for crop production. Hence, some weeds in *paira* crops are benificial.

10.4.4 Growing rainfed grasspea as *paira/utera* crop for retention of moisture, for the next crop

Grasspea (*Lathyrus sativus*), a winter, hardy, rainfed crop, utilises moisture for its growth and development and still it leaves more moisture in the soils after its harvest, where the seeds of next rainfed crops (greengram or jute) may be germinated well in the summer and those crops (greengram and jute) ultimately give higher yields due to presence of moisture in the soils. Therefore, under rainfed condition for planting either greengram or jute in summer, grasspea may be grown as a preceding *paira* crop in *rabi* season.

10.4.5 Depth of seeding

Depth of seeding is very helpful for germination of seeds of crops, under rainfed conditions. It has been observed that in lower depth, soil moisture is more. It has also been observed that wheat seeds germinate well when the crop is sown at a depth of 8-12 cm below soil, by making furrow with tine, under rainfed conditions.

Practical Class : Sowing rainfed wheat at greater depth (8-12 cm), growing *khesari,* for conserving soil moisture and growing green gram

Lesson 104 : Sow seeds of wheat in rainfed 'rice-fallows' at depths of 4, 8 and 12 cm) for observing the germination position of seeds at 15 DAS and plant count at 45 DAS

Introduction

Depth of seeding, is very important for seed germination, moisture retention, growth and yield of rainfed wheat, particularly in the eastern parts of the country, where heavy rainfall is available in *kharif* but not in *rabi*. The learners, therefore should be acquainted with this phenomenon. Seeds of wheat should be sown at a depth of 4 cm, 8 cm and 12 cm. depth by the middle of November, in rainfed rice-fallows, for observation on soil moisture condition (visually), germination of wheat seeds in one sq.m. area. It is necessary to count the plant population at 45 DAS.

Objective

Sowing rainfed wheat at different depth

- O to get more soil moisture for germination,
- O to facilitate the growth of wheat,
- O to get more grain yield under rainfed conditions.

Material required

❖ 'Rice-fallow' land under rainfed conditions, ❖ Bullock (2), ploughman (1) other ploughing arrangements, with *desi* plough, ❖ Ladder (1), ❖ Balance (1 or 2 kg. capacity), ❖ Wheat seeds (of any variety), as per requirement for 1000 sq.m area, @ 125 kg seed rate/ha, ❖ NPK-fertilizer, as required for the same land, @ 40-20-20 kg of N, P_2O_5 and K_2O kg/ha, ❖ Tine (4), ❖ String (nylon) – 500 gm, ❖ Pegs – 8, ❖ Labours (4), ❖ Note book, ❖ Pen/pencil.

Procedure

- Plough the area twice,
- Apply the entire required N, P_2O_5 and K_2O/ha, separately first urea (N) and by mixing P_2O_5 and K_2O in the second lot, by broadcasting on soil,
- Ladder the soils twice, length and cross-wise,
- Divide the land in three areas equally (A, B and C),
- Make furrows at depths of 4, 8 and 12 cm in each of the plots separately, at a line-to-line space of 25 cm.
- Sow seeds uniformly, on the furrows made in each plot, with the amount of seeds dividing in 3 portions,
- Cover the seeds manually in land, from the sides,
- Observe the germination (in one sq.m. area), 15 days after sowing the rainfed wheat,
- Count the plant number (per sq.m area) of wheat, at 45 DAS (days after sowing)

Assignments for observations

(a) 15 days after sowing, on germination (per sq.m), 4 samples

Sample	Plot A Depth : 4 cm.	Plot B Depth : 8 cm	Plot C Depth : 12
Sample 1			
Sample 2			
Sample 3			
Sample 4			
Total No. of samples			
Mean of samples			

Comments

(b) Records at 45 days after sowing, on wheat, tillers from one sq.m. area, in 4 samples

Sample	Plot A (4 cm depth)	Plot B (8 cm depth)	Plot C (12 cm depth)
Sample 1			
Sample 2			
Sample 3			
Sample 4			
Total No. of samples			
Mean of samples			

Conclusions

[Generally, under rainfed conditions wheat germination is about 40 to 50 per sq.m. area, where plant number (i.e. tillers) per sqm/area is about 150-200].

Lesson 105 : (a) Sowing rainfed grasspea (*khesari*) as *paira / utera* crop for moisture conservation

(b) Sow rainfed green gram, in pre-*kharif* season immediately after harvest of grasspea as a *paira* crop and after irrigated wheat. Observe the germination condition of rainfed *green gram* and record the plant numbers at 15 and 45 DAS and comment on them

[For sowing rainfed grasspea as in (a), consult lesson 49; for (b) consult the following].

Introduction

Due to frequent rainfall in pre-*kharif* season in eastern parts of the country, it is possible to grow rainfed green gram, if sown by broadcast just after rainfed *rabi paira* crop of grasspea or before the harvest of matured irrigated wheat (in *rabi*) under 'wheat-green gram-rice' cropping system [where only wheat is irrigated and other two crops *green gram* and rice are grown under rainfed conditions]. Therefore, the learners should know the techniques for raising rainfed green gram both after *paira* grasspea or wheat as rainfed crop.

Objective

(A) Rainfed green gram after irrigated wheat

- O to raise rainfed green gram after irrigated wheat, for good yield,
- O to increase the cropping intensity,
- O to encourage the growers for increasing yields of crops in the system, economically, (both rainfed and irrigated combination).

Material required

❖ Wheat field, under irrigation at matured condition, ❖ Sickles (4), ❖ Tieing rope/string/coir, ❖ Truck/van, ❖ Threshing floor, ❖ Godown, ❖ Green gram seed, as required for 1000 sq.m. area @ 60 kg seed rate/ha, ❖ Bullock (2), ploughman (1), *desi* plough (1) with accessories, ❖ Bida/ladder - 1, ❖ Note book, ❖ Pen/pencil.

Procedure

- Broadcast green gram seeds, just one day before wheat harvest,
- Harvest the wheat crop, one day after broadcasting green gram seeds, with the help of sickle,
- Remove the wheat with bundles through vans/truck from the field,
- Apply scratching type *bida*, followed by heavy laddering twice.

[**Note :** If there is enough moisture for germination green gram seeds (green gram seeds germinate under low moisture in the soils, if land is heavily pulverised), no need of applying *bida*; two heavy ladderings are enough].

Assignment : Observation :

(i) For germination, 15 days after sowing and for plant population 45 DAS, records should be taken (population 45 DAS, i.e. on plant number/sq.m area in 4 samples)

Sample	Plant number/sqm 15 DAS	45 DAS
Sample 1		
Sample 2		
Sample 3		
Sample 4		
Total No. of samples		
Mean of samples		

(B) Rainfed green gram after rainfed *rabi paira/utera* crop of grasspea (*khesari*)

Objective; **Material required** and **Procedure** as in (A) above.

Assignment and observations : (as in A and B under i)

(**Note :** Instead of previous crop wheat, it will be *khesari*)

(ii) For observation of records, on number of branches/plant, at 45 days after sowing

Sample	Branches/plant (average no. from 5 plant) at 45 DAS
Sample 1	
Sample 2	
Sample 3	
Sample 4	
Total No. of samples	
Mean of samples	

Conclusion

Sample	Remarks
Sample 1	
Sample 2	
Sample 3	
Sample 4	
Total No. of samples	
Mean of samples	

10.5 CROPS FOR RAINFED SYSTEMS

Choosing the crops under 'rainfed cropping systems', is very important for the success in crop-production. Hence, crops under (i) cereals, (ii) pulses, (iii) oil seeds, (iv) fibres, (v) fodder etc. should be selected after consideration of different things. For this, they have been discussed in short below :

10.5.1 Cereals

Among the *cereals*, oat, barley, *jowar*, *bajra*, millets etc. can be grown under rainfed conditions, as they adjust themselves with lower moisture content in soils. Therefore, under multiple cropping programmes, in rainfed system, these crops should be considered, according to situations.

10.5.2 Pulses

Important *pulse crops* like grasspea, gram, greengram, blackgram, lentil, pea etc. may be grown under rainfed conditions, as they can survive in a better way in soils which are deficient in proper moisture content. Hence, for multiple cropping programme, under rainfed condition, these crops should be considered according to situations.

10.5.3 Oilseeds

The crops like mustard/rape, sunflower, niger, safflower, sesame etc. under *'oilseed'* group, can withstand low moisture content in soil, under rainfed conditions. So, for any multiple cropping programme under rainfed conditions, these crops should be considered, according to situations.

10.5.4 Fibres

Bast fibre crops like jute, mesta, sunnhemp etc. could be grown under rainfed conditions. For any rainfed cropping programme in eastern part of the country, therefore, these crops should be considered, for different situations.

10.5.5 Fodders

Among the fodder crops, grasspea, blackgram, napier grass, diananath grass, ricebean, cowpea etc. could be grown under rainfed conditions. Hence it is necessary to consider these crops, for fitting in rainfed multiple cropping programmes.

Practical Class : Sowing of different crops under rainfed areas in lines in small plots (for observations)

Show the following crops in small lines :

(i) Barley, lentil, oat, mustard and grasspea, in *rabi* season,

(ii) Greengram, cowpea, sesame, ricebean, sunhemp (for fibre), mesta and thin napier in *kharif* season, and

(iii) Sunnhemp (for seed), mesta (for seed), maize (only in eastern part of India), napier and blackgram in *kharif* season, for observation from time-to-time. This is actually the museum.

CROP MUSEUM

'Crop Museum' is the area, by the side of the main paths in the farms of the university, for most of the crops, grown in different seasons (*rabi*, pre-*kharif* and *kharif*), in small areas, should be exhibited to the students, for increasing their knowledge. Though, the crops are grown in small areas, it has got tremendous impact / importance in learning, as because the students of all the years are observing them from germination to harvesting, while they are visiting the farms. In the crop museum, one name plate should be there where (i) Common name, (ii) Scientific name, (iii) Date of sowing and (iv) Type of crop, should be written.

[**Note :** An area of two hectares of land should be selected by both sides of the main path, like long strip, for 3 seasons and one for anothers. Four or more rows of each crop will be grown in each season, measuring about 15m length but with different spacing (as per crop)].

Lesson 106 : Sow barley, lentil, oat, mustard and grasspea in *rabi* season under rainfed conditions

Introduction

The entire area for sowing *rabi* crops should be ploughed 4 times, with laddering (4) for more pulverization. These crops should not be provided with irrigation facilities.

Objective :

The crops in *rabi* should be sown

O to improve knowledge of the student,

O to get an idea about crop,

O to utilize these knowledge in future.

Material required

✣ A good land by the side of path of University Farm, ✣ Bullock (2), Ploughman (1), ✣ Mouldboard *desi* plough, ✣ Ladder 1, ✣ Balance (5 kg capacity), ✣ NPK-fertilizer as per calculation) for 1000 sq m area @ 40-20-20, legumes and 80-40-40 for non-legumes of N, P_2O_5 and K_2O/ha (dividing by the number of crops), ✣ Thread/coir, ✣ Tine/wheel-hoe (for making furrow), ✣ Board/plate (with recognisable information as mentioned earlier), ✣ Seeds of different crops (as mentioned earlier), ✣ Note book, ✣ Pen/pencil,

Specific information, on crops like barley, lentil, oat, mustard and grasspea, under rainfed conditions

Crops	Sowing time	Seed rate (kg/ha)	Spacing in cm. for line	Crop type
1. Barley	Mid-November	100	25 cm (line)	Cereal
2. Lentil	Mid-November	60	25 cm (line)	Pulse
3. Oat	Mid-November	100	25 cm (line)	Cereal
4. Mustard	Mid-October	8	40 cm (line)	Oil seed
5. Grasspea	Mid-November	30	Broadcast	Pulse

Procedure

- Plough the land 2 times, followed by 2 ladderings,
- Plough the land again third time, without laddering,
- Give the general dose of N, P_2O_5 and K_2O kg/ha for required amount [@ 40-20-20 respectively for legumes and 60-30-30 for non-legumes],
- Ladder the land (after applying fertilizers),
- Give plough again for the 4th time (if necessary),
- Sow the crops (as above) by giving seed rate, spacing etc. in *rabi* (as mentioned), and

- Put the name plate/placards (as earlier) with information of the crops in each case.

Lesson 107 : Sow greengram, cowpea, sesame, sunflower, sunnhemp (for fibre), ricebean (for fodder), mesta (for fibre) and thin- napier (fodder) in 4 lines in pre-*kharif* season, under rainfed conditions

Introduction

In pre-*kharif* season, some crops like greengram, cowpea, sesame, sunflo- wer, sunnhemp, ricebean, mesta, thinnapier etc. can be grown under rainfed conditions, in areas where rainfall is available. It is possible in eastern part of the country, but not in other parts. Therefore, the learners must know the condition of these crop, by growing them in small areas in the season.

Objective

Sowing of crops in museum plot, in pre-*kharif* season, under rainfed conditions

O to observe the growing conditions/behaviour,

O to know the proper date of planting,

O to gather knowledge about types of crops.

Material required

✥ A good land by the sides of the University Farm main path, ✥ Bullock (2), Ploughman (1), ✥ Mouldboard or *desi* plough 1, ✥ Ladder 1, ✥ Balance (5 kg capacity), ✥ NPK-fertilizer (as per calculation for each of the crop area, with the same as mentioned earlier), ✥ Thread/coir, ✥ Tine/wheel-hoe (for making furrow), ✥ Board/ plate (name with recognisable information), ✥ Seeds of crops mentioned, ✥ Note book, ✥ Pen/pencil.

Specific information, on greengram, cowpea, sunflower, sunnhemp (for fibre), ricebean (for fodder), mesta (for fibre) and thinnapier (fodder) in 4 lines, in pre-*kharif* season, under rainfed conditions in pre-*kharif* season

Crop	Sowing time	Seed rate (kg/ha)	Spacing (cm) (in lines)	Crop-type (commercial)
Greengram	March/April	60	50	Pulse
Cowpea	March/April	60	60	Pulse & fodder
Sunflower	March/April	80	60	Oil seed
Sunnhemp	April/May	40	30	Fibre
Ricebean	April/May	60	60	Fodder
Mesta	April/May	80	30	Fibre
Thinnapier	April/May	15	40	Fodder

Procedure : (as in lesson 106)

Lesson 108 : Sow napiergrass, cowpea, ricebean, mesta, *jowar*, *bajra*, blackgram, sunnhemp (for seeds), maize for seeds (only in eastern India), under rainfed condition in *kharif* season, in each small plot

Introduction

In *kharif* season, some crops like napiergrass, cowpea, ricebean, mesta, *jowar*, *bajra*, blackgram, sunnhemp (for seeds), maize etc. are grown in eastern India, under rainfed condition, because of high rainfall in *kharif* season in small plots, each for observation by the learners.

Objective

The above mentioned crop may be grown

- O to observe the growing condition,
- O to know the proper date of planting,
- O to gather knowledge about types of crops.

Material required and **Procedure** as in lesson 106.

Specific information, on napier (fodder), cowpea (fodder), ricebean (fodder), mesta (seed), *jowar*, *bajra*, blackgram (fodder), sunhemp (seed), maize, under rainfed conditions, in eastern or other parts of the country in *kharif* season

Crop	Sowing time	Seed rate (kg/ha)	Spacing (cm) (row-to-row)	Crop type
Napier (F)	May/June	80,000 clumps	25	Fodder
Cowpea (F)	May/June	60	40 cm	Fodder
Ricebean (F)	June/July	60	40 cm	Fodder
Mesta (S)	June/July	80	30	Fibre
Jowar (S)	June/July	80	30	Cereal
Bajra (S)	June/July	80	30	Cereal
Blackgram (S)	July/August	60	30 cm	Pulse
Sunnhemp (F)	June/July	60	30 cm	Seed
Maize (S)	June/July	80	40 cm	Seed

Note : F = fodder; S = seed; F = fibre.

CHAPTER 11

MAJOR CROPS CULTIVATION (in short)

11.1 RABI SEASON

11.1.1 Cereal crops

Cereal food crops are highly needed as '*food crops* in India.

Wheat

1.	*Climatic requirement*	:	Temperatures range from 5° to 16° at tillering and 35-36°C at maturity; winter season
2.	*Soil requirement*	:	Alluvial/loamy/sandy loam/clay loam
3.	*Required land situation*	:	Flat/hilly land; flat land better
4.	*Importance of crop*	:	Major food in India
5.	*Fertilizer management*	:	N, P_2O_5 & K_2O @ 100, 50 & 50 kg/ha, respectively
6.	*Water management*	:	4 to 5 irrigations at 3 weeks interval
7.	*Duration of crop*	:	4 to 5 months in plains and 5 to 6 months in hills
8.	*Parts used*	:	Grain for human being and straw for cattle
9.	*By-product*	:	Straw and bran
10.	*Use of by-product*	:	Straw : Cattle feed/mulch/fuel/manure/paper

Maize

1.	*Climatic requirement*	:	Medium to high temperature, humidity & rainfall
2.	*Soil requirement*	:	Loamy / alluvial / sandy loam
3.	*Required land situation*	:	High land in plains and flat land in hill
4.	*Importance of crop*	:	Major food and industrial products
5.	*Fertilizer management*	:	N,P_2O_5 & K_2O @ 120, 60 & 60 kg/ha, respectively

6.	*Water management*	: 5 to 6 irrigations in winter, 3-4 in summer
7.	*Duration of crop*	: 4 to 5 months in plains and hills
8.	*Parts used*	: Seed for consumption, stover for cattle feed
9.	*By-product*	: Stover
10.	*Use of by-product*	: Cattle feed/trailing vine crop/fuel/peg

Rice (*boro*)

1.	*Climatic requirement*	: High temperature, high humidity and water stagna-tion needed for, transplantation, in cool conditions
2.	*Soil requirement*	: Clay/clayey loam/clayey/loamy
3.	*Required land situation*	: Low, medium and flat land
4.	*Importance of crop*	: Major food (for carbohydrate)
5.	*Fertilizer management*	: N, P_2O_5 & K_2O @ 80-100,40-50 & 40-50 kg/ha, respectively
6.	*Water management*	: Always water stagnation needed, hence 6-8 times irrigation
7.	*Duration of crop*	: 4-6 months
8.	*Parts used*	: Seeds for human beings and straw for cattle
9.	*By-product*	: Straw and bran/husk
10.	*Use of by-product*	: Cattle feed, roofing/mulching/organic matter/ fuel/ paper board

Oat

1.	*Climatic requirement*	: Low temperature, dry condition
2.	*Soil requirement*	: Alluvial / loamy / saline soil
3.	*Required land situation*	: Flat lands in plains and hills
4.	*Importance of crop*	: Can withstand salinity; food for human being
5.	*Fertilizer management*	: N, P_2O_5 & K_2O @ 80, 30 & 30 kg/ha, respectively
6.	*Water management*	: Normally rainfed, 1-2 irrigation for high yields
7.	*Duration of crop*	: 4 to 5 months
8.	*Parts used*	: Seeds for human beings/ straw for cattle
9.	*By-product*	: Stover or straw
10.	*Use of by-product*	: Cattle feed, roofing, organic matter, fuel, and mulching

Barley

1. *Climatic requirement* : Low, medium temperature, dry weather
2. *Soil requirement* : Loamy, sandy loam, loamy, clayey, saline soils
3. *Required land situation* : Flat lands in plains and hills
4. *Importance of crop* : Minor food for human beings and major fodder for cattle
5. *Fertilizer management* : N, P_2O_5 & K_2O @ 80, 40 & 40 kg/ha, respectively
6. *Water management* : Normally rainfed, 1-2 irrigation at flowering for higher production
7. *Duration of crop* : 4-5 months
8. *Parts used* : Seeds for human consumption and straw for cattle
9. *By-product* : Straw
10. *Use of by-product* : Cattle feed/fuel/organic matter/roofling

Jowar

1. *Climatic requirement* : Warm climate with high temperature
2. *Soil requirement* : Loamy/sandy loam/sandy
3. *Required land situation* : High flat lands
4. *Importance of crop* : Tall plant like sugarcane, good for fodder
5. *Fertilizer management* : N, P_2O_5 & K_2O @ 60, 30 & 30 kg/ha, respectively
6. *Water management* : Normally rainfed, 1-2 irrigations for higher production
7. *Duration of crop* : 4 to 5 months
8. *Parts used* : Seeds for human beings and straw for cattle
9. *By-product* : Straw/stover
10. *Use of by-product* : Cattle food/stick for supporting structure/ fuel/ organic matter/fencing

Bajra

1. *Climatic requirement* : Warm climate
2. *Soil requirement* : Loamy/sandy loam/sandy
3. *Required land situation* : Flat high land
4. *Importance of crop* : This is good food both for human beings and cattle in rainfed areas

5. *Fertilizer management* : N, P_2O_5 & K_2O @ 60, 30 & 30 kg/ha, respectively
6. *Water management* : Normally rainfed, 1-2 irrigations for higher production
7. *Duration of crop* : 4-5 months
8. *Parts used* : Seeds
9. *By-product* : Straw/stover
10. *Use of by-product* : Cattle feed/fuel/organic matter/fencing

11.1.2 Pulse crops

Pulse crops, under family *Leguminaceae* and sub-family *Papilionaceae*, contain high amount of proteins. This (pulse) is used as *dal* in India and soup in foreign countries.

Lentil

1. *Climatic requirement* : Medium or low temperature and dry condition
2. *Soil requirement* : Loamy, clayey loam, sandy loam
3. *Required land situation* : Flat lands of plains and hills
4. *Importance of crop* : Major pulse in northern, central and eastern India
5. *Fertilizer management* : N, P_2O_5 & K_2O @ 40, 60 & 30 kg/ha, respectively
6. *Water management* : Rainfed normally, 1-2 irrigations for higher yields
7. *Duration of crop* : 4 – 5 months
8. *Parts used* : Seeds
9. *By-product* : Stover/*bhusa*
10. *Use of by-product* : Cattle feed, fuel, organic matter and mulch

Gram

1. *Climatic requirement* : Low or medium temperature, dry conditions
2. *Soil requirement* : Loamy, sandy loam and clayey loam soils
3. *Required land situation* : Flat lands of plains and hills
4. *Importance of crop* : Major pulse in India except south India
5. *Fertilizer management* : N, P_2O_5 & K_2O @ 40, 60 & 60 kg/ha, respectively
6. *Water management* : Normally rainfed, 1-2 irrigations for higher production, at flowering

7. *Duration of crop* : 5-6 months
8. *Parts used* : Seeds for *dal*
9. *By-product* : Stover / *bhusa*
10. *Use of by-product* : Cattle feed, fuel, organic matter and mulch

Grasspea

1. *Climatic requirement* : Medium to low temperature, dry conditions
2. *Soil requirement* : Clay silt, clay, clayey, loam, alluvial soils
3. *Required land situation* : Low lands in plains and hills
4. *Importance of crop* : Minor *dal* (pulse), fodder/soil-moisture enriching crop / green manuring/cover crop
5. *Fertilizer management* : N, P_2O_5 & K_2O @ 40, 60 & 40 kg/ha (as *paira no fertilizer required*), respectively
6. *Water management* : Rainfed crop
7. *Duration of crop* : 4 to 5 months
8. *Parts used* : Seeds for man and plants for fodder
9. *By-product* : Stover/*bhusa*
10. *Use of by-product* : Cattle feed, organic matter, fuel and mulching

Pea

1. *Climatic requirement* : Low or medium temperature, dry conditions
2. *Soil requirement* : Sandy, loamy, clayey loam soils
3. *Required land situation* : Flat lands of plains and hills
4. *Importance of crop* : Major winter vegetable and pulse (*dal*)
5. *Fertilizer management* : N, P_2O_5 & K_2O @ 40, 60 & 40 kg/ha, respectively
6. *Water management* : Normally rainfed, 1-2 irrigations for higher yields
7. *Duration of crop* : 4-5 months
8. *Parts used* : Seeds and plants
9. *By-product* : Stover
10. *Use of by-product* : Cattle feed, fodder, fuel, organic matter and mulch

11.1.3. Oilseed crops

Oilseed crops give the 'oil and fats to the human diets. It is the medium of cooking for people, all over the country.

Mustard/rape

1. *Climatic requirement* : Low/medium temperature with dry conditions
2. *Soil requirement* : Loamy/alluvial soils
3. *Required land situation* : Flat lands in plains and hills
4. *Importance of crop* : Major mstard varieties are with yellow flowers
5. *Fertilizer management* : N, P_2O_5 & K_2O @ 80, 60, & 60 Kg/ha, respectively
6. *Water management* : Normally rainfed; 1-2 irrigations if irrigated
7. *Duration of crop* : 3-4 months
8. *Parts used* : Seeds for oils and cake for manures/cattle feed
9. *By-product* : Oilcake and stover
10. *Use of by-product* : Cattle feeds, manure, fuel and organic matter

Safflower

1. *Climatic requirement* : Low or medium temperature, dry weather
2. *Soil requirement* : Sandy loam, clayey loam, loamy soils etc.
3. *Required land situation* : Flat lands of plains and hills
4. *Importance of crop* : Minor edible oilseed, mostly used in perfume materials as base
5. *Fertilizer management* : N, P_2O_5 & K_2O @ 60-30-30 kg/ha, respectively
6. *Water management* : Normally rainfed, irrigation at flowering, for high yields (if irrigated)
7. *Duration of crop* : 4-5 months
8. *Parts used* : Seeds for oil/perfumes
9. *By-product* : Stover
10. *Use of by-product* : Fuel, organic matter, mulch

Niger

1. *Climatic requirement* : Low or medium temperature, dry conditions
2. *Soil requirement* : Loamy/alluvial/sandy loam soils
3. *Required land situation* : Flat lands of plains and hills

4.	*Importance of crop*	:	Minor oil, spines on leaves
5.	*Fertilizer management*	:	N, P_2O_5 & K_2O @ 60, 30 & 30 kg/ha, respectively
6.	*Water management*	:	Normally rainfed, irrigation at flowering, if irrigated
7.	*Duration of crop*	:	4-5 months
8.	*Parts used*	:	Seeds
9.	*By-product*	:	Stover, oilcake
10.	*Use of by-product*	:	Cattle feed, manures, fuel, organic matter, etc.

11.1.3 Fibre crops

Both bast and lint fibres are used in textile industries for clothing and manufacturing of bags/hessians etc.

Flax

1.	*Climatic requirement*	:	Low or medium temperature, dry weather
2.	*Soil requirement*	:	Sandy loam, loam and alluvial soils
3.	*Required land situation*	:	Flat lands of plains and hills
4.	*Importance of crop*	:	Minor fibre crop, having blue flowers, with less importance
5.	*Fertilizer management*	:	N, P_2O_5 & K_2O @ 60, 30 & 30 kg/ha, respectively
6.	*Water management*	:	Normally rainfed; irrigation at flowering for high yields, if irrigated.
7.	*Duration of crop*	:	4-5 months
8.	*Parts used*	:	Bast fibre
9.	*By-product*	:	Stover
10.	*Use of by-product*	:	Paper board, fuel, organic matter and mulch

11.1.4. Sugar and tuber crop

Sugars are obtained from sugar crops and starch/carbohydrates etc. are obtained from tuber/rhizome crops

Sugarbeet

1.	*Climatic requirement*	:	Low or medium temperature, dry weather
2.	*Soil requirement*	:	Loamy, alluvial, sandy loam and saline soils

3.	*Required land situation*	:	Flat lands of plains and hills
4.	*Importance of crop*	:	Major sugar crop, can withsand salinity
5.	*Fertilizer management*	:	N, P_2O_5 & K_2O @ 120, 60 & 60 kg/ha, respectively
6.	*Water management*	:	3-4 irrigations, 20-25 days interval
7.	*Duration of crop*	:	4-5 months
8.	*Parts used*	:	Swollen roots, for sugar
9.	*By-product*	:	Foliage (leaves and stalks)
10.	*Use of by-product*	:	Cattle feed, organic manure, fuel, mulch etc.

Potato

1.	*Climatic requirement*	:	Low or medium temperature, dry weather
2.	*Soil requirement*	:	Loamy, sandy loam, alluvial soils etc.
3.	*Required land situation*	:	Flat land of plains and hills
4.	*Importance of crop*	:	Major vegetable, tubers grow inside soils
5.	*Fertilizer management*	:	N, P_2O_5 & K_2O @ 120, 60, 60 kg/ha, respectively
6.	*Water management*	:	Earlier flash irrigation, later 5-8 channel irrigations
7.	*Duration of crop*	:	4-5 months
8.	*Parts used*	:	Tuber (potato) as vegetable
9.	*By-product*	:	Haulm and leaves
10.	*Use of by-product*	:	Cattle feed, manure, mulch, organic manure, fuel etc.

11.1.5 Fodder crops

Fodders/forages are required for cattle, goats, buffaloes, horses etc. as foods/feeds.

Oat

1.	*Climatic requirement*	:	Low or medium temperature, dry weather
2.	*Soil requirement*	:	Loamy, clayey loam and sandy loam soils
3.	*Required land situation*	:	Flat lands of plains and hills
4.	*Importance of crop*	:	Main fodder crop, long ear-head
5.	*Fertilizer management*	:	N, P_2O_5 & K_2O @ 80, 50 & 50 kg/ha

6.	*Water management*	: 1-2 irrigate at flowering, can be grown as rainfed crop
7.	*Duration of crop*	: 4-5 months
8.	*Parts used*	: Plant (whole) as fodder
9.	*By-product*	: No by-product, if it is cut for fodder
10.	*Use of by-product*	: No by-product, no use.

Barley

1.	*Climatic requirement*	: Low or medium temperature with dry weather
2.	*Soil requirement*	: Loamy, clayey loam and alluvial soils
3.	*Required land situation*	: Flat land of plains and hills
4.	*Importance of crop*	: Major fodder crop (entire plant), can be grown in saline soils also
5.	*Fertilizer management*	: N, P_2O_5 & K_2O @ 60, 30, 30 kg/ha, respectively
6.	*Water management*	: 1-2 irrigations at flowering, can be grown as rainfed crop
7.	*Duration of crop*	: 4-5 months
8.	*Parts used*	: Entire plant for fodder
9.	*By-product*	: No by-product
10.	*Use of by-product*	: No, utilization, as no by-product.

Berseem

1.	*Climatic requirement*	: Low or medium temperature, dry weather
2.	*Soil requirement*	: Loamy, clayey loam and alluvial soils
3.	*Required land situation*	: Flat lands of plains and hills
4.	*Importance of crop*	: Major legume fodder, having trilobed leaves
5.	*Fertilizer management*	: N, P_2O_5 & K_2O @ 40, 80 & 60 kg/ha, respectively
6.	*Water management*	: 3-4 irrigations at 25-30 days interval
7.	*Duration of crop*	: 4-5 months
8.	*Parts used*	: Entire plant for fodder/hay
9.	*By-product*	: No by-product
10.	*Use of by-product*	: No by-product, no use.

Lucern

1.	*Climatic requirement*	:	Low or medium temperature, dry weather
2.	*Soil requirement*	:	Loamy, clayey and clayey loam soils
3.	*Required land situation*	:	Flat lands of plains and hills
4.	*Importance of crop*	:	Major legume fodder crop, having trilobed leaves
5.	*Fertilizer management*	:	N, P_2O_5 & K_2O @ 40, 80 & 80 kg/ha, respectively
6.	*Water management*	:	3 to 4 irrigations at 25-30 days interval
7.	*Duration of crop*	:	4 to 5 months
8.	*Parts used*	:	Entire plant as fodder/hay
9.	*By-product*	:	No by-product, as entire plant is cut for fodder
10.	*Use of by-product*	:	No by-product, no use.

Grasspea

1.	*Climatic requirement*	:	Low or medium temperature, dry weather
2.	*Soil requirement*	:	Clayey, clayey loam and loamy soils
3.	*Required land situation*	:	Flat land of plains and hills
4.	*Importance of crop*	:	Winter fodder with blue/red flowers; *paira* crops
5.	*Fertilizer management*	:	N, P_2O_5 & K_2O @ 20, 40 & 40 kg/ha, respectively (No fertilizer required as *paira* crop)
6.	*Water management*	:	Rainfed crop
7.	*Duration of crop*	:	4 to 5 months
8.	*Parts used*	:	Entire plant is used as fodder/hay
9.	*By-product*	:	No by-product
10.	*Use of by-product*	:	As no by-product, no use

11.1.6 Green manuring crops

The crops, which are used for green manuring purposes in order to improve soil fertility and soil health, are called the green manure crops.

Grasspea

1.	*Climatic requirement*	:	Low or medium temperature, dry weather
2.	*Soil requirement*	:	Clayey, clayey loam, and loamy soils
3.	*Required land situation*	:	Flat lands of plains and hills

4.	*Importance of crop*	:	May be grown as fodder, green manuring or pulse or *paira* crop
5.	*Fertilizer management*	:	N, P_2O_5 & K_2O @ 20, 40 & 40 kg/ha, respectively (practically no fertilizer required as *paira* crop)
6.	*Water management*	:	Rain fed crop
7.	*Duration of crop*	:	4-5 months
8.	*Parts used*	:	Entire plant is ploughed under
9.	*By-product*	:	No by-product
10.	*Use of by-product*	:	No by-product, no use

Practical Class : Recording observations for the changes in growth at different stages and yields of different crops in the respective seasons

Lesson 109 : Recording observations, on the changes in growth at different stages of crops (Oat, mustard, wheat, potato and lentil) at 45 DAS in *rabi* season

Introduction

Recording observations, on the change in growth of crops at different stages particularly at earlier stage (45 DAS), is very important, in respect of growth behaviour of crops.

Objective

- O to study the growth behaviour of crops,
- O to implement further programmes in the cropping system,

Material required

✣ Standing crops (as mentioned) in the field, ✣ Scales (small and big),
✣ Pegs – 10, ✣ Balance (1, 2 kg capacity), ✣ Note book, ✣ Pen/pencil.

Procedure

Pegging should be done in 1 sq.m area in each crop, for recording plant height and tillers number from 10 plants.

Observations on changes of growth of crops at 45 DAS should be recorded.

Special characters of *rabi* crops

Growth characters	Crops in *rabi* / season				
	Oat	Mustard	Wheat	Potato	Lentil
Plant type	Annual	Annual	Annual	Annual	Annual
Plant character (cm)	Erect	Erect	Erect	Vine	Vine
Height/length	30-40	30-40	30-40	20-25	20-25
Tillers/branches/ sq.m)	100-150	5-6	150-200	3-4	2-3

Assignment

Observation at 45 DAS

Plant characters	Crops in *rabi*				
	Oat	Mustard	Wheat	Potato	Lentil
Leaf type					
Root type					
Stem type					
Green matter (g/sqm)					

[**Note** : Destructive sample from 25 cm x 25 cm area. Pullout the entire plants and weigh them, multiply this weight (g) by 16 to get 1 sq.m area].

Lesson 110. Recording observations, on the growth changes of sample *rabi* crops (as in lesson 109), at 60 days after sowing)

Introduction; **Objective**; **Material required** and **Procedure** : All as in lesson 109.

Observations

Observations on the changes of growth characters of the said crops at 60 days after sowing

Growth characters	Crops in *rabi*				
	Oat	Mustard	Wheat	Potato	Lentil
Plant height/length (cm)	40-50	40-50	40-50	25-30	25-30
Tillers/branch per plant (sqm)	150-200	4-5	200-350	4-5	4-5

Assignment on green matter (g/m^2)

Sample	Crops in *rabi*				
	Oat	Mustard	Wheat	Potato	Lentil
Sample 1					
Sample 2					
Sample 3					
Sample 4					
Total					
Mean					

Note : Destructive samples for green matter (as in lesson 109).

Comments

Lesson 111. Recording observations, on growth changes of the same *rabi* crops (as in lesson 109), at 90 days after sowing

Introduction; **Objective**; **Material required** and **Procedure** : All as in lesson 109.

Observations : Recording observations, on changes of growth characters of the said crops at 90 days after sowing :

Growth characters	Crops in *rabi*				
	Oat	Mustard	Wheat	Potato	Lentil
Plant height/length (cm)	50-55	50-55	50-60	35-40	30-35
Tillers /sqm or branch/plant	200-250	5-6	250-350	5-6	405
Effective tiller/ sqm or branch/ plant	180-240	5-6	220-340	5-6	4-5

Assignment : (as in lesson 110, instead of 60 DAS; it should be 90 DAS)

Lesson 112. Recording observations, on changes of growth of the same crops (as in lesson 109) at harvest

Introduction; **Objective**; **Material required** and **Procedure** : All as in lesson 109.

Observations

Recording observations on changes of growth and reproductive characters of the said crops in *rabi* at harvest.

Growth and reproductive characters	Crops in *rabi*				
	Oat	Mustard	Wheat	Potato	Lentil
(a) Growth characters					
1. Plant height / length (cm)	50-80	60-80	80-90	50-60	40-45
2. Tillers (sqm) or branch/plant	300-350	7-8	350-400	6-7	5-6
(b) Reproductive characters					
1. Effective tillers / sqm or pod number / branch	250-300	7-8	300-350	10-15 (Tuber/sqm)	2-3
2. Grain number/ earhead or per pod	25-30	7-8	30-40	-	2-3
3. 1000-seed weight (g)	20-25	8-9	35-40	-	15-20

Assignment : Recording yields (kg/sqm)

Sample	Crops in *rabi*				
	Oat	Mustard	Wheat	Potato	Lentil
Sample 1					
Sample 2					
Sample 3					
Sample 4					
Total					
Mean					

Note : Potato, a tuber crop, grows in underground; reproductive character is stolen; and tuber per plant or per sqm. area, is measured. In other crops seed-yields should be estimated.

11.2 PRE-*KHARIF* SEASON

11.2.1 Cereal crops

Aus (direct-seeded) rice

1. *Climatic requirement* : Warm weather with rainfall
2. *Soil requirement* : Clay, clayey, alluvial and loamy soils
3. *Required land situation* : Flat lands of plains and hills

4. *Importance of crop* : Major food-crop (carbohydrate), water stagnation needed

5. *Fertilizer management* : N, P_2O_5 & K_2O @ 100, 50 & 50 kg/ha, respectively

6. *Water management* : 3-4 irrigations in dry areas, but no irrigation in high rainfall areas

7. *Duration of crop* : 4-5 months (sometiems 3-4 months)

8. *Parts used* : Seeds for consumption and straw for cattle

9. *By-product* : Straw

10. *Use of by-product* : Cattle feed, fuel, mulch, organic matter and roofing materials

Maize

1. *Climatic requirement* : Medium temperature with dry/wet weather

2. *Soil requirement* : Loamy, alluvial and clayey loam

3. *Required land situation* : Flat lands of plains and hills

4. *Importance of crop* : Cereal (carbohydrate); silk and tussel present; seeds available from cob

5. *Fertilizer management* : N, P_2O_5 & K_2O @ 120, 60, 60 kg/ha, respectively

6. *Water management* : 3-4 irrigations at 25 days intervals

7. *Duration of crop* : 4-5 months

8. *Parts used* : Cobs for roasting and grains for food

9. *By-product* : Straw/stover (with leaves)

10. *Use of by-product* : Cattle food, fuel, stick, base for trailing crop, organic matter, pegs and, mulch

11.2.2 Pulse crops

All pulse crops are under *Leguminoceae*, sub-family *Papilionaceae* and they content high amount of protein. They fix nitrogen in soils from air.

Greengram

1. *Climatic requirement* : Hot weather with rainfall

2. *Soil requirement* : Loamy, clayey and sandy loam soils

3. *Required land situation* : High flat land of plains and hills

4. *Importance of crop* : Grain for major pulse (*dal*); in roots nodules present; N-fixation 50 kg/ha

5.	*Fertilizer management*	:	N, P_2O_5 & K_2O @ 40, 60 & 40 kg/ha, respectively
6.	*Water management*	:	Mainly rainfed, 1-2 irrigations at flowering, for good yields
7.	*Duration of crop*	:	85-95 days after sowing
8.	*Parts used*	:	Grain as *dal* (pulse) or soup
9.	*By-product*	:	Stover
10.	*Use of by-product*	:	Cattle feed, fuel, green manure, mulch, organic matter etc.

Arahar/Pigeonpea

1.	*Climatic requirement*	:	High temperature with rainfall [some variety (*rabi arahar*) may be grown in winter also]
2.	*Soil requirement*	:	Loamy, sandy loam, alluvial and clayey loam
3.	*Required land situation*	:	High lands in plains and hills
4.	*Importance of crop*	:	Seed used as pulse, big plants with yellow flowers
5.	*Fertilizer management*	:	N, P_2O_5 & K_2O @ 40, 60, 30 kg/ha, respectively
6.	*Water management*	:	Rainfed crop, 1-2 irrigations, if irrigated
7.	*Duration of crop*	:	100-120 days or more
8.	*Parts used*	:	Seeds/*dal*
9.	*By-product*	:	Stalk/stover/*bhusa*
10.	*Use of by-product*	:	Fuel, cattle feed, stick for trailing vegetables/ fodder crops (cowpea, *rajma*, ricebean).

11.2.3 Oilseed crops

Sesame

1.	*Climatic requirement*	:	Warm climate, with rainfall
2.	*Soil requirement*	:	Loamy, sandy loam and clayey loam
3.	*Required land situation*	:	High flat and medium lands of plains and hills
4.	*Importance of crop*	:	Seeds used as oil and confectionery; having white flowers
5.	*Fertilizer management*	:	N, P_2O_5 & K_2O @ 60, 30 & 30 kg/ha, respectively
6.	*Water management*	:	Rainfed; 1-2 irrigations for good yields, if irrigated
7.	*Duration of crop*	:	100-120 days

8. *Parts used* : Seeds for oil
9. *By-product* : Stalk / leaves
10. *Use of by-product* : Fuel, organic matter, peg, and mulch

Groundnut

1. *Climatic requirement* : Warm climate, with rainfall
2. *Soil requirement* : Loamy, sandy loam, alluvial and clayey loam
3. *Required land situation* : High and flat lands in plains and hills
4. *Importance of crop* : Seeds are used for oil, butter/confectionary
5. *Fertilizer management* : N, P_2O_5 & K_2O @ 40, 60 & 40 kg/ha, respectively
6. *Water management* : Normally rainfed; 1-2 irrigations at flowering (before & after), for higher yields
7. *Duration of crop* : 100-120 days (sometimes more)
8. *Parts used* : Seeds for oils
9. *By-product* : Haulm/*bhusa* / oilcake
10. *Use of by-product* : Fuel, mulch, manure, fodder, organic matter etc.

[**Note** : In Bengal, groundnut (short duration variety ICGV 86124) may be grown in pre-*kharif* season and haulm may be utilized for fodder or green manuring crop in the *kharif* rice]

Sunflower

1. *Climatic requirement* : Warm climate, rainfed
2. *Soil requirement* : Loamy, sandy loam, sandy and alluvial soils
3. *Required land situation* : High flat lands in plains/hills
4. *Importance of crop* : Seeds for oil/margarine preparation, big head
5. *Fertilizer management* : N, P_2O_5 & K_2O @ 60, 30 & 30 kg/ha, respectively
6. *Water management* : Normally raifned, 1-2 irrigations at flowering for higher yields
7. *Duration of crop* : 100-120 days
8. *Parts used* : Seeds for oil/beautification/margarine
9. *By-product* : Stalk / stover
10. *Use of by-product* : Fuel, organic matter, mulch and pegs

11.2.4 Fibre crops

Bast fibres are obtained from jute, mesta, sunnhemp, etc. and lint fibre from cotton.

Jute

1.	*Climatic requirement*	: High rainfall with high temperature and humidity
2.	*Soil requirement*	: Alluvial, loamy and clayey loam soils
3.	*Required land situation*	: Low lands for *capsularis* and high lands for *olitorius*
4.	*Importance of crop*	: Long pod for *olitorius* and round pod for *capsularis*
5.	*Fertilizer management*	: N, P_2O_5 & K_2O @ 40-80, 30-40 & 50-60 kg/ha, respectively
6.	*Water management*	: Normally rainfed; 1-2 irrigations, if irrigated
7.	*Duration of crop*	: 120-140 days
8.	*Parts used*	: Firbes (bast) and leaves (as manure)
9.	*By-product*	: Stick/*pakati* and leaves with twigs
10.	*Use of by-product*	: Fencing/stick in *boroj*/fuel, paper pulp, organic matter and manure

Mesta :

1.	*Climatic requirement*	: High temperature, high rainfall and humidity
2.	*Soil requirement*	: Alluvial, loamy, sandy and sandy loam soils
3.	*Required land situation*	: Low, medium and high lands in plains
4.	*Importance of crop*	: Bast fibre, fleshy flower with red fruits
5.	*Fertilizer management*	: N, P_2O_5 & K_2O @ 60, 40 & 40 kg/ha, respectively
6.	*Water management*	: Normally rainfed, 1-2 irrigations, if required
7.	*Duration of crop*	: 100-140 days
8.	*Parts used*	: Bark (fibre) on the stem
9.	*By-product*	: Stick (*pakati*) and leaves
10.	*Use of by-product*	: Indoor fencing, paper pulp, fuel, mulch, organic matter etc.

Sunnhemp

1.	*Climate requirements*	: High temperature, high rainfall and humidity
2.	*Soil requirement*	: Alluvial, loamy and clayey loam soil

3. *Required land situation* : Low, medium, high lands in plains
4. *Importance of crop* : Fibre from bark, used in cigarette paper/ tissue paper (for seeds, in *rabi*)
5. *Fertilizer management* : N, P_2O_5 & K_2O @ 60, 40 & 40 kg/ha, respectively
6. *Water management* : Normally rainfed; 1-2 irrigation, if irrigated
7. *Duration of crop* : 120-140 days
8. *Parts used* : Stems for fibre and leaves for manures
9. *By-product* : Stick and leaves
10. *Use of by-product* : Paper pulp, fuel, stacking for vegetable crops, organic matter and pegs

11.2.5 Sugar crops

Sugarcane

1. *Climatic requirement* : High temperature with humidity and rainfall
2. *Soil requirement* : Clayey, alluvial, loam, loamy and black-cotton soils
3. *Required land situation* : High flat and medium land
4. *Importance of crop* : For sugar (from stem) production, tall plants
5. *Fertilizer management* : N, P_2O_5 & K_2O @ 120, 60 & 60 kg/ha, respectively
6. *Water management* : 7-10 times irrigations (some times more)
7. *Duration of crop* : 365 to 540 days (*Eksali*/*Adsali* crops)
8. *Parts used* : Stems for juice/sugar/*gur*/propagation
9. *By-product* : Bagasse and leaves
10. *Use of by-product* : Bag/wrapping/printing papers, fuel, card-board, soil ameliorant, cattle feed and mulch

11.2.6 Rhizomes

Elephant-foot-yam [*Amorphophallus (ol)*]

1. *Climatic requirement* : High temperature, high rainfall and humidity
2. *Soil requirement* : Loamy, clayey, sandy loam and loamy soils
3. *Required land situation* : High lands in plains and hills
4. *Importance of crop* : Only one pseudo-stem, variegated in nature
5. *Fertilizer management* : N, P_2O_5 & K_2O @ 120, 60 & 80 kg/ha, respectively

6.	*Water management*	:	Normally rainfed, 1-2 irrigations, if irrigated
7.	*Duration of crop*	:	140-180 days
8.	*Parts used*	:	Rhizome (corm) and stem for vegetables
9.	*By-product*	:	Leaves and stem (pseudo)
10.	*Use of by-product*	:	Vegetable, organic matter, fuel, mulch etc.

Arum / taro

1.	*Climatic requirement*	:	Moderate to high rainfall, high temperature and humidity
2.	*Soil requirement*	:	Clayey, loamy, sandy loam and clayey loam soils
3.	*Required land situation*	:	High, medium and low lands (according to varieties)
4.	*Importance of crop*	:	Sap is itching type, vegetable, (different types).
5.	*Fertilizer management*	:	N, P_2O_5 & K_2O @ 100, 50 & 50 kg/ha, respectively
6.	*Water management*	:	Normally rainfed; but 1-2 irrigations if grown in highland or under transplanted conditions
7.	*Duration of crop*	:	150-160 days
8.	*Parts used*	:	Rhizomes, petioles and runners for vegetables
9.	*By-product*	:	Leaf
10.	*Use of by-product*	:	For fuel, organic matter and mulch

11.2.7 Fodder crops

Both legumes and non-legume fodders are grown under different situations and fed to the cattle.

Cowpea

1.	*Climatic requirement*	:	High temperature, with moderate rainfall
2.	*Soil requirement*	:	Alluvial, loamy, sand and sandy loam soils
3.	*Required land situation*	:	High or medium lands
4.	*Importance of crop*	:	Used as vegetable, pulse, cover and fodder; used for reclaiming degraded soils
5.	*Fertilizer management*	:	N, P_2O_5 & K_2O @ 40, 60 & 60 kg/ha, respectively
6.	*Water management*	:	Normally rainfed; 1-2 irrigations, if irrigated
7.	*Duration of crop*	:	110-130 days

8. *Parts used* : Entire crop cut for fodder; pods for vegetable
9. *By-product* : No by-product, when cut for fodder
10. *Use of by-product* : No use.

Ricebean

1. *Climatic requirement* : Moderate rainfall with high temperature and humidity
2. *Soil requirement* : Alluvial, loamy and sandy loam soils
3. *Required land situation* : High and medium lands
4. *Importance of crop* : Can be used as fodder, pulse, G.M. and fuel
5. *Fertilizer management* : N, P_2O_5 & K_2O @ 50, 60 & 50 kg/ha, respectively
6. *Water management* : Normally rainfed; 1-2 irrigations at flowering for good yield, if irrigated
7. *Duration of crop* : 110-140 days
8. *Parts used* : Whole plant is fodder
9. *By-product* : No by-product
10. *Use of by-product* : No use, as no by-product

Blackgram

1. *Climatic requirement* : High temperature, with high and moderate rainfall
2. *Soil requirement* : Loamy, clayey, alluvial, sandy and sandy loam soils
3. *Required land situation* : High medium lands
4. *Importance of crop* : Can be used as *dal*, fodder, G.M. and fuel
5. *Fertilizer management* : N, P_2O_5 & K_2O @ 40, 60 and 40 kg/ha, respectively
6. *Water management* : Normally rainfed; 1-2 irrigations, if irrigated
7. *Duration of crop* : 100-130 days
8. *Parts used* : Whole plants is used for fodder
9. *By-product* : No by-product
10. *Use of by-product* : As no by-product, no use

Napier

1. *Climatic requirement* : High temperature with moderate rainfall
2. *Soil requirement* : Alluvial, loamy and clayey loam soils

3.	*Required land situation*	:	High and medium lands
4.	*Importance of crop*	:	Long stem with narrow leaves
5.	*Fertilizer management*	:	N, P_2O_5 & K_2O @ 100, 50 & 50 kg/ha, respectively
6.	*Water management*	:	4-5 irrigations at 25-30 days interval
7.	*Duration of crop*	:	110-140 days
8.	*Parts used*	:	Whole plants for fodder are harvested
9.	*By-product*	:	No by-product
10.	*Use of by-product*	:	No use, as no by-product

11.2.8 Green manure crops

Dhaincha / baghadhaincha

1.	*Climatic requirement*	:	High temperature with rainfall and humidity
2.	*Soil requirement*	:	Alluvial, clayey, clay, loamy and sandy loam soils
3.	*Required land situation*	:	Low and medium lands
4.	*Importance of crop*	:	Big plants with yellow flowers, can reclaim saline-alkali soils
5.	*Fertilizer management*	:	N, P_2O_5 & K_2O @ 40, 60 & 40 kg/ha, respectively
6.	*Water management*	:	Normally rainfed
7.	*Duration of crop*	:	110-130 days
8.	*Parts used*	:	Entire plant is ploughed-under, when tender, for G.M.
9.	*By-product*	:	No by-product
10.	*Use of by-product*	:	No use, as no by-product

Blackgram

1.	*Climatic requirement*	:	Moderate temperature with high rainfall
2.	*Soil requirement*	:	Alluvial, loamy, clayey and sandy loam soils
3.	*Required land situation*	:	High or medium lands
4.	*Importance of crop*	:	Pulse crop, can be used for GM & cover crop and fodder crop
5.	*Fertilizer management*	:	N, P_2O_5 & K_2O @ 40, 60 & 40 kg/ha, respectively
6.	*Water management*	:	Normally rainfed

7.	*Duration of crop*	:	110-130 days
8.	*Parts used*	:	Entire plant is ploughed - under
9.	*By-product*	:	No by-product
10.	*Use of by-product*	:	No use, as no by-product

Sunnhemp

1.	*Climatic requirement*	:	High rainfall, temperature and humidity
2.	*Soil requirement*	:	Loamy, alluvial and sandy loam soils
3.	*Required land situation*	:	High and medium lands
4.	*Importance of crop*	:	Can be grown as fibre crop; *paira* crop (for seed) in *rabi*, after transplanted *kharif* rice
5.	*Fertilizer management*	:	N, P_2O_5 & K_2O @ 40, 60 & 60 kg/ha, respectively
6.	*Water management*	:	Normally rainfed
7.	*Duration of crop*	:	110-130 days
8.	*Parts used*	:	Entire plant is ploughed-under, when tender
9.	*By-product*	:	No by-product
10.	*Use of by-product*	:	No use, as no by-product

***Guar* (Cluster bean)**

1.	*Climatic requirement*	:	Moderate rainfall with high temperature
2.	*Soil requirement*	:	Loamy, alluvial and sandy loam soils
3.	*Required land situation*	:	High or medium lands
4.	*Importance of crop*	:	Can be used as vegetable, beside G.M.
5.	*Fertilizer management*	:	N, P_2O_5 & K_2O @ 40, 60 & 50 kg/ha, respectively
6.	*Water management*	:	Normally rainfed
7.	*Duration of crop*	:	90-110 days
8.	*Parts used*	:	Whole plant is ploughed – under for G.M.
9.	*By-product*	:	No by-product
10.	*Use of by-product*	:	No use, as no by-product

Practical Class : Observations should be recorded from time-to-time, on growth changes different at stages of the crops [jute, rice (*aus*), sesame, maize, green gram in pre-*kharif* season]

Lesson 113 : Recording observations, on changes of growth of crops (as above) at 45 DAS in pre-*kharif* season

Introduction; **Objective** and **Material required** as in lesson 109.

Procedure

- Select 4 areas (samples) in different places, under different crops (as mentioned) in pre-*kharif* season.
- Put 4 pegs in 4 corners, measuring 1 m length and breadth, in each sample.
- In the area of 1 m^2, select some (say 10) plants for recording observations under different categories, in 4 samples.
- Make them average, and put them in record book.
- Destructive sample (green matter) should be taken from 25 cm × 25 cm and that should be multiplied by 16 for 1 sqm area.

Observations on changes of growth of crops (as mentioned), at pre-*kharif* season, at 45 DAS

Growth characters	Crops in pre-*kharif*				
	Jute	Rice (*aus*)	Sesame	Maize	Green gram
Plant type	Annual	Annual	Annual	Annual	Annual
Plant character	Erect	Erect	Erect	Erect	Vine
Height/length (cm)	30-40	20-30	20-30	30-40	10-20
Tillers/sqm or branches per plant (1 m^2)	0	3-4	0	0	2

Assignments (at 45 DAS)

Plant characters	Crops in pre-*kharif*				
	Jute	Rice (*aus*)	Sesame	Maize	Green gram
Root type					
Leaf type					
Stem type					
Green matter (g/sqm)					

Comments

Lesson 114. Recording observations, on changes of growth of pre-*kharif* crops (as in lesson 113) at 60 days after sowing

Introduction; **Objective**; **Material required** and **Procedure** as in lesson 113.

Observation (at full growth)

Growth characters	Crops in pre-*kharif* season				
	Jute	Rice (*aus*)	Sesame	Maize	Green gram
Plant height / length (cm)	100-150	60-80	60-80	100-180	60-80
Tillers/m^2 or branches/plant	0	100-120	2-3	0	2-3

Assignments (At 60 DAS)

Observations to be recorded at 60 DAS

Plant characters	Crops in pre-*kharif* season				
	Jute	Rice (*aus*)	Sesame	Maize	Green gram
Root type					
Leaf type					
Stem type					
Green matter (g/m^2)					

Comments

Lesson 115. Recording observations, on changes of growth of pre-*kharif* crops (as in lesson 113), at 90 DAS

Introduction; **Objective**; **Material required** and **Procedure** as in lesson 109.

Observations on growth characters of the said crops at 90 DAS

Growth characters	Crops in pre-*kharif* season				
	Jute	Rice (*aus*)	Sesame	Maize	Green gram
Plant height (cm)	150-200	70-85	70-90	140-180	70-75
Tillers/m^2 or branch/plant	0	120-150	2-3	0	2-3
Effective tillers / sqm or branch / plant	0	115-125	3-4	0	3-4
Grains/ear-head or pod/branch	15-16	40-45	5-8	400-500	6-8

Lesson 116. Recording observations, on changes of growth of pre-*kharif* crops (as in lesson 113), at harvest

Introduction; Objective; Material required and **Procedure** as in lesson 109.

Observations

Recording observations on changes of (a) growth and (b) reprod-uctive characters of the said crops in pre-*kharif* season at harvest.

Growth and reproductive characters	Crops in pre-*kharif* season				
	Jute	Rice (*aus*)	Sesame	Maize	Green gram
(a) Growth characters :					
1. Plant height/ length (cm)	180-225	80-90	80-95	150-200	75-80
2. Tillers (sqm) or branch/plant	0	180-200	3-4	0	3-4
(b) Reproductive characters :					
1. Effective tillers / sqm or pod number / branch	0	115-125	3-4	0	3-4
2. Grain no. ear or seed no. / branch	15-16	40-50	5-8	400-500	6-8
3. 1000-seed weight (g)	10-12	20-25	10-15	80-100	25-35

11.3 *KHARIF* SEASON

11.3.1 Cereal crops

Rice (aman)

1.	*Climatic requirement*	:	High/moderate temperature with high rain- fall and humidity
2.	*Soil requirement*	:	Clayey, loamy, silty loam and clayey loam soils
3.	*Required land situation*	:	Low and medium lands, having waterlogged
4.	*Importance of crop*	:	Major food (carbohydrate) in India
5.	*Fertilizer management*	:	N, P_2O_5 & K_2O @ 100, 50 & 50 kg/ha, respectively
6.	*Water management*	:	4-5 irrigations, if needed, otherwise rainfed, water stagnation needed
7.	*Duration of crop*	:	120 to 150 days
8.	*Parts used*	:	Grains for human beings, straw for cattle
9.	*By-product*	:	Straw

10.	*Use of by-product*	:	Cattle feed, fuel, bedding for cattle/poultry, organic matter, mulch and roof materials.

Maize

1.	*Climatic requirement*	:	Moderate temperature, moderate rainfall
2.	*Soil requirement*	:	Loamy, alluvial, clayey loam and sandy soils
3.	*Required land situation*	:	Flat high lands of plains/hills
4.	*Importance of crop*	:	Cereal for both human beings, cattle & poultry
5.	*Fertilizer management*	:	N, P_2O_5 & K_2O @ 120, 60, 60 kg/ha, respectively
6.	*Water management*	:	3-4 irrigations (if rainfall is less)
7.	*Duration of crop*	:	120-140 days
8.	*Parts used*	:	Cobs (for roasting)/ grain for man, cattle and poultry
9.	*By-product available*	:	Stalk with leaves
10.	*Use of by-product*	:	Cattle feed, fuel, organic manure, pegs, used in trailing vine crops

11.3.2 Pulse crops

Greengram (Green gram)

1.	*Climatic requirement*	:	Warm climate, with rainfall
2.	*Soil requirement*	:	Loamy, alluvial, sany loam soils
3.	*Required land situation*	:	High and medium lands
4.	*Importance of crop*	:	Big leaves, trilobed; star-like inflorescence
5.	*Fertilizer management*	:	N, P_2O_5 & K_2O @ 40, 60 & 60 kg/ha, respectively
6.	*Water management*	:	Normally rainfed, 1-2 irrigations if irrigated
7.	*Duration of crop*	:	110-130 days
8.	*Parts used*	:	Seeds for pulse (*dal*) or soup
9.	*By-product*	:	Stalk/stover
10.	*Use of by-product*	:	Cattle feed, fuel, organic matter, mulch, residues

Cowpea

1.	*Climatic requirement*	:	High hot temperature with rainfall
2.	*Soil requirement*	:	Alluvial, loamy and clayey loam soils

3.	*Required land situation*	:	High and medium lands
4.	*Importance of crop*	:	Used as vegetable, G.M., cover crop, pulse
5.	*Fertilizer management*	:	N, P_2O_5 & K_2O @ 40, 60, 60 kg/ha, respectively
6.	*Water management*	:	Normally rainfed, 1-2 irrigations, if irrigated
7.	*Duration of crop*	:	120-140 days
8.	*Parts used*	:	Pods/seeds (vine used as fodder)
9.	*By-product*	:	Stalk or stover (when used as vegetable or pulse)
10.	*Use of by-product*	:	Fuel, organic matter, mulch, cattle feed

Ricebean

1.	*Climate requirements*	:	High temperature with rainfall and humidity
2.	*Soil requirement*	:	Loamy, alluvial and sandy loam soils
3.	*Required land situation*	:	High and medium lands
4.	*Importance of crop*	:	Trailing habits, controlling weeds, improving soil fertility (magic-crop for barren lands)
5.	*Fertilizer management*	:	N, P_2O_5 & K_2O @ 40, 60 & 40 kg/ha, respectively
6.	*Water management*	:	Normally rainfed; 1-2 irrigations, if irrigated
7.	*Duration of crop*	:	120-140 days
8.	*Parts used*	:	Seeds for *dal*, whole plant for fodder
9.	*By-product*	:	Stalk / stover (if seeds taken) / *bhusa*
10.	*Use of by-product*	:	Cattle feed, fuel, organic matter, G.M. cover crop.

11.3.3 Oilseed crops

[Different oilseed crops are obtained from different families, some of which are edible and some are not, but required for other purposes. Some non-edible oil crops may be utilized for bio-fuel purposes. These oilseeds are as follows] :

Sesame

1.	*Climatic requirement*	:	High temperature with rainfall and humidity
2.	*Soil requirement*	:	Alluvial, loamy and sandy loam soils
3.	*Required land situation*	:	High and medium lands
4.	*Importance of crop*	:	C4-type-crop, can be grown all times, white flower
5.	*Fertilizer management*	:	N, P_2O_5 & K_2O @ 60, 30 & 30 kg/ha, respectively

6.	*Water management*	:	Normally rainfed, 1-2 irrigations, if irrigated
7.	*Duration of crop*	:	110-130 days
8.	*Parts used*	:	Seeds for edible oil
9.	*By-product*	:	Stalk/stover
10.	*Use of by-product*	:	Fuel, organic matter, pegs (for vegetable staking)

Sunflower

1.	*Climatic requirement*	:	High temperature with rainfall
2.	*Soil requirement*	:	Loamy, sandy loam and alluvial soils
3.	*Required land situation*	:	High or medium lands
4.	*Importance of crop*	:	Head (big) moves according to sun's direction
5.	*Fertilizer management*	:	N, P_2O_5 & K_2O @ 60, 30 & 30 kg/ha, respectively
6.	*Water management*	:	Normally rainfed, 1-2 irrigations, if irrigated
7.	*Duration of crop*	:	110-120 days
8.	*Parts used*	:	Seeds for edible oil
9.	*By-product*	:	Stalk and leaves
10.	*Use of by-product*	:	Fuel, organic manure, pegs, mulch, cattle feed

Groundnut

1.	*Climatic requirement*	:	Hot, High temperature with rainfall
2.	*Soil requirement*	:	Loamy, sandy loam, clayey loam & sandy soils
3.	*Required land situation*	:	High and medium lands
4.	*Importance of crop*	:	Pods are developed below the ground with yellow flowers that develop above ground
5.	*Fertilizer management*	:	N, P_2O_5 & K_2O @ 40, 60 & 60 kg/ha, respectively
6.	*Water management*	:	3-4 irrigations, rainfed in eastern India
7.	*Duration of crop*	:	120-160 days
8.	*Parts used*	:	Nuts for edible oils/roasted food
9.	*By-product*	:	Haulms and leaves
10.	*Use of by-product*	:	Fuel, green manure, fodder, cattle feeds and mulch

[**Note** : In eastern parts, it can be grown in pre-*kharif* and soft twigs can be used as fodder or G.M.]

Jatropha/Physic nut

1.	*Climatic requirement*	:	Warm weather/climate with rainfall
2.	*Soil requirement*	:	Loamy, sandy loam, alluvium and sandy soils
3.	*Required land situation*	:	High and medium lands
4.	*Importance of crop*	:	Bio-fuel, petrol to be substituted to some extent
5.	*Fertilizer management*	:	N, P_2O_5 & K_2O @ 80, 60 & 30 kg/ha, respectively
6.	*Water management*	:	Normally rainfeed; 2-3 irrigations, if irrigated
7.	*Duration of crop*	:	2-3 years
8.	*Parts used*	:	Seeds for bio-fuel/fuel
9.	*By-product*	:	Plant (shrubs) and leaves
10.	*Use of by-product*	:	Fuel, cattle feed, organic manure, pegs and logs

Castor

1.	*Climatic requirement*	:	Moderate hot climate with rainfall
2.	*Soil requirement*	:	Loamy, sandy loam, alluvial and sandy soils
3.	*Required land situation*	:	High and medium lands
4.	*Importance of crop*	:	Lubricant (for cars) for industries, purgative for man/cattle, fuel
5.	*Fertilizer management*	:	N, P_2O_5 & K_2O @ 60, 30 & 30 kg/ha, respectively
6.	*Water management*	:	Normally rainfed; 1-2 irrigations before and after flowering, if irrigated
7.	*Duration of crop*	:	2-3 years
8.	*Parts used*	:	Seeds for biofuel / fuel
9.	*By-product*	:	Stalk and leaves
10.	*Use of by-product*	:	Fuel, organic matter, peg, cattle feed etc.

Banbherenda (*Jatropha glandulifera* Roxb.)

1.	*Climatic requirement*	:	Warm weather with rainfall
2.	*Soil requirement*	:	Loamy, sandy loam, alluvial and sandy soils
3.	*Required land situation*	:	High or medium lands
4.	*Importance of crop*	:	Some engines may be run with oils instead of petrol, diesel

5.	*Fertilizer management*	:	N, P_2O_5 & K_2O @ 80, 60 & 40 kg/ha, respectively
6.	*Water management*	:	2-3 irrigations annually, if irrigated
7.	*Duration of crop*	:	2-3 years
8.	*Parts used*	:	Seeds for non-edible oils
9.	*By-product*	:	Stem and leaves
10.	*Use of by-product*	:	Fuel, peg, organic matter, cattle feed etc.

11.3.4 Fibre crops

Jute (for seed)

1.	*Climatic requirement*	:	Hot, High temperature, high humidiy and rainfall
2.	*Soil requirement*	:	Alluvial, loamy and sandy loam soils
3.	*Required land situation*	:	High, medium and low land
4.	*Importance of crop*	:	Bast fibre from bark/stem, seeds small
5.	*Fertilizer management*	:	N, P_2O_5 & K_2O @ 80, 30 & 60 kg/ha, respectively
6.	*Water management*	:	2-3 irrigations, if irrigated; normally rainfed
7.	*Duration of crop*	:	120-130 days
8.	*Parts used*	:	Seeds/for seed-purpose and stems for fibres
9.	*By-product*	:	Stick (*pakati*)/ leaves
10.	*Use of by-product*	:	Stick for paper boards, fench, paper and leaves for manure (leaf contains about 0.5% N)

Mesta (for seed)

1.	*Climatic requirement*	:	High temperature with high and moderate rainfall and humidity
2.	*Soil requirement*	:	Loamy, alluvial, clayey loam and sandy loam soils
3.	*Required land situation*	:	High, medium and low lands
4.	*Importance of crop*	:	Bast fibre from stem; big coloured flowers
5.	*Fertilizer management*	:	N, P_2O_5 & K_2O @ 60, 30, 60 kg/ha, respectively
6.	*Water management*	:	Normally rainfed; 1-2 irrigations, if irrigated
7.	*Duration of crop*	:	110-140 days

8.	*Parts used*	:	Fibre from stem and seeds from pods
9.	*By-product*	:	Stalk/stick and leaves/*bhusa*
10.	*Use of by-product*	:	Stick for fuel, pegs, fence, paper board, paper; leaves for manures, mulch, cattle feed and vegetable

Sunnhemp (for seed)

1.	*Climatic requirement*	:	High temperature; high rainfall & humidity
2.	*Soil requirement*	:	Alluvial, loamy, sandy loam and sandy soils
3.	*Required land situation*	:	High, medium and low lands
4.	*Importance of crop*	:	Bast fibre from stem/bark; roots from stems
5.	*Fertilizer management*	:	N, P_2O_5 & K_2O @ 40, 40 & 60 kg/ha, respectively
6.	*Water management*	:	Normally rainfed; 1-2 irrigations, if irrigated
7.	*Duration of crop*	:	110-130 days
8.	*Parts used*	:	Fibre from stem and seeds from pods
9.	*By-product*	:	Stick/stalk and leaves
10.	*Use of by-product*	:	Stick as fuel, pegs fence, and leaves for organic manure, cattle feed etc.

1.3.5 Rhizomes

Asparagus (Satamuli)

1.	*Climatic requirement*	:	Dry condition, warm weather, with rainfall
2.	*Soil requirement*	:	Sandy, rocky, sandy loam and loamy soils
3.	*Required land situation*	:	High and medium lands
4.	*Importance of crop*	:	Swollen roots, spinous shrub and large plants
5.	*Fertilizer management*	:	N, P_2O_5 & K_2O @ 60, 30, 30 kg/ha, respectively
6.	*Water management*	:	Normally rainfed; 1-2 irrigations, if irrigated
7.	*Duration of crop*	:	2-3 years
8.	*Parts used*	:	Roots for food (starch and carbohydrate)
9.	*By-product*	:	Stem and twigs
10.	*Use of by-product*	:	Fuel, pegs, organic matter, veterinary medicine, cattle feeds etc.

Amorphophallus (ol)

1.	*Climatic requirement*	:	High temperateure, rainfall and humidity
2.	*Soil requirement*	:	All types of soil except heavy clay
3.	*Required land situation*	:	High and medium lands
4.	*Importance of crop*	:	Big corm, used as vegetable, variegated stones
5.	*Fertilizer management*	:	N, P_2O_5 & K_2O @ 100, 50 & 60 kg/ha, respectively
6.	*Water management*	:	Normally rainfed; 1-2 irrigations, if irrigated
7.	*Duration of crop*	:	120-150 days
8.	*Parts used*	:	Big corm used as vegetable; pseudo-stem used as vegetable sometimes.
9.	*By-product*	:	Stalk and leaves
10.	*Use of by-product*	:	Fuel, organic matter, vegetable etc.

Arum / taro

1.	*Climatic requirement*	:	High temperature with high rainfall and humidity
2.	*Soil requirement*	:	All types of soils except sandy soils
3.	*Required land situation*	:	Low, medium and high lands (for different varieties)
4.	*Importance of crop*	:	Some varieties are grown in water; runners available and edible
5.	*Fertilizer management*	:	N, P_2O_5 & K_2O @ 100, 40, 60 kg/ha, respectively
6.	*Water management*	:	Grown in rainfed/transplanted or dry conditions
7.	*Duration of crop*	:	150-160 (extended with runner)
8.	*Parts used*	:	Rhizomes, petiole and runners used as vegetable
9.	*By-product*	:	Leaves and roots
10.	*Use of by-product*	:	Fuel, mulch, organic matter, plate

11.3.6 Fodder crops

Napier or thinnapier

1.	*Climatic requirement*	:	High temperature, moderate rainfall and humidity
2.	*Soil requirement*	:	Alluvial, loamy, sandy loam and sandy soils
3.	*Required land situation*		High and medium lands
4.	*Importance of crop*	:	Long stem with long and narrow leaves

5.	*Fertilizer management*	:	N, P_2O_5 & K_2O @ 100, 50 & 50 kg/ha, respectively
6.	*Water management*	:	3-4 irrigations, at 20 days interval
7.	*Duration of crop*	:	110-140 days
8.	*Parts used*	:	Whole plants are cut at flowering for fodder
9.	*By-product*	:	No by-product
10.	*Use of by-product*	:	No use, as no by-product

Teosinte :

1.	*Climatic requirement*	:	Moderate temperature and moderate rainfall
2.	*Soil requirement*	:	Loamy, stony, alluvial, sandy loam and sandy soils
3.	*Required land situation*	:	High and medium lands
4.	*Importance of crop*	:	Sparing (not compact) cobs, yellow colour
5.	*Fertilizer management*	:	N, P_2O_5 & K_2O @ 100, 50, 50 kg/ha, respectively
6.	*Water management*	:	4-5 irrigations, 20 days interval
7.	*Duration of crop*	:	110-140 days
8.	*Parts used*	:	Entire crop plants are used as fodder
9.	*By-product*	:	No by-product
10.	*Use of by-product*	:	No use of by-product, as no by-product

Cowpea

1.	*Climatic requirement*	:	Moderate rainfall with high temperature
2.	*Soil requirement*	:	Alluvial, loamy, sandy loam & sandy soils
3.	*Required land situation*	:	High and medium lands
4.	*Importance of crop*	:	The crop may be used as cover crop, vegetable, G.M., pulse, fodder etc.
5.	*Fertilizer management*	:	N, P_2O_5 & K_2O @ 40, 60 & 50 kg/ha, respectively
6.	*Water management*	:	Normally rainfed; 1-2 irrigations, if irrigated
7.	*Duration of crop*	:	110-130 days
8.	*Parts used*	:	Entire plants cut, at flowering for fodder
9.	*By-product*	:	No by-product
10.	*Use of by-product*	:	No use, as no by-product

Blackgram

1.	*Climatic requirement*	:	High/moderate rainfall with high tempeature
2.	*Soil requirement*	:	Loamy, alluvial, sandy and sandy soils
3.	*Required land situation*	:	High and medium lands
4.	*Importance of crop*	:	Used as pulse, G.M., fodder & N-fixation
5.	*Fertilizer management*	:	N, P_2O_5 & K_2O @ 40, 60, 40 kg/ha, respectively
6.	*Water management*	:	Rain-fed, 1-2 irrigations, if irrigated
7.	*Duration of crop*	:	100-130 days
8.	*Parts used*	:	Whole plant cut, for fodder at flowering
9.	*By-product*	:	No by-product
10.	*Use of by-product*	:	As no by-product, no use.

Ricebean

1.	*Climatic requirement*	:	Moderate rainfall with high temperature
2.	*Soil requirement*	:	Alluvial, loamy, sandy and clayey loam soil
3.	*Required land situation*	:	High and medium lands
4.	*Importance of crop*	:	Used as fodder, G.M., pulse, pod, weed- control
5.	*Fertilizer management*	:	N, P_2O_5 & K_2O @ 50, 60 & 50 kg/ha, respectively
6.	*Water management*	:	Normally rainfed; 1-2 irrigations, if irrigated
7.	*Duration of crop*	:	110-140 days
8.	*Parts used*	:	Whole plants are cut for fodders
9.	*By-product*	:	No by-product
10.	*Use of by-product*	:	No use, as no by-product

11.3.7 Green manure crops

Dhaincha / baghadhaincha

1.	*Climatic requirement*	:	High rainfall with high temperature and humidity
2.	*Soil requirement*	:	Alluvial, clayey loam and loamy clay soils
3.	*Required land situation*	:	Low and medium lands
4.	*Importance of crop*	:	This crop may be used to reclaim saline-alkali soils; nodules available in roots/stems of baghadhaincha

5.	*Fertilizer management*	:	N, P_2O_5 & K_2O @ 40, 60 & 40 kg/ha, respectively, nodule available in roots only of *dhaincha,* but nodules are in both roots and stems in *baghadhaincha*
6.	*Water management*	:	Normally rainfed (1-2 irrigation, if needed)
7.	*Duration of crop*	:	110-130 days, but 45-50 days crop is good for GM
8.	*Parts used*	:	Entire plants are ploughed-under as G.M., at 45-50 DAS
9.	*By-product*	:	No by-product
10.	*Use of by-product*	:	As no by-product, no use

Cowpea

1.	*Climatic requirement*	:	Moderate rainfall with high temperature
2.	*Soil requirement*	:	Alluvial, loamy, clayey loam and sandy loam soils
3.	*Required land situation*	:	High and medium lands
4.	*Importance of crop*	:	Improves soil organic matter and pods used as vegetable
5.	*Fertilizer management*	:	N, P_2O_5 & K_2O @ 40, 60 & 50 kg/ha, respectively
6.	*Water management*	:	Normally rainfed; 1-2 irrigations, if irrigated
7.	*Duration of crop*	:	110-130 days, for GM, 45-50 DAS
8.	*Parts used*	:	Entire plants are ploughed-under as GM, at 45-50 DAS
9.	*By-product*	:	No by-product
10.	*Use of by-product*	:	No use, as no by-product

Guar/Cluster bean

1.	*Climatic requirement*	:	Moderate rainfall with high temperature
2.	*Soil requirement*	:	Loamy, clayey loam and alluvial, soils
3.	*Required land situation*	:	High and medium lands
4.	*Importance of crop*	:	Can be used as vegetable (pod)
5.	*Fertilizer management*	:	N, P_2O_5 & K_2O @ 40, 60, 50 kg/ha, respectively
6.	*Water management*	:	Normally rainfed; 1-2 irrigations, if irrigated
7.	*Duration of crop*	:	90-110 days, but for GM, 45-50 DAS
8.	*Parts used*	:	Entire plants are ploughed-under as GM, at 40-50 days after sowing

9.	*By-product*	:	No by-product
10.	*Use of by-product*	:	As no by-product, no use

Sunnhemp

1.	*Climatic requirement*	:	High rainfall, temperature and humidity
2.	*Soil requirement*	:	Loamy, alluvial and sandy loam soils
3.	*Required land situation*	:	High and medium lands
4.	*Importance of crop*	:	Can also be grown as fibre crop; for seed, it can be grown in *rabi* as *paira* crop
5.	*Fertilizer management*	:	N, P_2O_5 & K_2O @ 40, 60 & 60 kg/ha, respectively
6.	*Water management*	:	Normally rainfed; 1-2 irrigations, if irrigated
7.	*Duration of crop*	:	110-130 days, but for GM 45-50 DAS
8.	*Parts used*	:	Entire plants are ploughed-under, at 45-50 DAS
9.	*By-product*	:	No by-product
10.	*Use of by-product*	:	As no by-product, no use

Blackgram

1.	*Climatic requirement*	:	Moderate rainfall, high temperature and humidity
2.	*Soil requirement*	:	Alluvial, loamy, sandy loam and sandy soils
3.	*Required land situation*	:	High and medium lands after flood in low lands
4.	*Importance of crop*	:	Can be used as pulse, G.M. & fodder crops
5.	*Fertilizer management*	:	N, P_2O_5 & K_2O @ 40, 60 & 50 kg/ha, respectively
6.	*Water management*	:	Normally rainfed; 1-2 irrigations, if irrigated
7.	*Duration of crop*	:	110-130 days but for G.M. 45-50 DAS
8.	*Parts used*	:	Entire plants ploughed-under at 45-50 DAS
9.	*By-product*	:	No by-product
10.	*Use of by-product*	:	As no by-product, no use.

Practical Class : Recording observations from time-to-time, on morp- hological changes of growth at stages and yields of crops in *kharif* season

Lesson 117. Recording observations, on changes of growth of crops (Rice, maize, green gram, cowpea, ricebean, sesame, sunflower, groundnut, jatropha and jute (for seed), at 45 DAS, in *kharif* season

Introduction; **Objective**; **Material required** and **Procedure :** All as in lesson 109.

Observations on changes of growth characters of *kharif* crop, at 45 DAS

Kharif crops	Plant type	Plant characters	Height/ length (cm)	Tillers plants/sqm
Rice (*aman*)	Annual	Erect	10-15	15-16
Maize	Annual	Erect	15-20	16-20
Green gram	Annual	Vine	5-7	30-40
Cowpea	Annual	Vine	8-10	30-40
Ricebean	Annual	Vine	8-10	30-40
Sesame	Annual	Erect	5-7	30-40
Sunflower	Annual	Erect	6-8	30-40
Groundnut	Annual	Vine	5-7	30-40
Jatropha	Perennial	Erect	6-8	20-25
Jute (seed)	Annual	Erect	5-6	30-40

Assignment for green matter and others, at 45 DAS (to be recorded)

Crops in *kharif*	Plant characters			
	Leaf type	Root type	Stem type	Green matter (g/sqm)
Rice (*aman*)				
Maize				
Green gram				
Cowpea				
Rice bean				
Sesame				
Sunflower				
Groundnut				
Jatropha				
Jute (for seed)				

[**Note**: Destructive samples for green matters – method as per the note in assignment table in lesson 109].

Comments :

Lesson 118 : Recording observations, on changes of growth of same *kharif* crops (as in lesson 117), at 60 DAS

Introduction; Objective; Material required; Procedure : All as in lesson 109.

Observations on changes of growth characters of crops (as in lesson 117) of same *kharif* crops at 60 DAS

Kharif crops	Plant type (cm)	Branch	Crop type	Plant type
Rice (*aman*)	60-80	70-80	Cereal	Annual
Maize	80-90	25-30	Cereal	Annual
Green gram	60-70	30-40	Pulse	Annual
Cowpea	60-70	30-40	Fodder	Annual
Rice bean	60-70	30-40	Fodder	Annual
Sesame	60-80	40-60	Oilseed	Annual
Sunflower	60-80	30-40	Oilseed	Annual
Groundnut	60-70	30-40	Oilseed	Annual
Jatropha	60-80	30-40	Oilseed	Annual
Jute (seed)	80-90	30-40	Fibre seed	Annual

Assignment at 60 DAS (to be recorded)

Crops in *kharif*	Plant characters			
	Root type	Leaf type	Stem type	Green matter (g/sqm)
Rice (*aman*)				
Maize				
Green gram				
Cowpea				
Ricebean				
Sesame				
Sunflower				
Groundnut				
Jatropha				
Jute (for seed)				

Lesson 119 : Recording observations, on changes of growth of the same *kharif* crops (as in lesson 117), at 90 DAS

Introduction; Objective; Material required; Procedure : All as in lesson 109.

Observations of growth characters of crops (as mentioned earlier) in *kharif* season at 60 to 90 DAS

Kharif crops	Plant height length (cm)	Tiller/m^2 or branch/plant	Eff. tillers/m^2 or branch plant	Grains/ear or pod/branch
Rice (*aman*)	80-90	120-150	118-140	70-80
Maize	180-200	1-2	1-2	400-500
Green gram	70-80	6-8	6-8	5-6
Cowpea	70-80	6-8	6-8	7-8
Rice bean	70-80	6-8	6-8	6-8
Sesame	70-80	6-8	6-8	15-20
Sunflower	70-80	6-8	6-8	100-110
Groundnut	70-80	3-4	3-4	2-3
Jatropha	70-80	6-8	6-8	4-6
Jute (seed)	150-200	6-8	6-8	14-16

Note: Eff = effective; m^2 = square meter; cm= centimetre

Assignment at 60-90 DAS (to be recorded)

Crops in *kharif*	Plant characters			
	Root type	Leaf type	Stem type	Green matter (g/sqm)
Rice (*aman*)				
Maize				
Green gram				
Cowpea				
Rice bean				
Sesame				
Sunflower				
Groundnut				
Jatropha				
Jute (for seed)				

Comments :

Lesson 120 : Recording observations, on changes of growth of *kharif* crops (as in lesson 117), at harvest

Introduction; Objective; Material required; Procedure: All as in lesson 109.

Observations on growth characters of the same crops (as in 117) in *kharif* season at harvest

Kharif season crops	Growth/production charactrs	
	Plant height/length (cm)	Tillers/sqm or branch/plant
(a) Growth characters :		
Rice (*aman*)	90-100	125-140
Maize	200-225	1-2
Green gram	75-85	6-10
Cowpea	75-85	6-8
Rice bean	75-85	6-8
Sesame	75-85	6-8
Sunflower	75-85	6-8
Groundnut	70-80	3-5
Jatropha	75-80	6-10
Jute (for seed)	140-180	6-10

(b) Reproductive characters

Kharif crops	Yield-components		
	Effective tillers/ m^2 or pod/branch	Grains/ear head or seeds/pod	1000-seed weight (gm)
Rice (*aman*)	200-250	120-130	20-25
Maize	2-3	400-500	10-15 (100-seeds)
Green gram	20-25	8-9	22-30
Cowpea	20-25	8-12	25-30
Rice bean	15-20	8-10	30-40
Sesame	8-10	80-90	10-15
Sunflower	100-125	110-120	20-35
Groundnut	10-25	2-4	10-15 (100-seeds)
Jatropha	10-25	8-10	35-40
Jute (seed)	10-25	6-8	5-10

CHAPTER 12

CROPPING SYSTEM

'Cropping system' plays a vital role, on the economic conditions of the farming community of the particular area in India.

For this cropping system, natural resources should be exploited heavily as far as possible. Of course, in all the places, it will not be possible, because of adverse resources facilities. But under these circumstances, additional efforts should be made to implement some cropping systems.

Criteria for 'intensive cropping' system should be based on:

(i) Climatic conditions, (ii) Land situations, (iii) Soil types, (iv) Good (fertile) lands, (v) Irrigation facilities availability, (vi) Availability of manures / fertilizers, (vii) Infra-structure for storage seeds, fibres, tubers, fodders etc. (viii) Good threshing floors, (ix) Proper machinaries (Tructor/power tiller etc.) for quick soil preparation, (x) Facilities for processing harvested products, (xi) Facilities for marketing for good price, (xii) Transport facilities for moving the produce, besides social structure.

12.1 UNDER RAINFED CONDITIONS

12.1.1 *Paira/Utera* crops

Contradictory to the common thinking, it is possible to go for multiple cropping programmes, in eastern parts of the country, even under rainfed conditions, because of (i) high rainfall, (ii) longer duration of the rainy season, (iii) good quality soils (alluvial), (iv) more flat lands in some areas, (v) land situations (flat plains), and (vi) seasonal variation with 3 crop-seasons (*rabi*, pre-*kharif* and *kharif*).

Under these conditions, *paira* or *utera* crop is possible under rainfed conditions and it has been found that the following rainfed cropping systems (3-crop) are possible, with good crop yields and for maintaining soil fertility :

1. Grasspea (P) - green gram- transplanted *kharif* rice, 2. Oat (P)- green gram – transplanted *kharif* rice, 3. Sunnhemp (for seed, P)- green gram- transplanted *kharif* rice, 4. Grasspea- mesta- transplanted *kharif* rice, 5. Oat (P)-mesta- transplanted *kharif* rice, 6. Sunnhemp (P)- mesta- transplanted *kharif* rice, 7. Grasspea (P)- groundnut- transplanted *kharif* rice, 8. Lentil (P)- groundnut- transplanted *kharif* rice, 9. Oat (P)-

groundnut- transplanted *kharif* rice, 10. Grasspea (P)- jute- transplanted *kharif* rice, 11. Lentil (P)- jute- transplanted *kharif* rice.

There are 5 eco-systems in India, such as (1) Arid, (2) Semi-arid, (3) Sub-humid, (4) Humid and (5) Coastal. Therefore, cropping system varies from place to place. Therefore, it has been observed that 3 crops may be raised even under rainfed conditions, provided one *paira/utera* crop is raised in the system. In eastern India, a good amount of water is present at the time of harvest of transplanted *kharif* rice; with these waters *paira* or *utera* crop may be germinated, later with residual moisture that *paira* crop may be raised. In pre-*kharif* and *kharif* seasons, no irrigation is necessary for any crop, because of high amount of rainfall and long duration of rainfall, Under these condition *paira / utera* (no-tillage relay) crop is a must, to exploit the situations.

Note : P = *Paira / Utera* crop.

12.1.2 Rainfed crops, in general

Rainfed cropping systems in India, are as follows

			Kharif	*Rabi*	Summer
1.	Arid ecosystem :	(i)	Rice -	Gobisarson -	–
2.	Semi-arid – ecosystem :				
		(i)	Rice -	mustard -	sunflower
		(ii)	Greengram -	mustard -	sunflower
		(iii)	Soybean -	sunflower -	greengram
		(iv)	Clusterbean -	barley -	–
		(v)	Sesame -	barley	green gram
3.	Sub-humid ecosystem :	(i)	Maize -	–	gobisarson
		(ii)	Rice -	sunflower -	greengram
		(iii)	Rice -	mustard -	sesame
		(iv)	Soybean -	gram -	–
		(v)	Blackgram -	gram -	–
4.	Humid ecosystem :	(i)	Rice -	–	rice
		(ii)	Rice -	mustard -	rice
		(iii)	Rice -	Frenchbean -	rice
		(iv)	Rice -	mustard -	–
		(v)	Rice -	grasspea (*paira*) -	jute
5.	Coastal ecosystem	(i)	Rice -	pigeonpea -	–
		(ii)	Rice -	–	groundnut
		(iii)	Rice -	gram -	sorghum
		(iv)	Rice -	sorghum -	sesame

[**Note:**Adopted from **Annual Report (1998-99) of All-India** Co-ordinated Research Project on Cropping System (ICAR), PDCSR, Modipuram, Meerut 250110, UP].

Special rainfed 3-crop systems in West Bengal (under humid- eco-system) with *rabi paira/utera* crops after *kharif* transplanted rice:

Kharif		*Rabi (paira / utera)*		Pre-*kharif*
Transplanted rice	–	grasspea	–	jute
Transplanted rice	–	lentil	–	jute
Transplanted rice	–	sunnhemp (seed)	–	jute
Transplanted rice	–	grasspea	–	groudnut
Transplanted rice	–	lentil	–	groundnut
Transplanted rice	–	sunnhemp (seed)	–	groundnut
Transplanted rice	–	lentil	–	green gram
Transplanted rice	–	sunnhemp (seed)	–	green gram

Source: Das, N.R. (2006).

Practical Class : Chalkout the programmes for rainfed cropping systems in high rainfall and some dry regions of the country

Lesson 121: Prepare the rainfed cropping systems for (i) High rain-fall areas, (ii) Coastal areas, and (iii) Semi-dry areas of the country

Introduction

'Cropping system' is the system where different crops (two more) are grown in different seasons in the same year, having maximum crop production in the same piece of land. For this system, inputs like fertilizers, water etc. are needed. But under rainfed condition also, multiple cropping systems are followed. Programme may be chalked out for (rainfed areas) the multiple cropping system both under high rainfall areas (like eastern part) and dry regions (like Rajasthan /Gujarat etc) from the information.

Objective

Multiple cropping under the existing system, is due to under the existing weather conditions,

O to get maximum crop production,

O to increase the number of crops in the system.

Material required

✥ The lands under different crops, ✥ Book or other materials for information on cropping system, ✥ Recorded data of crops, ✥ Note book, ✥ Pen/pencil.

Procedure

Collect the data regarding crops/varieties/sowing time/ production etc. then frame the rainfed cropping system under (i) high rainfall areas, (ii) coastal areas, and (iii) semi-dry areas of the country.

Rainfed cropping system in high rainfall areas

Following are the cropping systems under rainfed conditions in high rainfall areas (like, West Bengal, Assam etc) :

(i)	Rice	rice,	-
(ii)	Rice	mustard	rice,
(iii)	Rice	frenchbean	rice,
(iv)	Rice	mustard	- ,
(v)	Maize	-	gobisorson,
(vi)	Rice	mustard	sesame,
(vii)	Soyabean	-	gram,
(viii)	Blackgram	-	gram

Coastal areas

Following are the cropping systems in coastal areas, under rainfed conditions, particularly in West Bengal, Orissa, Andhra Pradesh etc :

(i)	Rice -	pigeon pea	-
(ii)	Rice -	-	groundnut,
(iii)	Rice -	gram -	sorghum,
(iv)	Rice -	sorghum -	sesame – etc.

Semi-dry areas

Following are the main cropping systems in semi-dry and rainfed areas, particularly in Gujarat, Rajasthan, U.P., M. P. etc.

(i)	Rice -	mustard -	sunflower,
(ii)	Greengram -	mustard -	sunflower
(iii)	Sesame -	barley	greengram,
(iv)	Soyabean -	sunflower –	greengram
(v)	Cluster bean -	barley -	etc.

Lesson 122 : Prepare the `special rainfed 3-crops system', with *paira-/utera* crops, based on eastern region (West Bengal), describing the conditions required for *paira/utera* crops.

Introduction; **Objective**; **Material required** (All as in lesson 121).

Procedure

In rice (*aman*) at maturity stage, under marshy/muddy and rainfed conditions, without tillage, grow the *paira/utera* crops (p) are grown in *rabi* season. Thereafter, pre-*kharif* and *kharif* season crops could be grown with natural rainfall with normal tillage, having the special 3-crop system such as:

Special rainfed 3-crop system with *paira/utera* crops

(1) Grasspea (p) – green gram – transplanted *kharif* rice,

(2) Sunnhemp (p) (for seed) – green gram-transplanted *kharif* rice,

(3) Oat (p) – green gram – transplanted *kharif* rice,

(4) Grasspea (p) – mesta - transplanted *kharif* rice,

(5) Oat (p) – mesta - transplanted *kharif* rice,

(6) Sunnhemp (p) – mesta - transplanted *kharif* rice,

(7) Grasspea (p) – groundnut - transplanted *kharif* rice,

(8) Lentil (p) – groundnut - transplanted *kharif* rice,

(9) Oat (p) – groundnut - transplanted *kharif* rice,

(10) Grasspea (p) – jute - transplanted *kharif* rice,

(11) Lentil (p) – jute - transplanted *kharif* rice etc.

Conditions required for *paira/utera* crops

(1) Low land with small amount of water at the end of *kharif*, trans-planted rice, (2) High rainfall, in rainy season, (3) Flat land, (4) Alluvial soils, (5) In different seasons with different climatic conditions, (6) Good transp-orting system and (7) Availability of adequate labourers.

[**Note**: *Paira/utera* cropping systems may be followed mostly in eastern parts of the country (West Bengal, Assam, Tripura, and some parts of Bihar, UP. and Orissa), particularly in low lying areas, where rainfall is high and continues for 6-7 months].

12.2 UNDER IRRIGATED CONDITIONS

12.2.1 Cropping systems for eastern parts of the country (Humid ecosystem)

In eastern parts of the country, generally *rabi* crops are irrigated; other crops in the systems are rainfed or irrigated, if needed (when rainfall is less). This eastern part of the country is in the humid-ecosystem. The cropping system in this ecosystem are as follows:

Kharif -		Rabi -	Summer (pre-kharif)
(i)	Rice -	rice (boro) -	G. M. (green manure)
(ii)	Rice -	potato -	sesame
(iii)	Rice -	potato -	groundnut

(iv)	Rice -	potato -	aus rice (direct-seeded)
(v)	Rice -	wheat -	jute
(vi)	Rice -	cabbage -	rice (boro)

12.2.2 Cropping systems for other regions of India

In other parts of the country i.e. in all the ecosystems (except humid- ecosystem, which has been stated earlier), the cropping systems are as follow:

	Kharif	*Rabi*	Summer
(A) Arid ecosystem :			
(i)	Pearl millet -	mustard -	greengram
(ii)	Soybean -	wheat -	cowpea
(iii)	Pearlmillet -	potato -	greengram
(B) Semiarid ecosystem :			
(i)	Maize -	potato -	green gram -
(ii)	Rice -	maize -	G.M.
(iii)	Maize -	wheat -	sunflower
(iv)	Groundnut -	wheat -	greengram
(v)	Soybean -	wheat -	greengram
(C) Sub-humid ecosystem :			
(i)	Maize-	*toria* mustard -	potato
(ii)	Maize -	pea -	potato
(iii)	Maize -	soybean -	frenchbean
(iv)	Rice -	wheat -	fodder
(v)	Rice -	potato -	wheat
(D) Coastal ecosystem :			
(i)	Rice -	cabbage -	lady's finger
(ii)	Rice -	tomato -	leafy vegetable
(iii)	Rice -	potato -	sesame
(iv)	Rice -	wheat -	greengram
(v)	Rice -	rice -	pigeonpea
(vi)	Rice -	rice -	sesame

Source : Das, N.R. (2006)

Practical Class : Chalkout the programmes for irrigated cropping systems, in different ecosystems in India

Lesson 123 : Prepare the irrigated cropping systems in different eco-systems in India, describing the weather conditions therein

Introduction; Objective; Material required (All as in lesson 121)

Procedure (see again)

Prepare the irrigated cropping systems for different ecosystems in India and describe the weather conditions in brief there-in (as in 12.2.2)

	Kharif	*Rabi-*	Summer
(i) Cropping system under Arid-zone ecosystem :			
(1)	Pearlmillet -	mustard -	greengram
(2)	Soybean -	wheat -	cowpea,
(3)	Pearlmillet -	potato -	greengram.
(ii) Cropping system under Semi-arid-zone ecosystem :			
(1)	Maize -	potato -	greengram
(2)	Rice -	maize -	greengram
(3)	Maize -	wheat -	sunflower
(4)	Groundnut -	wheat -	greengram
(5)	Soybean -	wheat -	greengram
(iii) Cropping system under Sub-humid zone ecosystem :			
(1)	Maize -	toria -	potato
(2)	Maize -	pea -	potato
(3)	Maize -	soybean -	frenchbean
(4)	Rice	wheat	fodder (cowpea)
(5)	Rice	potato	wheat
(iv) Cropping system under Humid-zone ecosystem :			
(1)	Rice -	wheat -	jute
(2)	Rice -	maize -	jute
(3)	Rice -	mustard -	greengram
(4)	Maize -	wheat -	greengram
(5)	Maize	potato	jute
(v) Cropping system under Coastal-zone ecosystem :			
(1)	Rice -	cabbage -	lady's finger
(2)	Rice -	tomato -	leafy vegetables
(3)	Rice -	potato -	sesame

(4)	Rice -	rice -	pigeonpea
(5)	Rice -	rice -	sesame
(6)	Rice -	potato -	greengram
(7)	Rice -	rice -	greengram

Weather conditions in different ecosystems

(i) Arid-zone ecosystem

This area practically is very dry; most of the crops are grown with irrigation. Low rainfall in rainy season, winter and summer are severe and dry. These areas are mostly in north-western parts of India.

(ii) Semi-arid-zone ecosystem

The area is also practically dry; and most of the crops are grown with irrigations. Rainfall is also low, like arid-zone in rainy season; winter and summer are severe. Most of the areas are in north-west and central parts of the country.

(iii) Sub-humid-zone ecosystem

Practically this area is semi-humid consisting of the middle and northern areas of the country. Here rainfall in rainy season is higher; winter is prolonged and severe but not so severe as in *arid* or *semiarid* regions. Here summer is also severe.

(iv) Humid-zone ecosystem

This area is highly humid, due to too much rainfall and also long duration of rainfall (April to October). Distinct crop-seasons are observed as *kharif*, *rabi* and pre-*kharif*. Winter is short and mild. Practically all crops are grown. This area is in the eastern and north-eastern parts of India. Actually this area consists of both the wet low lands and high hills. Good vegetations are available. Summer is not so severe here.

(vi) Coastal-zone ecosystem

These are the coastal areas of West Bengal, Orissa, Andhra Pradesh, Tamil Nadu, Maharastra, Karnataka and Gajarat, having saline soil and sultry weather. Winter and summer are not so severe. Most of the crops cannot be grown, only some specific crops are grown. There is a potential of growing cashew-nuts, coconuts etc. as plantation crops, besides other crops.

12.3 PARTIAL RAINFED AND PARTIAL IRRIGATED CONDITIONS (IN EASTERN PARTS OF INDIA)

12.3.1 Irrigation in *rabi* season plus rainfed in pre-*kharif* and *kharif* seasons

Generally, irrigation is given in *rabi* season which is a dry period, in the eastern parts of the country, particularly in West Bengal, Assam, Tripura, Meghalaya, Manipur etc. Here rainfall is high nearly 1400 to 3000 mm annually and duration is about 7 to 8 months (starting from March to October). Therefore, multiple cropping programme is very common, even in rainfed areas.

In irrigated areas, the *rabi* crops are fully irrigated and both in pre- *kharif* and *kharif* seasons practically no irrigation is given except some low-rainfall-year.

The major cropping systems are as follows :

Kharif		*Rabi*	Summer
(i)	Rice -	wheat -	jute,
(ii)	Rice -	maize -	jute,
(iii)	Rice -	wheat -	green gram,
(iv)	Maize -	wheat -	jute,
(v)	Rice (*aus*) -	potato -	jute,
(vi)	Rice -	wheat -	groundnut etc.

12.3.2. Irrigation in *rabi* season plus partially rainfed and irrigated crops in pre-*kharif* and *kharif* seasons

The same cropping systems, stated earlier may be grown under irrigated conditions in some cases. Water shortage may be there either in pre-*kharif* /*kharif* seasons, then only irrigations should be given there. In these cases, *kharif* (mostly transplanted rice) crop suffers a lot, because of lack of water- stagnation which is needed for this crop, all the time from transplantation to harvest.

Practical Class : Chalkout the plan for irrigated + rainfed intensive cropping programmes

Lesson 124: Prepare the ecologically combined (irrigated and rain-fed) 3-crop systems in eastern parts of the country

Introduction

In the eastern parts of the country, particularly in West Bengal, Assam, Tripura, Meghalaya etc., rainfall is high, along with its duration, and persists for long time, ranging from March/April to October. In the rainy season, there is plenty of water; though in *rabi* season, rainfall is very low. More moisture is available in soils. In most of the crops, during pre-*kharif* and *kharif*, seasons no irrigation is practically needed, however, only in *rabi* season irrigation is needed. Therefore, practically 3 to 4 crops are raised with combined effects of irrigation and rainfed conditions.

Objective

Preparation of plan for combined water sources (irrigation and rainfed) of some cropping systems in eastern parts of India is

O to get the maximum crop yield of the system, under existing climatic conditions,

- O to utilize both rain-water and irrigation-water for the intensive cropping systems,
- O to economise cost of irrigation,
- O to protect the crop against the excessive water from irrigation.

Material required

❖ Irrigated high/medium lands, ❖ Irrigation arrangements, ❖ Crops (seeds) for the multiple cropping programmes, ❖ Spade – 4, ❖ Rope / threads – 500 gm, ❖ Rain gauge, ❖ Note book, ❖ Pen/pencil.

Procedure

- Grow *rabi* crops with the facilities of irrigations at different phases (during *rabi* season wheat, potato, maize etc. could be grown)
- Grow the pre-*kharif* crops like jute, green gram, rice (direct-seeded or *aus*), maize etc. under natural rainfall conditions. If necessary, apply irrigation, as facilities of irrigations are available,
- Similarly, as above, grow the *kharif* crops like transplanted *kharif* rice, maize, mesta (for seed), jute (for seeds), with natural precipit-ation. If necessary, apply irrigation water, as irrigation facilities are available here.

Thus the intensive cropping systems, with combined water sources (irrigation + natural rainfall), the following system may be followed in wet or humid ecosystem :

Rabi (irrigated)		Pre-*kharif* (rainfed)	*Kharif* (rainfed)
(1)	Wheat -	jute -	*transplanted* rice
(2)	Potato -	*aus* rice -	jute (for seed)
(3)	Wheat -	green gram -	transplanted rice
(4)	Maize -	jute -	transplanted rice
(5)	Maize -	green gram -	mesta (for seed)
(6)	Wheat -	groundnut -	transplanted rice

[**Note :** In eastern parts of India, this is possible because of dry conditions in *rabi* and wet conditions in both pre-*kharif* and *kharif* seasons. In most of the low-lying areas, excessive rainfall water is the root cause of low yield of *kharif* crops. Transplanted *kharif* rice, requires huge amount of water and is grown with accumulated water from rainfall].

■■■■

CHAPTER 13

HARVESTING OF CROPS, GRAIN-YIELDS AND THEIR STORAGE

13.1 HARVESTING OF DIFFERENT CROPS

'*Harvesting*' is a process of collecting the main products, alongwith the by-products of the crops, grown in the fields. Therefore, different methods are adopted, for harvesting different crops under different conditions and situations. Different crops, under 'cereals, pulses, oilseeds, fibres, sugars, tubers and fodders, are stated below in brief : For all crops, harvesting criteria should be identified first, before harvesting the crops.

13.1.1 Cereal crops (wheat, rice, maize, *bajra*, oat, barley, *ragi*, *jowar*, *etc.*)

Wheat : Harvesting criteria for wheat is (i) about cent percent of drying of stem and leaves, (ii) soil should be free from moisture. (iii) grain should be with about18-20% moisture. Too much of drying the stem and leaves, breaks and there is loss of ear head also sometimes.

Normally, cutting of stem at soil base, is completed with the help of sickles. Of course, under mechanised cultivation, wheat is harvested by the combined harvester, where grains are separated in the harvester itself. Generally, wheat that is harvested with sickles, are carried to the thereshing floor, then grains are separated with thresher or by trampling with bullocks under high temperature conditions.

Rice : (a) Pre-*kharif* (*aus*) : When nearly 75 to 80% of plants are ripe, is indicated by yellowing of leaves; *aus* rice is harvested by cutting stem with sickle, at the base of the plants and bundles are prepared and threshing is completed by paddy thresher after 2-3 days.

(b) *Kharif* (*aman*) and (c) *rabi* (*boro*) : Both *kharif* (*aman*) and *rabi* (*boro*) rices are harvested in the same manner. The criteria for harvesting these types of rices (*aman* and *boro*) will be about 90-100% of dry leaves and stem. The plants are cut at the base of the stem with sickles and kept on the field itself, for sun drying for 3 to 4 days and then bundles are prepared and after bringing the rice bundles in the threshing floor, they are threshed with the help of paddy thresher. In the same manner, barley, *ragi* and oat are harvested and threshed. Of course, in some places different methods are adopted for threshing rice.

Maize : Maize is a big plant, under cereals, but it shows two types of reproductive systems i.e. silk and tussel. Silks are on the head of the cob (female part) and the tussels at the top (male part). Symptoms at harvest will be about 80-90% dry leaves, tussel and cobs, but 50-60% dry stem. Then the cobs are separately harvested but the plant will remain as such. Then cobs are decorticated to get the seeds, with the help of *decorticator*. If decorticator is not available, then seeds are separated manually. Of course green cobs are sold when the they are in dough-stage. Now-a-days, very small cobs are sold at earlier stage with high price, for the fine smell. The soup is prepared with *'tender cobs'* for emitting fine smell.

***Jowar* and *bajra* :** *Jowar* and *bajra* are the tall plants like maize, but are harvested separately. The ripe inflorescence (where seeds are stored), are taken away by cutting from the top of the plants and harvest the stem individually, by cutting the stems from the base, with sickles. The seeds of both *jowar* and *bajra* are separated by beating or trampling by bullocks under high hot weather conditions.

(N.B. The plants of maize, *jowar* and *bajra* may also be utilized as fodder, if harvested latter. Maize plants sometimes, may be utilized for trailing purpose of vine type crops like ricebean, *rajma,* cowpea etc.).

13.1.2 Pulse crops (green gram, lentil, pea, gram, grasspea and *arahar*)

Greengram : Greengram requires plucking by hands for some ripe (first- time) pods which are black in colour, but at the same time the same plants are green in colour. But at second time harvesting, entire upper portion of plants is cut with all pod-bearing portions by sickles. The rest of the plant- portions remain in the field, as such, as `crop-residue' for organic matter. Then after bringing them (cut portion of pods) on the threshing floor, seeds are removed by beating or trampling by bullock.

Lentil, pea, gram and grasspea : Regarding harvesting of lentil, pea, gram and grasspea, the same method is used (as in greengram) as they are in the same category. They are pulled manually and spread over the fields for 2-3 days, for sun drying. After drying, all the materials are taken to the threshing floors and grains are removed by beating. Of course, sometimes the plants of gram are cut with the help of sickle, as the stem is very strong to pull out. The remaining plant portion i.e. *bhusha* coming from threshing, is a good cattle feed, of course.

***Arahar* :** *Arahar* is a big plant with hard stem. When most of the leaves are shed the pods are harvested by cutting the top portion with *hasua*. The top portion is dried and threshed by beating and the seeds are collected by winnowing. When amount is high, then the top portion is trampled by bullocks after sun drying for 2-3 days, for collection of seeds.

13.1.3 Oilseed crops (mustard/rape, sunflower, sesame, safflower and groundnut)

Mustard/rape, sunflower, sesame and safflower : The crops are harvested by sickle by cutting the stem at the top level, after most of the leaves and stems are dry. Then they are removed immediately to the threshing floor where these crops are threshed by beating to collect the seeds. All harvests should be done in the morning, to minimise seed loss by shattering. The plants then are spread over the floor for drying for 2-3 days and seeds are collected by beating or by trampling of bullocks. Caution should be taken about safflower which is spiny in nature. Precaution should be taken for this crop.

Groundnut : Groundnut is an oilseed crop, under *Leguminaceae* family-sub family *Papilionaceae*. It is cultivated for nuts which grow inside the soils, after fertilization above the soil. Harvesting is done when leaves and stems are dry. Harvesting is done first by uprooting the plants manually, then by spading cautiously to get the nuts inside the soils. Nuts then are collected by removing soils.

13.1.4 Fibre crops (jute, mesta, sunnhemp, flax and cotton)

Jute, mesta and sunnhemp : Jute, mesta and sunnhemp are all summer - and tall crops, having bast fibres and sticks. They are harvested in the same manner as jute. There are 5 steps for fibre extraction i.e. (i) Harvesting, (ii) Leaf shedding, (iii) Retting, (iv) Fibre extraction, (v) Drying of fibres and (vi) Bundling in *lachis*.

Harvesting is done by cutting individual plant at the base, with the help of sickle/*hasua*, then they are bundled up. The bundles are then made stand at slanting position to **shed leaves**. After shedding leaves and for retting, the bundles are fitted in '***Jak***' by placing the bundles on water surface, one-after another in a long structure ('*jak*'). About 20-25 days after setting 'jak', in water, the bundles are brought on the landside and manually the fibres are **extracted** individually and washed in clean and running water. Then the washed fibres are **dried** on bamboo poles, for 2-3 days. This dried fibre is actually the jute product. After drying, the jute fibres (long) are bundled in small form (*lachi*) for further big bundles.

Flax : The flax, a small *rabi* bast fibre crop, after harvesting by pulling the plants, is put into ditch/pond, for retting. The fibre is extracted by beating them after 20-25 days. The fibre is then dried on bamboo pole and stored after making *lachis*. Of course, this is the minor crop but valuable for quality and use.

Cotton : After plucking bolls from the cotton plant in winter (immedia-tely after *kharif*), white fibres are separated from the caly-x. This raw- white-fibre is the cotton. Of course, further processing is done in the factory for use in different purposes.

13.1.5 Sugar crops (sugarcane, sugarbeet and yambean)

Sugarcane : Normally, matured sugarcane should have partially dry leaves. After removing leaves the nodes and internodes are open and they are filled up fully with juice and particular colour (according to colour of the varieties) are prominent. At this stage, the brix value is high, then sugarcane should be harvested. But above all, '*hand refractometer*' is the best instrument, (though small handy and portable), for determining the harvesting time. 'Refractometer Brix – RB' is main for determining the actual harvesting time, because at that time sucrose percentage exceeds 16% and juice purity increases over 85%, then it should be assumed that the cane is matured. Then the sugarcane is harvested.

Sugarbeet : Sugarbeet, the small heighted and surface-ball-type crop, (when full grown) and leaves are pale in colour, when it matures. The sucrose percentage may be determined by RB. When sucrose percentage is about 16%, it should be harvested. Generally the sugarbeet is pulled out manually and collect and the leaves are fed to cattle. The ball-like structure, the 'sugarbeet' is then processed by crushing, for manufacturing sugar for use.

Yambean (*Sakalu*) : It is a vine crop and swollen root portion is under the soils. This 'yambean' is harvested from the soils with help of spade, when the seeds above are about to ripe. Of course, the 'yambean' is eaten raw as such, because of low content of sugar.

13.1.6 Tuber crops (potato and *amorphophallus*)

Potato : Potato grows in the underground. When the tuber is hard and skin is rough, by the time haulm above the soils will be half-dead, i.e. the haulm will be mostly with leaves or less number of leaves or yellow leaves, then it will be assumed that tuber is mature, beneath the plants and inside the soils.

Either with man-drawn small plough or with the spade, the potato is normally harvested. Of course, by mechanical way it can be harvested and graded for storage and sale.

Amorphophallus (*ol*) : *Amorphophallus*, a big flat ball-like rhizome (corm), beneath the soil, is harvested with the help of spade by removing soils around, leaving the pseudo-stem-stalk and leaves. At any time, it can be harvested for vegetable, if it is bigger in size after 3 months of sowing. Of course, after 120 days it becomes bigger fully.

13.1.7 Fodder crops (Napier, berseem, lucern, blackgram and cowpea)

Napier and thinnapier : Napier and thinnapier are more or less the same type of tall grass, are cut at the base of individual plants, when green, with the help of sickles, at flowering stage to get more and nutritive fodder for the cattle. These green plants are then fed to cattle, buffaloes etc.

Lucern and berseem: Lucern and berseem, two trilobed-leaved legumes, are harvested by cutting the entire crop (or cutting), with the help of sickle, when green, keeping the lower stem about 6 cm above the soils. Generally 3-4 cuttings are taken, where long duration winter is there, but in short duration–winter like eastern parts of the country, only 2-3 or 1-2 cuttings are taken. Sometimes, only one cutting is taken, keeping crop for seed there. The entire green mass (fodder) is then taken into the cowshed for consumption by the cattle etc.

Cowpea and blackgram : Cowpea and blackgram, big trilobed-leaved legumes, are harvested when entire crop is green, with the help of sickle, at flowering stage, where nutritive value is high. The whole green mass (fodder) is taken into cowshed; for consumption by the cattle.

Practical Class: Harvesting of crops of wheat, rice, green gram, mustard, jute, maize, sugarcane and potato, in different seasons with their harvesting criteria

Lesson 125 : What should be the harvesting criteria of (a) wheat, (b) *boro* paddy, (c) potato and (d) mustard in *rabi* season; (i) *aus* rice, (ii) green gram and (iii) jute in pre-*kharif* season and (A) sugarcane, (B) maize and (C) *aman* rice in *kharif* season ? Write them in detail.

Harvesting criteria for crops in different seasons

I. *Rabi* season

(a) Wheat : (as in 13.1.1); **(b) *Boro* paddy :** [as in 13.1.1 rice (b) (*boro*)]; **(c) Potato :** (as in 13.1.6);

II. Pre-*kharif* season

(i) *Aus* rice (direct-seeded rice): [as in 13.1.1. rice (a)]; **(ii) Greengram :** (as in 13.1.2); **(iii) *Jute* :** (as in 13.1.4);

III. *Kharif* season

(A) Sugarcane : (as in 13.1.5); **(B) Maize** : (as in 13.1.1); **(C) *Aman* transplanted rice** : [as in 13.1.1 rice (b)].

Lesson 126: Practise *harvesting* of wheat after describing the procedure

Introduction

Wheat, the winter loving crop, ripens with the rise of temperature in summer. Therefore, moisture content in both the plant or soil, is less. From the dry crop when harvested, it is easy to remove grain from it and for storing maintenance of low moisture in the grain is easy to regulate. Then immediately after harvest, threshing may be done in most of the areas.

Objective

Harvesting of wheat is done -

O to get yields (grain or straw),

O to vacate the land for the next crop.

Material required

✤ The field with matured wheat crop, ✤ Sickles, ✤ Rope/string, ✤ Labourers, ✤ Truck/van, ✤ Threshing floor, ✤ Pair of bullocks, ✤ Bullock cart, ✤ Bullock operator (all these are for ordinary harvesting) but for mechanised harvesting or harvester/combined harvester is needed along with its accessories, ✤ Note book and ✤ Pen/pencil.

Procedure

- Cut the wheat plants at, the soil surface, with sickles,
- Tie-them-up with rope/string in the form of small bundles,
- Carry the bundles either in the head-load or in van/truck/bullock cart to the threshing floors,
- Separate wheat grain with wheat-thresher or by beating.
- Dry them in the sun for 2 more days, for removing grains from the plants,
- Note down the date of harvest and date of separation of grains, along with the area harvested.

Practise harvesting of wheat in the filed

(teacher will lead them to harvest wheat)

Lesson 127 : Practise harvesting of *boro* paddy, after describing the procedure

Introduction

Though *boro* paddy is transplanted in winter under puddled conditions, it is harvested in summer in dry conditions of the field. Of course, the *boro* rice plant is still pale in colour, but not so dry as in wheat. It is kept in the field after harvesting and bundlings for drying, removed and threshed on the floors with help of paddy thresher. This is the ordinary process of harvesting *boro* paddy but now-a-days harvester is also used, for its harvest.

Objective : (as in lesson 126).

Material required

❖ The field with matured *boro* rice crop (in *winter* season), ❖ Sickles – 2, ❖ Labourers–5, ❖ Truck/van/cart–1, ❖ Threshing floor–1, ❖ Paddy thresher with operators–1, ❖ Note book and ❖ Pen/pencil.

Procedure

- Harvest paddy, when the land is dry and rice plant is pale yellow, with the help of sickles just at ground level,
- Place the rice plants in rows, for drying in the field, in a systematic manner,
- Keep the rice plants for 2-3 days as such, in the field,
- After 2-3 days, rice plants should be tied with rice plant itself, in small bundles,
- Carry the bundles either on van/truck/cart or head load to the threshing floor for separating grains,
- Dry the bundles in floor again before threshing,
- Thresh paddy with paddy thresher,

- Record in note book the date of harvest, days for keep them in the field, carrying date to the threshing floor and date of final processing of grains.

Practise harvesting of *boro* paddy in the field:

(the teacher would lead the learners to harvest *boro* paddy)

The learners should participate in –

(1) *Boro* paddy cutting/harvesting operation, (2) Tieing up of bundles of *boro* paddy, (3) Carrying the bundles with the help of trucks/van to the threshing floors and (4) Threshing programme of *boro* paddy.

Lesson 128 : Practise harvesting of potato, after describing the procedure

Introduction

Potato, the winter crop, though sown in dry condition, but the tubers are saleable to the market according to the price of potato. Otherwise, it is harvested at earlier part of the summer, when the land is dry and upper skin is hard. As tubers are grown in the underground, hence its harvest is also done by digging soils and thereafter sorting is done for market as such, or to place the potato in the cold storage.

Objective

Timely harvesting of potato is done -

O to get the tubers (potato) for marketing or for keeping in the cold storage,

O to get ready for the next crop,

O to avoid spoilage due to disease/insects/other pest etc.

Materials required

✥ Field with matured potato crop, ✥ Spades – 2, ✥ Man-drawn small plough – 2, ✥ Baskets – 4, ✥ Labourers – 5, ✥ Sacks – 4, ✥ Sickles – 2, ✥ Balance (50 kg capacity) – 1, ✥ Truck/van/cart – 1, ✥ Note book and ✥ Pen/ pencil.

Procedure

- Remove the dry plants from the ridges, by cutting them with sickles,
- Run the small plough deeply over the ridges cautiously to avail cutting the tubers (or remove soils from the ridges to get potato),
- Use the spades for removing the obstruction, if any,
- Collect the potato/tubers (matured) in a basket,
- Sort them out into 3 categories, such as, big, medium and small size potato,
- Fill up the sacks with potato from the baskets,
- Carry these sacks into the trucks/van/cart to the market place or cold storage, after further necessary sorting.

Practise harvesting of potato

The learners should participate in the harvesting programme of potato for

1. Detecting harvesting time and to observe the conditions for maturity,
2. Running the spade or plough or mechanised harvester,
3. Collection of potato tubers from the field,
4. Sorting-out or grading the potato,
5. Collection of potato, into sacks,
6. Participation in carrying potato to the cold storage.
7. Observe the cold storage conditions and record temperature for the maintenance of storage upto the selling-time.

Lesson 129 : Practise harvesting of mustard, after describing the procedure

Introduction

Mustard, the oilseed crop, is used for mustard-oil, for cooking mostly in eastern parts of India. Out of so many spp., *'rai'* group is cultivated in a mass scale. The siliqua (pod) of mustard comes from the flower of the long inflorescence. The lower siliquas mature early, while that on the top most portion, only flowers are observed during that time. Hence, harvesting should be done when 75% of siliquas are matured, by showing whitish or brownish in colour.

Objective

Harvesting of mustard is done timely -

O to get higher yields (both grain and stover),

O to release the land for the next crop.

Materials required

✤ A field with matured mustard crop, ✤ Sickles – 2, ✤ Rope/string/ coir (for tie-up the bundles), ✤ Truck/van cart, trolley, ✤ Note book ✤ Pen/ pencil.

Procedure

- Observe the mustard fields, whether it is fit for harvesting or not,
- Cut the upper portion of the mustard plants upto lower last siliqua observed,
- Tie-up the cut-portion of mustard plants with string/coir in small bundles,
- Carry these bundles into the truck/van/trolley/cart etc. to carry again to the farm threshing floor,.
- Get seeds by bleating the bundles and dry them,
- Note down the records like :

  (i) Date of harvest, height of the cut (above soil surface),

(ii) Tieing-up the materials,

(iii) Carry the materials to the threshing floor.

[**Note :** (1) Best seeds/grains of mustard are shattered in the field when ripe, hence it should be harvested early in the morning, just at ripening,

(2) for most of the varieties of mustard mature early (within 70-85 days) after sowing. Hence land should be released earlier for next crop.

Practise harvesting of mustard in the field :

Participate in the following programmes for harvesting of mustard : (1) Cutting the plants, (2) Tieing-up the bundles, (3) Carrying the bundles to the threshing floor and (4) separating the grains.

Lesson 130 : Practise harvesting of *aus* (direct-seeded) rice, after describing the procedure

Introduction

Aus rice (direct-seeded) is categorised as (i) low land *aus* and (ii) high land *aus* grown under rainfed condition in summer, in eastern India where rainfall is high. We are concerned herewith high land *aus*. Low land *aus* is sown in dry condition in February/March in low lands by broadcasting, seeds immediately after rainfall, then flood-water comes and rice is harvested in water. But in high land, *aus* rice is broadcast immediately after rainfall in April/May and harvested in July/August.

Objective

High land *aus* rice (direct-seeded) is harvested -

O to get the yields (grains/straw),

O to release the land for sowing other next crops.

Materials required

✤ A field with matured *aus* rice, ✤ Sickles – 2, ✤ Coir/string/rope – 500 g, ✤ Truck/trolley/van / cart – 1, ✤ Note book and ✤ Pen/pencil.

Procedure

- Cut the *aus* rice at the base, when the plants are yellowish in colour,
- Place the cut crops in lines, for drying in the sun,
- After 2-3 days, tie-up the small bundles with the rice plant itself,
- Remove the small bundles in big bundles with coirs/string etc.
- Carry the big bundles with tractors/van/carts, to the threshing floor,
- Note down the harvest date of *aus* rice and also the threshing date.

Practise the harvesting of *aus* (direct-seeded) rice in the field

Participate in the following *aus* rice harvesting programme :

(1) Cutting the *aus* rice, (2) Tieing-up the harvested materials, (3) Carrying the big bundles of plant-materials to the threshing floor, for threshing, (4) Thresh the *aus* rice and dry it, after winnowing.

Lesson 131 : Practise harvesting of green gram, after describing the procedure

Introduction

Green gram, the pulse crop, ripens periodically, the pods of which are black in colour at ripening stage. Generally, picking of black coloured pods is done twice. First time, the black pods only and at second time entire top portion is harvested with sickles and carried to the threshing floor, for further processing. The first lot gives the best seeds. There is a chance of shattering big pods, if matures earlier and not harvested in time.

Objective

The harvesting of green gram is done -

O to get the yields of grain/seed and stover or crop residue,

O to release the land for sowing the next crop.

Materials required

❖ A field with matured green gram, ❖ Sacks – 2, ❖ Coir/string – 500 gm, ❖ Tubs (plastic)/buskets – 2, ❖ Truck/trolley/van/cart, ❖ Note book and Pen/ pencil.

Procedure

- Pick-up manually the ripened pod which are black in colour first, keeping the plants intact,
- Pods should be kept in tubs,
- After about 7-10 days of first picking, cut the top portion of the green gram plants and put them into sacks,
- Keep the sacks by the sides of the field,
- After completion of cutting the tops, remove the sacks and tubs to the threshing floor,
- Dry the top portion,
- Separate the grains from *bhusa,* by winnowing.
- Incorporate the lower portion of the plants as crop-residue, into the soil, for maintaining soil fertility.
- Note down the dates of harvest of first picking and harvest the other alongwith the date for incorporation of lower portion into the soils.

[Note : The entire plant materials are then ploughed-under, for using *crop-residues* for maintaining soil fertility of the land].

Practise harvesting of green gram in the field

Participate in the following programmes in harvesting of green gram :

(1) First picking of matured pods, (2) Cutting the top portion of green gram plants and sacking, (3) Carry seed materials to the threshing floor. (4) Collection of seeds from the pods, and (5) Incorporation of crop-residue (lower portion) into the soils.

Lesson 132: Practise harvesting of jute, after describing the procedure

Introduction

Jute, the tall bast fibre crop, is grown in eastern parts of India, where temperature, humidity and rainfall are high in pre-*kharif* season. Its harvest and processing are quite different, from other crops. With sharp serrated sickles tall plants of jute are cut, with shedded leaves in the fields the bundles are prepared and carried to the water source for steeping, *'jak'* is prepared for retting and then stripping of fibre is done. Ultimately drying is completed. All these are done for processing of fibre.

Objective

Harvesting of jute crop is done -

O to get fibre and stick yields of jute,

O to get leaves for maintaining soil fertility,

O to release the land for transplantation of *kharif* rice.

Material required

✥ The field with full grown jute crop of about 120 days, ✥ Sharp serrated sickles or *hasua*, ✥ Truck/trolley/van/cart etc. (to carry the jute bundles from the fields to the source of water), ✥ Note book with pen/pencil.

Procedure

- Cut the individual plants of jute with sickles/*hasuas* at the ground level,
- Spread the plants on the ground in lines systematically,
- Tie-up some plants into a bundle of about 6-8 cm radius with jute plants,
- Make the bundles standing slantingly in some places in the field, to remove leaves (to be spread later on the surface of soil, for maintaining soil fertility).
- Remove the bundles to the source of water for steeping.
- Prepare the *jak* (accumulated jute - bundles in water)
- Remove the bundles after about 3 weeks, for stripping fibres,
- Strip the fibre, individually from each plant,
- Dry the fibre and form the *lachhi*, after adequate drying.

Practise harvesting of jute crop in the field

The learners should participate in the following programmes for harvesting of jute :

(1) Cutting the jute plants, (2) Tieing-up the bundles, (3) Heaping the bundles (for shedding leaves), (4) Steeping, (5) Stripping programme, and (6) Drying programmes.

Lesson 133 : Practise harvesting of sugarcane, after describing the procedure

Introduction

Sugarcane, the sugar crop, is a long duration crop (*eksali* & *adshsali* crop) ranging from 11 to 20 months. Sugar or *gur* are obtained from the juice, which is again obtained by crushing canes of sugarcane, the tall crop of about 3 metres long. The long canes are cut at the ground level, leaves are removed, rest of the cleaned canes after removing the top, are sent to the factory/ consumption place, with trolley/van/trucks/cart, etc.

Objective

Harvesting of sugarcane crop is done -

- O to get sugarcanes (stems) for *gur* or sugars,
- O to release the land either for keeping ratoon crop or for sowing next crops,
- O to utilize the leaves and bagasse either for fuel or for organic matter / manures.

Material required

✤ A field of matured sugarcane, ✤ Refractometer (for testing sucrose in the cane), ✤ Sharp serrated sickles/*hasua*, ✤ Tieing-up the bundles with materials like coir/rope/ string etc. ✤ Truck/trolley/van/cart, ✤ Note book and ✤ Pen/pencil.

Procedure

- Cut the individual cane with sharp serrated sickles/*hasua*, at the ground level and place them (canes) on the ground in rows systematically,
- Remove the leaves from the canes cautiously (so that eyes are not damaged),
- Remove the top portion with leaves,
- Bundle the canes (with 10-12 number) by tieing them up with coir/string or with the leaves of sugarcane, at both ends.
- Remove the bundles in truck/van/cart etc. to the consumption places like factories, market, or other places.

Practise harvesting of sugarcane crop in the field

Participate in the following programmes for harvesting of sugarcane :

(1) Cutting the canes, (2) Removing the leaves from the sugarcane-body, (3) Carrying the bundles to the factories.

Lesson 134: Practise harvesting of maize, after describing the procedure

Introduction

Maize, the tall cereal crops are grown throughout the year. This is cultivated either for fodder, for grains, for cobs, for pop-corn or for some essence. It is harvested at different stages at different times. It is harvested as fodder when it is in flowering stage; as grain at maturity; as cob at dough-stage; as pop-corn at maturity (with specific variety), and as essence 3-5 days after silkings. We would consider it here, as grain crop for our studies.

Objective

The harvesting of maize is done -

- O to get yields in different forms (such as fodder, grains, cobs, for essence or as pop corn),
- O to vacate the land for sowing the next crop.

Material required

✣ A field with matured crop of maize, ✣ Sickle (sharpened and serrated) – 4, ✣ *Hasua* – 4, ✣ Baskets/big tub – 4, ✣ Sacks – 4, ✣ Truck/trolley/ van/ cart – 1, ✣ Maize sheller – 1, ✣ Balance – 2, (spring balance, 1 or 2 kg and 30 kg capacity), ✣ Note book and ✣ Pen/pencil.

Procedure

- Pick up the ripe cobs from each plant by hand,
- Put the cobs into the tubs/sacks/buckets,
- Cut the individual plant (without cob) at ground levels for fuel (or it may be used as support base of some crops),
- Bring the basket/sacks etc. containing cobs, to the threshing floor,
- Dry them in the floor for two days,
- Remove the grains from the cobs with the help of maize shellers,
- Put the seeds into the tub/basket and put them into sun for drying,
- Note down the date of harvest, shelling the seeds and drying dates and all methodologies used.

Practise harvesting of maize crop in the field

The learners should participate in the following harvesting programmes :

(1) Picking-up the cobs from the plants, (2) Cutting the plants, (3) Maize shelling and (4) Drying processes.

Lesson 135 : Practise harvesting of *aman* (transplanted *kharif*) rice, after describing the procedure

Introduction

Out of three types of rice (*aus*, *aman*, and *boro*), *aman*, i.e. transplanted *kharif* rice is very important in eastern parts of the country, because of large area coverage, due to huge amount of rain-water availability in the rainy (*kharif*) season. Nearly eighty percent of area under *kharif* rice, is grown under rainfed conditions, with huge amount of standing water received from rainfall only, though at the time of harvest the medium land is more or less dry. The learners should know the harvesting technique of *aman* rice.

Objective

Harvesting of *kharif* transplanted rice is done —

O to get the yields (grains and straw) of rice,

O to release the land for sowing other crops,

O to bring pliability (good tilth) of the soils.

[due to alternate drying and wetting (for rain, if any)].

Material required

✤ A field with matured transplanted *kharif* rice, ✤ Sickles (sharp serrated) – 4, ✤ String/rope/coir – 500 g, ✤ Truck/trolley/van/cart –1, ✤ Note book and ✤ Pen/pencil.

Procedure

- Cut the rice plants in the 'hill' just at the ground level and spread them on the land in line systematically,
- After two or three days, those plants are tied-up in the form of small bundles (*guchi*) of about 3-4 cm. in radius, with the help of rice plants in the morning,
- Keep the bundles for 2 days for further drying,
- After two days, tie-up the small bundles into big bundles with coir/ rope/string.
- Carry the big bundles in truck/trolley/van/cart, to the good threshing floor for threshing.

Practise harvesting of *kharif* (*aman*) rice

Participate in the following harvesting programmes :

(1) Cutting the plants and spreading them on surface of the soil, (2) Tieing-up the small and big bundles, (3) Carrying the materials to the threshing floor.

Lesson 136: Give a full picture of symptoms of harvesting of (a) wheat, (b) *boro* paddy, (c) potato, (d) mustard, (e) green gram and (f) jute

Full picture of symptoms for harvesting of

(a) Wheat : (1) Both plants and soils should be completely dry, [but in eastern parts, this situation may not be possible to get all the times. Wheat should be harvested earlier to avoid the storm (*Kalbaisakhi*) in April]. (2) Atmospheric temperature should be high (more than 34°C), moisture (humidity) should be less than 45% and without rainfall, (not even cloudy). All these conditions prevail in March/April/May.

(b) *Boro* paddy: (1) Bright sunshine, high temperature and less humidity, are required for harvesting, (2) Land should be dry, (3) Plants are 75-80% dry, (4) Pale yellow colour of the stem and leaves. (Harvesting is done in March/April/May).

(c) Potato : (1) Most of the leaves are partially dry, though stems are still green, (2) The land should be dry, crackings should be visible on the ridges, (3) Skin of the tuber (potato) should be hard and rough, (4) Size of the potato should be bigger, (5) Colour of the tubers should be attractive and sometimes light/ pale yellow (not always), (Of course potato can be harvested at any time because of high price in the market).

(d) Mustard : (1) The land should be dry, (2) Plants should be about 70-80% dry, with pale yellow colour of the leaves, (3) Shattering takes place from lower siliqua.

(**Note:** Because of shattering nature, mustard, after harvest, should be placed on polythene sheet, if harvested in proper maturity. To avoid shattering therefore, mustard should be harvested earlier, when siliquas are partially pale).

(e) Green gram: [Ripening does not synchronise at one time. Hence pods should be picked-up twice].

(1) Pods should be black in colour (earlier picking), (2) Plants should be partially dry particularly the leaves, (3) Land should be dry. (Rain free, bright sunshine day needed for harvest, but in May/June or July this situation is not available all the times, hence some seeds shatter).

(f) Jute (*Olitorius*) : (Though jute can be harvested at any tine, but for good amount of yield and good quality of fibre, it should be harvested at about 120-130 days after sowing).

(1) Pod formation time is the best time for harvesting of jute for having good yield and for quality fibre of jute, [i.e. at 120-130 DAS], (2) Lower leaves fall rapidly on the ground, (3) Growth ceased, [All these characters appear in the months of July/August i.e. about 130 days after sowing].

[*Note*: Generally *capsularis* jute is grown in low lands. Hence harvesting techniques are quite different but its area is less] and (4) Just at flowering or pod-formation stage is the best time for harvesting *olitorius* jute.

Lesson 137 : Describe the symptoms of harvesting of (a) mesta, (b) grasspea, (c) ricebean and (d) jute for seed

Description of symptoms of harvesting of

(a) Mesta

(i) For fibre : Mesta, the bast fibre crop, is grown in areas where rainfall, temperature and humidity are moderate. It is generally grown in poor soils, with poor management hence it is called the 'poor-man's-crop', for fibre it is grown in pre-*kharif* (April/May) season and harvested in July/ August, like jute crop. Generally 120-130 day crop is better for both fibre yield and quality. At this stage the fibre is strong, though the plants are green. Of course, like jute it can be harvested earlier also but with poor yield and with good quality. It can also be harvested at later stage of growth (beyond 130 DAS) for good yields but with bad quality.

(ii) For seeds : For seed purpose, mesta should be sown in the month of June/July and is harvested in October/Novemebr. Mesta should be harvested when the plants and pods are dry, along with the land.

(b) Grasspea

(i) For seed : Grasspea (*khesari*), the pulse crop, is grown in winter season in low lying areas, particularly as *paira / utera* crop in *kharif* rice field. Some of the varieties contain BOAA [B - (N) – Oxalyl amino-alanine], The toxic substance, probably responsible for lathyrism disease in human beings, hence it was banned. Some varieties are free from this toxic substance. Now ban has been lifted. Now, it is grown in most of the low lying lands for seeds. In West Bengal, in rice-fallows, it is one of the best pulse crops. This crop is grown for seed in West Bengal, Assam, M.P. Bihar etc. It should be harvested drying and when it is fully dry, in the morning and kept in the field for drying and collection of seeds is done by beating the dried materials.

(ii) For fodder : This *khesari* is also a good fodder crop, during *rabi* season when it is green. Its herbage production is also good, particularly in rice field. Its herbage productivity is high after sunnhemp, grown as *paira* crop for seed.

(c) Ricebean

(i) For seeds : Ricebean, the minor pulse crop, is grown in pre-*kharif* (summer)/ *kharif* seasons, for seeds. It is harvested when the plants are dry in the month of January/February. It is a quick growing crop, but flowers late and seed is harvested in the later part of *kharif* or earlier part of *rabi* season. This can be grown with minimal or zero-tillage. For seed of course, this crop may be sown in July/August and harvested in November/December or even in January when it is fully dry.

(ii) For fodder : Ricebean crop can also be grown for fodder purpose when it is in green stage. For this, it should be sown in the months of April/ May. Either it can be cut in the months of June/July for fodder when it is fully green and kept for seed, or as such, it can be harvested totally for fodder in July/August when it is flowering or pre-flowering stage and vacated the lands for sowing other crops like *toria* or mustard in *rabi* or blackgram, maize etc. in pre-*kharif*.

(d) Jute for seeds

Jute for seed is sown with the same date as it is done in mesta for seed. It is sown in June/July and harvested in October/November or December, when plants, seed etc. are fully dry. After harvesting the top portion of the plants, seeds are collected by beating the tops and dried.

Lesson 138 : Describe the symptoms of harvesting of (A) maize, (B) sugarcane and (C) transplanted *kharif* rice

Description of symptoms for harvesting of

(A) Maize

(i) For cobs : Maize is cultivated for different purposes like, grain, cobs, fodder, essence, support-base etc. In some urban areas, cobs (earhead of maize, at dough stage), are eaten either partially by heating or by boiling with water. The crop which is used as cobs, require less time, only 2 to 2.5 months. After tacking cobs, the entire plants may be utilized as fodder, as it is green.

(ii) For seeds : Maize, the cereal crop, may be cultivated and harvested for seeds both for consumption by human beings, cattle and poultry. In that case its duration is about 4 months or so and it can be grown at any time of the year, as it C_4 - type crop. The dry cobs containing seeds, numbering about 200-400, are harvested when the plants and cobs are pale yellow. These seeds are shelled from the cobs. Only one or two cobs are, generally, available from each plant. Maize, when harvested for seeds, is fully dry (both leaves and stems).

(B) Sugarcane

In sugarcane, only the stems are utilized for juice from which *gur* or sugars are produced. It is either *eksali* or *adhsali* crop. As because it is a long duration crop, and cumbersome cultivation, now-a-days it is being replaced by other crops in the multiple cropping systems. The sugarcane when matured, the lower leaves are dry, but not the top leaves), eyes are swollen, the internodes are full of juice. At that time, the brix value is about 18% (indicating glucose percentage) and juice is very sweet.

(C) Transplanted *kharif* rice

Symptoms for harvesting of transplanted *kharif* rice are as follows : (1) Leaves about 70-80% are dry, (2) Stems are also dry in about 70-80%, (3) Soils are dry (except some lowlands), (4) Entire field is pale yellow in colour. All these symptoms indicate the time for harvesting of transplanted *kharif* or *aman* rice.

Lesson 139 : Describe the harvesting criteria of (a) napier grass, (b) berseem, (c) lucern and (d) blackgram for fodder

Harvesting criteria :

(a) Napier grass (fodder) : Generally, napier grass is cultivated in April/May and 2-3 cuts are taken, first one after 2 months after sowing and later at one month interval. At that time the plants are succulents and green. When the plants become hard, cutting for fodder should be stopped. For this grass fodder, normally irrigation is needed; but in the eastern part, irrigation is not needed, because of high rainfall from the months of April/May to September/October. Either this grass is fed to the cattle in stalls or it may be prepared as silage or hay (at later stage). For harvesting, for fooder, main criteria should be succulence and greenness. Production, succulence, and greenness increase if nitrogenous fertilizers are given in sufficient amount.

(b) Berseem (fodder): Berseem, the leguminous fodder, is grown in *rabi* season, under irrigated conditions. Generally, 2-3 cuttings are taken; first cutting is taken 2 months after sowing and subsequently, at later stage, one month intervals, if properly irrigated with high dose of phosphatic fertilizers. Generally, *succulence* and *greenness* are the major criteria for harvesting/ cutting of berseem, beside the duration of cutting. This legume fodder is generally fed to the cattle in the stall by mixing with napier grass or any other grass fodder, to avoid `loose-motion' of the cattle, because of high content of protein in legumes. The ratio of mixing legume and grass should be 1 : 1 under normal conditions. Production, succulency and greenness of the fodder increase, if phosphatic fertilizer, along with other fertilizers and micro- nutrients (particularly molybdenum) are applied properly.

(c) Lucern (fodder) : Lucern, the legume fodder, is also cultivated in *rabi*, under irrigated conditions. All other aspects are like 'berseem' as mentioned earlier above.

(d) Blackgram (fodder) : Blackgram, the leguminous fodder crop, is cultivated during pre-*kharif* season for fodder, but during *kharif* season for pulse (seeds) crop. This may also be grown for green manure crop in pre-*kharif* season in eastern part of the country. Only one cut (total harvest) at 1-1½ months after sowing, is the best to harvest blackgram, as at this stage maximum production, succulent and green fodders are achieved/obtained. This harvested fodder is then fed to the cattle in the stall with the mixture of some grassy fodders or straws. As this is leguminous fodder, hence high content of protein is present and hence it should be fed to the cattle by mixing with other grass fodder positively, for proper digestion. Greenishness and succulence are two main criteria for harvesting of blackgram, along with the production and for this duration of harvest (from sowing to harvest) is the best criteria. In blackgram, it is harvested at 45 – 50 days after sowing. Total harvest is done at that time.

13.2 CROP-YIELDS (PRODUCE) AND PROCESSING

Generally, after harvest of crops, yields (grain/straw or stover, fibre/stick, juice or sugar/stover etc.) are processed for consumption or storage. The crop yields are different and therefore, they are processed differently.

13.1.1 Cereal crops (Wheat, rice, maize, *bajra*, oat, barley and *jowar*)

Wheat : After harvesting wheat, it is carried to the threshing floor for drying further for 2 to 3 days. The grain is separated either by treading the materials under the feet of bullock or by wheat thresher or by beating the materials with sticks or by tractor. Of course, grain is separated by the combined harvester under mechanised cultivation, and dried after winnowing.

Rice : Rice, after harvesting, material is kept in the field and brought back to threshing floor for further drying in the sun. The dried materials (straw + grain), the grain is separated by beating with stick, or treading the materials under the feet of human beings, or by paddy thresher, or by beating on the bamboo mat or by combined harvester. Dried materials are separated after winnowing against the wind.

Maize : Either cobs are harvested at ripening of cobs or by total harvest of the plant and cobs are separated the on threshing floor. For both the cases, the cobs are dried and grains are separated from the rachies/stocks/ears/ heads with the 'maize sheller'; the shelled grains then are dried and packed in bags/bins, etc.

***Bajra* and *Jowar* :** The top portions of *bajra and jowar* are taken in the threshing floor, after the total harvest and dried. Grains are separated either by beating with stick, or by treading the materials under the feet of bullocks or by running the tractor over the materials. Seeds are dried after winnowing and put into bags/bins.

***Oat* and *barley* :** Total harvest is done and the materials are brought back to threshing floor; kept them for some days (2-3 days) and after drying, grains are separated either by beating with stick, or by treading the materials under the feet of bullocks, or by running the tractor. The separated grains are dried after winnowing and put into bags/sack/bins etc.

13.2.2 Pulse crops (Green gram, lentil, pea, gram, grasspea and *arahar*)

Greengram : Greengram/*moong*/mungbean, is grown in summer and is harvested in dry days and the top portion is brought to the threshing floor, where it is kept for 2-3 days for sun-drying. After drying the materials, seeds are separated by beating with stick or by treading the materials under the feet of the bullocks or by running the tractor. Seeds then again are dried, after winnowing. The dried seeds then put into bag/sack/bins.

Lentil : Lentil, is harvested totally in the morning and the material is brought to the threshing floor for further drying for 2-3 days. After drying, seeds are separated from the *bhusa*, by beating with stick, or treading the materials under the feet of the bullocks or by runnning the tractor. After winnowing, the seeds are dried again and kept in bag/sack/bins.

Pea : (Same as in lentil); **Gram :** (Same as in lentil); **Grasspea :** (Same as in lentil); ***Arahar* /*red gram* :** (Same as in mungbean).

13.2.3 Oilseed crops (Mustard/rape, sunflower, groundnut, sesame & safflower)

Mustard/rape : After harvesting the top portion of mustard/rape, the materials are taken to the threshing floor in the polythene lined bags, where they are heaped in methodical order and kept for 5-6 days. The heaped material is then dried in the sun. In the floor itself; the seeds shatter and from the materials further seeds are separated by beating with stick, or by treading the materials with bullocks or by running the tractor over the materials. The seeds are again dried after winnowing and kept in the sack/ bag/bins.

Sunflower : (as in mustard)

Groundnut : Harvesting of groundnut is quite different from that of other cereal or pulse crops, because of nuts which are obtained from the under ground. Groundnut plants in matured conditions, are either uprooted or picked-up by spade or small man-drawn plough and the nuts are bagged/ sacked after separating from soils, and the bags/sacks are then brought to the threshing floor for drying. After 2-3 days drying in the sun, the nuts are again put into bags/sacked and kept into bins, etc. as such (with shells).

Sesame : (as in mustard)

13.2.4 Fibre crops (Jute, mesta, sunnhemp, flax and cotton)

Jute : Jute, the tall bast fibre crop, is harvested individually and plants are kept the plants under bundles in the field, for some few days (3-4 days). After leaf-shedding, the bundles are carried to the water sources. Steeping is done and after about 3 weeks, stripping is done. The stripped fibre is then washed in clean water and dried. After thoroughly drying, the fibre is put into *lachis* which are the dried again and kept in bundle either for sale or for keeping in the big bundles into the godown for further use.

Mesta : (as in jute); **Sunnhemp :** (as in jute).

Flax : Flax, (for both seeds and fibre) crop grows in *rabi* season, the plants are harvested by uprooting or cutting the plants at base and taken to the water source for steeping. After 3 weeks the plants are stripped, washed and dried the fibre. The fibre is then tied into small *lachis* for further drying. After adequate drying in the sun the *lachis* are tied for bundles and put into godown for further use.

Cotton : In cotton, the lint fibre crop, only the bolls are harvested/picked up, at maturity when they are open in full form individually and manually. The fibres from bolls, as harvested, are taken to the threshing floor, for final picking the white fibre, avoiding the calyx and other foreign materials. The white fibre is then packed into bales/bags/sacks. Then the bales are used for different purpose in the factory/centres.

13.2.5 Sugar crops, rhizome and tuber crops (Sugarcane, sugarbeet, potato and rhizome)

Sugarcane : In sugarcane, the sugar crop, processing is different from other crops, as from cane (sugarcane stem) juice is obtained; from juice *gur* or sugars are produced. Therefore, processing is done in different forms, such as (i) sugarcane, (ii) crushing sugarcane, (iii) juice manipulation for (a) *gur* making and (b) sugar manufacturing.

(a) Gur : After harvesting sugarcane, the canes are cleaned by removing the leaves and other debris and the top. The cleaned sugarcane are kept in the field itself for *gur* making. The sugarcanes, in the field, are crushed for juice. Juice then is boiled (with the help of sugarcane leaves, left after cleaning the canes) in a special big flattened container and frothings are removed with some chemicals or *bherenda / bhendi* plants. After removing water from the juice, *gur* is produced. This *gur* is kept in different containers for further uses.

(b) Sugars : For sugar manufacturing, the cleaned canes are carried to the sugar manufacturing unit and crushed for the juice by some mechanical process. From juice, sugars are obtained and then sugars are packed in polythene lined bags/sacks etc. for marketing to different places.

Sugarbeet: Sugarbeet, the sugar crop, grows in *rabi* in underground, like vegetable beet, though in bigger form commercially. The sugarbeets, the ball-like underground roots, are uprooted individually and manually and immediately transported to the sugar factory, after cleaning. Then the cleaned sugarbeets are crushed for juice and from juice, sugar is manu-factured in the factories. For obtaining sugars, different new machineries are required. The produced sugars are then packed in bags/sacks etc. for further marketing.

Potato : After harvesting potato, it is cleaned (should be free from muds/ soils and other debris) and sorted according to the grades and put into bags/ sacks etc. The bags/sacks are carried to the market places or cold storages. Potato is the underground stem and is used as vegetable, hence no need of further processing like other seeds or so.

Rhizomes : *Amorphophallus* (*ol*)/*arum* etc. are in the *rhizome* group, the edible portion is used as vegetable. It is obtained from the modified stem and grows underground, in pre-*kharif* season. *'Ol'* will be discussed here shortly. After harvesting *ol* i.e. corm from the ground individually with the help of spade manually, the corms are cleaned (as because soil/muds/debris etc. are in corms) and sacked and directly sent to the market for selling after sorting. Of course, some good corms are kept for cold storage, to be used as future vegetable and some medium sized *ols* are kept for 'seed materials for future use.

13.2.6 Fodder crops (Napier, berseem, lucern blackgram and cowpea)

Fodder crops, after harvest (it is harvested in succulent and green conditions), are fed to the cattle/milch cow or buffaloes etc.

Napier grass : Napier grass, the tall grass, is cut at the base when it is around 90 days after sowing and when it is full green and succulent. These harvested materials are directly carried to the feeding-stall, through trolleys/ vans/carts etc. The harvested materials (stem and leaves) are chopped into pieces and fed to the milch cows/ buffaloes, with the mixture of some leguminous fodder like blackgram/cowpea etc. Due to huge amount of protein in legumes, this napier or thinnapier, are added to the feeds to increase the digestibility of the cattle and also for good health of animal and also for milk production.

Berseem : Berseem is a leguminous fodder crop, grows in *rabi* season. Unlike other crops, it can be cut several times and fed to the cattle. Generally 2-3 cuttings are given at 45-50, 75 and 105 days after sowing. Productions increase where low temperature persists for a long time. After cutting, the green materials (leaves and soft stem), it is loaded to the trolley/vans/carts etc. and carried to the feeding stall. As such, the materials are fed to the cattle/buffaloes with the addition of oat/barley etc.

Lucern : (as in berseem).

Blackgram : Blackgram, the leguminous crop, grown in pre-*kharif* / *kharif* season may be used as fodder crop, after total harvest. Harvesting is done either by uprooting or by cutting the plants at soil level, when the crop is green and succulent, nearly 40-50 days after sowing. After harvest, the material is shifted to the feeding-stall and directly fed to the cattle, after mixing with rice straw or other grassy fodders like napier/thinnapier etc.

Cowpea : (as in blackgram)

Practical Class: Processing of yields, after harvest of some field crops

Lesson 140 : Write the procedures of processing for getting the final yields of cereal crops (Wheat, rice, maize, *bajra*, oat, barley and *jowar*)

Procedure for processing : In getting the final yields of cereal crops (Wheat, rice, maize, *bajra*, oat, barley and *jowar*), the processings are described below.

(i) Wheat : After the deposition of wheat materials (grain and straw) after harvest, on the threshing floor, they are spread over the floor for further sunning for 2-3 days. Thereafter the wheat grain and straw are separated either by wheat thresher, or by treading the materials under the feet of cattle or by running tractor over the heaps of the harvested crops or by beating with the stick manually and by winnowing the materials.

(ii) Rice : In floor, the harvested crop (with grain and straw) is then threshed under the paddy thresher or it may be treaded under the feet of the cattle or by beating on the bamboo-mat. After separation of grains the materials are then winnowed after separation from straw. The bamboo-mat (made up of pieces of bamboo), measuring about 2m × 2m. area for separation and winnowing of grains. Grains are dried in the sun for 2 days again.

(iii) Maize : After the harvest of maize i.e. the cobs, the brackets are peeled-off and the yellow grains are separated from the rachis or ear-head by the maize shellers and kept on tub/busket. The separated grains then are dried in the sun for 2-more days and the final grains are kept in the bags/ sacks/bins, etc.

(iv) *Bajra* : *Bajra*, the cereal crop of dry areas, is sown in the month of June/July during rainy season. It can be utilized both for fodder and grain crop. As fodder this crop should be harvested; the entire plants at about 40-45 days after sowing, when the crop is green and succulent. But for seeds, it should be harvested when plants with leaves are dry. Normally entire plants are harvested by sickle as soil/ground level and carried to the threshing floor. There the top portion of the plants are cut and grains are separated by treading in materials under feet of the cattle or by beating with stick. Seeds then are dried and put into sacks/bins etc.

(v) Oat : (as in lesson 140 i); **(vi) Barley** : (as in lesson 140 i); **(vii) *Jowar* :** (as in lesson 140 iv).

Lesson 141 : Participation in the grain-processing of wheat and *kharif* rice in farms

The learners should participiple in the following grain-processing- programmes of (i) wheat and (ii) transplanted *kharif* rice in the farm with presence of the teachers/ demonstrators.

(a) Wheat: (i) Harvesting, (ii) shifting the materials to the threshing floor, (iii) drying the materials, (iv) threshing/separating the grains, (v) winnowing of the materials, (vi) drying the seeds, (vii) sacking or putting the produce into bins, etc. These are the different steps for processing of wheat.

(b) *Kharif* rice : (i) Harvesting, (ii) keeping the materials in the field for drying, (iii) shifting the materials to the threshing floor and drying, (iv) drying again the materials, (v) threshing/separating grain, (vi) winnowing, (vii) drying seeds, (viii) sacking, are the different steps for processing of *kharif* rice.

Lesson 142 : Describe the procedure for processing in getting final yields of pulse crops (greengram, blackgram, lentil, pea, gram, grasspea and red gram *arahar*)

(i) Greengram (pulse): Greengram (mungbean), the pulse crop, is harvested the top portion after first plucking and the pods of the top portions are taken to the threshing floor, where it is dried for 2-3 days in the sun. After proper sunning, grains are separated by treading the materials under the feet of the cattle or by tractor driving or by beating with bamboo sticks then stover is removed by hand and rake and seeds are winnowed and dried. There after it is sacked.

(ii) Blackgram (pulse) : (as in lesson 142 i).

(iii) Lentil (pulse) : After pulling the entire ripe plants of lentil, after maturity, the leguminous plants (i.e. pulse) are obtained after separating (as in lesson 142 i) and drying the seeds, it is sacked or put into air-tight bins.

(iv) Pea; (v) Gram; (vi) Grasspea : All as in lesson 142 iii.

(vii) *Arahar* (redgram/pigeonpea) (pulse) : *Arahar*/red gram/ pigeon-pea, is a pulse crop, grown normally in pre-*kharif* /*kharif* seasons and harvested at the latter part of the rainy season, though there are some varieties which can be grown in *rabi* season also. It is a big plant and its plants are used as fuel. The entire plants are harvested with sharp separated sickles/*hasua* at the base and shifted the plants to the threshing floor and the top portion (upto pods present) is cut and separated out and dried. After drying tread the materials under the feet of the cattle and separated grains should be winnowed, dried and sacked.

Lesson 143 : Practise the grain processing of blackgram and *arahar* in the farms, describing the procedures of processing

(i) Description of the procedures of processing of

(a) Blackgram: (as in lesson 142 ii); **(b) *Arahar* (Red gram/ pigeonpea)** : (as in lesson 142 iii).

(ii) Participation in the farm-programmes for seed-processing

The learners should participate in the seed-processing programmes (with the teachers/demonstrators).

(a) Blackgram : (1) Plucking/picking the earlier pods, (2) Cutting the top portion for other pods, (3) Shifting the top portion and pods to the threshing floor, (4) Drying the materials, (5) Separation of the grains/seeds, (6) Winnowing the materials after removing the stover, (7) Drying the seeds in the sun, (8) Packing/sacking seeds or putting seeds into bins.

(b) *Arahar* (Red gram/pigeonpea): (1) Cutting the *arahar* plants, (2) Shifting the harvested plants to the threshing floor, (3) Cutting the top por- tion (upto pods), (4) Drying the 'seed and stover' materials, (5) Separation of seeds by beating, (6) Drying the seeds, (7) Sacking or putting seeds into bins.

Lesson 144: Describe, in short, the procedure for getting the final yields of oilseed crops (Mustard, sunflower, groundnut, sesame, safflower and jatropha)

Description of the procedure for processing of oilseed crops (Mustard, sunflower, groundnut, sesame, safflower and jatropha)

(i) Mustard : Mustard, the oilseed crop, is grown in *rabi*, sown in October/November and harvested in Feb/March. It is harvested (only the top portion of the plants in the morning and shifted to threshing floor and kept in a heap, where from seeds are separated by treading the materials under the feet of cattle or by beating the materials with bamboo stick. The threshed materials are then winnowed and grains/seeds are separated The grains are then dried and kept in the sacks or put into bins.

(ii) Sunflower : (as in lesson 144i).

(iii) Groundnut/peanut : Groundnut/peanut, the oilseed crop under legumes, is grown in pre-*kharif*/*kharif* season and harvested in July/August/ November-December. The individual plant is either uprooted by pulling it or by removing the soils with the spade. As pods are grown in the soils, these are collected and dried. Under optimum moisture conditions, the plants are uprooted. The pods from the soils are collected manually and put into the basket/sack/tubs etc. and dried on the floor and as such, the pods are put into sacks or into bins.

(iv) Sesame; **(v) Safflower** and **(vi) Jatropha :** All as in lesson 144i.

Lesson 145 : Practise the processing for getting seeds of mustard and sesame in the farms

Practice for processing of getting seeds of mustard and sesame in the farms

The learners must participate in the seed-processing-programmes of mustard and sesame in the farm with the teacher/demonstrators.

(i) Mustard : (1) Cutting the top portion of the plants (upto the last siliqua), (2) Remove the cut portion from the field to the threshing floor, (3) Heap the cut portion on the threshing floor, (4) Separate the grain/seeds from stover, (5) Drying the seeds, (6) Put the seeds into sacks/bins.

(ii) Sesame : (as in mustard above).

Lesson 146 : Write, in short, the procedures for getting final yields of fibre crops of jute, mesta, sunnhemp, flax and cotton

(i) Jute : After cutting the individual plant, the crop should be kept for bundling with the jute plant itself. Bundles should be kept standing slantingly with top above, for shedding leaves. After shedding leaves, the bundles should be brought nearer to the water source, for putting into 'jak' for steeping. After sometimes (say 3 weeks) while the plants are partially retted, should be taken on the land for stripping the fibre from the stem. It (fibre) is then dried thoroughly on the *'ar'* and bundles are formed into *'lachi'*. The *lachis* are dried further. These *lachis* are then bundled together for big bundles and kept in godown in arranged manner.

(ii) Mesta; (iii) Sunnhemp and **(iv) Flask :** as in lesson 146i.

(v) Cotton : After blooming the cotton boll into full form, it (white lint fibre) is picked up by hand individually and kept into some basket. In similar way, the second or third time (if any) pickings are done. These bolls are made free from seeds in the threshing floor. Then these 'white fibres' (the cotton fibre) are used or stored.

Lesson 147: Practise in the processing for getting final yields of jute and mesta in the farms

Participate in the processing programmes for getting final fibre yields of jute and mesta in the farms

[**Note** : The learners will participate in the following programmes under the guidance of teachers]

(i) Jute : (a) Harvesting and plant-bundling, (b) Stacking the bundles for leaf-shedding in the field, (c) Removal of bundles from the field to water source, (d) Steeping : Setting the 'jak' or 'jag' in water for one and-half- months, (e) Retting : Removing partially decomposed 'jag' materials for stripping fibre, (f) Extraction : Stripping fibres from the stem of jute, (g) Washing the fibre in (running) water, (h) Drying the fibre on *'ar'* (bamboo poles), (i) Binding the *'lachis'* (j) Fibre further (for big bundles) bundling and (k) Storing.

(ii) Mesta : (as in lesson 147i).

Lesson 148: Describe the procedure for getting final yields of sugars from sugarcane

(i) Harvesting of cane: Individual canes are cut with sharp serrated sickles, at the ground level.

(ii) Processing of cane: Harvested canes are cleaned with sickle, leaves and adventitious roots are removed. The top portion of the cane is also removed.

(iii) Transportation of cane: The cleaned canes are then removed from the field to the market/factories etc. by carrying them in trolley/van/carts etc.

(iv) Crushing the cane for juice: The cleaned canes are crushed in the factories for juice.

(v) Bringing out '*gur*' or sugars from juice: From juice obtained from canes, *gur* or sugars are prepared.

(a) *Gur* preparation : *Gur* is prepared in the field it self, by putting juice into a special big flat rectangular iron container, after removing froths with the addition of *verenda*/chemicals etc.

(b) Sugar preparation : Sugars are manufactured in the factory by passing juice through the different chemical processes.

(vi) Sacking *gur*/sugar: The production of granular sugar bids are then packed in some sacks etc. for removing, distribution and selling.

Lesson 149: Participate in the *gur* /sugar manufacturing unit to know the processing in detail, for obtaining the produce

The learners should participate in the '*gur*' or sugar manufacturing process in the centres

(i) '*Gur*': (1) Harvesting the cane, (2) Processing of cane, (3) Transpor-tation of cane, (4) Crushing cane for juice, (5) Bringing out *gur* by boiling in the vat after cleaning froth with *verenda* plants or some chemicals, (6) Preparation of *gur* in containers and (7) Storing the containers into storage or sending for markets.

Assignment : (according to number from 1 to 6 as above)

(ii) Sugar manufacturing unit Assignment : [According to steps 1 to 6 (lesson 149) plus the notes from sugar factories].

Lesson 150 : Describe the procedure for getting final yields of potato

Description of procedure for getting potato yields

(a) Harvesting : Harvesting of potato is quite different from that of other crops. Potato tubers are grown in underground, hence these are lifted/ harvested with the help of spade or small man-drawn plough before picking up.

(b) Cleaning the field : The potato tubers which are adhered to muds and soils are cleaned with the help of hands; for drying the soils, last irrigation should be applied at least 3 weeks before the harvest of the crop.

(c) Sacking in field : After cleaning and grading, the potato tubers are put into sacks/gunny bags etc. for shifting from the field.

(d) Shifting the *sacks* from the field into the cold storage : The bags/sacks are shifted from the field to the markets or cold storage.

(e) Stacking the *sacks* in cold storage : In the cold storage, the sacks full of potatoes, are stacked upto a height of about 10-15m. The sacks/bags are generally carried to the godown either by lorries/van/trolleys etc. and the sacks/ bags are lifted with the help of big ladders to a greater height.

Lesson 151: Practise (a) for processing and (b) storage of potato in cold storage

(a) Practising for processing programmes of potato in the field

The learners should participate in the following programmes :

(1) Harvesting of potatoes by opening the ridges, (2) Picking up potatoes from the fields, (3) Cleaning potato, (4) Grading potato, (5) Putting graded potato into basket/bags/sacks, (6) Carrying bags/sacks to markets/cold storage.

(b) Practising for further processing of potato in cold storage

(1) Lifting the bags to greater height, (2) Maintenance of low temperature.

Assignment (for the cold storage)

(1) Arrangement of the sacks in the cold storage, (2) Height, length and width of cold storage, (3) Accommodation of cold storage, (4) Ventilation arrangement, (5) Climbing up arrangement, (6) Temperature-keeping and regulation arrangement, (7) Sprouting of tubers, (8) Relationship of sprouting and temperature and age of potato.

Lesson 152: Describe the procedures for processing of legume fodders (berseem, lucern, blackgram, rice-bean and grasspea)

Description of procedures for processing of berseem, lucern, black-gram, ricebean and grasspea

(i) Berseem (cutting): Berseem, the leguminous fodder, is grown in *rabi*. First cutting is done about 2- 2½ months after sowing, second cut and other cuts are completed again after 1-1½ month interval. All these harvested materials are brought from the fields to the cow-shed for feeding to the cattle. Then the crop is retained for seeds after last cut.

(ii) Lucern (cutting) : (as in lesson 152 i).

(iii) Blackgram (total harvest) : Blackgram, the legume fodder, is grown in pre-*kharif*. It is generally harvested two months after sowing, when the crop is green and succulent. After total harvest with the help of sickle, it should be carried to the stall feeding area and fed to the cattle, mixing with straws or grass-group fodder, as it contains high content of protein.

(iv) Ricebean (total harvest), **(v) Grass pea :** (total harvest) : both as in lesson 152 iii.

Lesson 153 : Participate in the `cutting-process' and harvesting of berseem, lucern, blackgram, ricebean and grasspea in the dairy areas

The learners should participate in cutting/harvesting programmes of

(i) Berseem (cutting) : (1) Cutting with sickles, (2) Accumulation of harvested materials in some places, (3) Putting the harvested materials into lorries/trolley/van etc. and sending to the feeding stalls, (4) Give the harvested berseem to the cattle after addition of grasses or straw of grass-group fodder/rice or wheat straw.

(ii) Lucern (cuttings) : (as in lesson 153 i).

(iii) Blackgram : (1) Cutting blackgram for fodder, (2) Collection and loading the materials in the trolley/van/cart, (3) Carrying this fodder to the feeding-stall, (4) Give the fodder materials to the cattle.

(iv) Ricebean, (v) Grasspea : both as in lesson 153 (iii).

Lesson 154 : Describe the procedure for processing of non-legume fodder crops (Napier, teosinte and fodder-mustard)

Description of procedures for processing/harvesting/cutting of non-legume fodders (Napier, teosinte and fodder-mustard)

(i) Napier : Napier the tall grass fodder, is grown in pre-*kharif* and *kharif* seasons. First cutting is given when the crop is 2-2½ months old, but subsequent cutting in given at about 45 days intervals. The criteria of fodder should be greenishness and succulency. After cutting, the materials are loaded in trolley/van/carts etc. and brought from the field to the feeding-stall without any delay. Succulency and greenishness decrease, delayed it. It is, generally, given to cattle immedi-ately, if possible with the mixture of some leguminous fodder, (like black gram/rice bean/ cowpea etc.).

(ii) Teosinte : Teosinte, the tall grass fodder, is also grown during pre-*kharif* and *kharif* seasons. Instead of several cuttings, it is generally cut once 2-2½ months after sowing. Of course, after 3 months also, it can be cut (whole) while the plant flowers. This fodder is chopped into pieces and fed to the cattle, mixing with some leguminous fodders like blackgram, ricebean etc. Here also succulency and greenish-ness is the criteria for fodder.

(iii) Fodder-mustard : One type of mustard, having big leaves with less pungeney, is used as fodder for cattle, under stall-feeding. This is grown in *rabi* season and when it is green and succulent, it is fed to the cattle after total harvest of crop.

Lesson 155 : Participate in the harvesting process of napier/thin napier in the dairy areas

The learners should participate in the harvesting/cutting process of napier in dairy farm areas (the teacher should be present) :

Napier harvesting/cutting process : (1) Cutting the individual napier plant, (2) Spreading over the soil surface, (3) Collection of plants and loading the material, in the trolley/van, (4) Carrying to the feeding stall directly, (5) Feed the cattle teosinte fodder by mixing some legume fodders (blackgram, ricebean or cowpea).

13.3 STORING OF SEEDS, FIBRE, STRAW AND BY-PRODUCTS

13.3.1 Cereal crops (Wheat, rice, maize, *bajra*, oat, barley and *jowar*)

After processing, the **grains** are stored either in rubber-lined bags or airtight bins, after drying fully so that moisture does not remain above 12% or so. This is true in case of other cereals (rice, maize, *bajra*, oat, barley and *jowar*), pulses (greengram, blackgram, lentil, pea gram, grasspea and *arahar*) and oilseeds (mustard, sunflower, groundnut, sesame, safflower and jatrapha).

Of course, by-products are used and stored differently. The stover or straw of cereal crops are stacked with methodical way under shed for further use, after proper sunning, except some crops like maize etc. Maize stover, after drying, may be utilized as fuel or for preparation of organic manures or for supporting sticks.

13.3.2 Pulse crops (Greengram, lentil, pea, gram, grasspea and *pigeon pea*)

Grains/seeds of pulse crops such as greengram, blackgram, lentil, pea, gram, grasspea and *pigeonpea,* are stored as in cereals (13.3.1). Of course the stover/*bhushi*/*bhusha*, obtained from all the crops, are utilized in different ways. The stover/*bhusha* of greengram, blackgram, lentil, pea, gram and grasspea are used for cattle feeds as it contains high amount of protein. The stover or sticks are used as fuel after drying. Jute sticks are used for paper manufacturing also.

13.3.3 Oilseed crops (Mustard/rape, groundnut, sunflower, sesame, safflower and jatropha)

Grains of oilseeds of mustard/rape, sunflower, groundnut, sunflower, sesame, safflower and jatropha are stored after drying seeds in the sun, so that moisture percentage does not remain above 12%, in the rubber-lined bags/ sacks or in airtight bins.

Of course, stover/sticks etc. of oilseed crops are used differently. Stovers yields of mustard/rape, sunflower and sesame safflower and groundnut and jatropha are used as fuels, but in some areas (eastern parts of India) ground- nut plants may be used as green manure or fodder, because of greenish-ness and succulency, as it is sown in pre-*kharif* when green plants are available, after getting pods.

13.3.4 Fibre crops (Jute, mesta, sunnhemp, flax and cotton)

Among the fibre crops, bast fibre crops (jute, mesta, sunnhemp and flax) are dealt with in one form and the lint fibre crop is considered separately, as their storing processes are quite different.

Bast fibre : After processing of bast fibres crops, fibres are made into bundles with *lachis* and the big bundles are then stacked one after another in the godown, devoid of high moisture. It should be ensured that fibre is dried properly in the sun before putting into bundles, otherwise there is a chance of fading of the colour of the product, because it will earn less money.

The by-products of jute, mesta, sunnhemp and flax, are the sticks (stover). The sticks are then dried in the sun for using as fuel. This stick may also be used as part of basic materials of paper, board etc.

Lint fibre : In case of lint fibre of cotton, the separated cotton (from the calyx and seeds of the bolls) fibre is kept in bundles (bales) under hydraulic pressure, and kept in the moisture-free area, in the factory, where different materials will be prepared from it. In case of cotton, the by-products (calyx and seeds) are used differently. The seeds are separated from the calyx. The calyx is used as fuel, whereas the seeds are kept, after drying, for future use (as in cereal/pulses/oilseeds) as seeds.

13.3.5 Sugar/Potato and rhizome (*amorphophallus*) crops

Sugarcane : Sugarcane, is grown as either *eksali* or *adhsali* crop, mostly in warm climate. The juice from which *gur* or sugars, are prepared, is obtained from the stems by crushing. The *gur* that is produced, is kept in the bags/or earthen pots for storing in stores/godown. The sugars that are produced in the sugar factories, are put into rubber-lined bags/sacks or other containers. The bags/sacks etc. with sugars, are sent to the markets for selling or kept into the stores/ godowns.

Potato : Potato, the underground stem and the important widely used vegetable crop, is cultivated in *rabi* season under ridge and furrow method. The harvested crop from the ridges, the tubers are sorted out according to grade, after cleaning them. The cleaned and graded potatoes are then put into jute-sacks, as such, in the field, where from the bags are carried to the market places or to the cold storage. The bags in the cold storage, are kept methodically one after another, at a greater height. Particular low temperature is then maintained for future use of potato.

Rhizomes (*amorphophallus*) : *Amorphophalls* (*ol*), the rhizome i.e. underground soft stem, i.s. now one of the best vegetables of Indian diet, because of availability of *amorphophallus* variety free from itching materials. This *ol* is cultivated in Pre-*kharif* /*kharif* season, under ridge and furrows. After harvesting *ol,* with the help of spade individually and manually, it is cleaned (as roots, soils, muds, debris etc. are present after harvesting) and the '*corms*' (individual head-like *ols*) are put into sacks/bags etc. These sacks are directly taken to the markets or placed in cold storage for future use.

13.3.6 Fodder crops (Legume and non-legume)

Generally, fodder crops particularly legumes that can not be stored for a long time. The crops, that are harvested green and succulent, are fed to cattle immediately. But in case of crops, under *Gramineae* group, can be *stored* in the form of hay or silage. Napier, thin napier etc. can be stored as hay also, if they are cut at the middle age when greenishness/greenness and succulency still are available.

Of course, straw/stover etc. of any cereal crop, can be stored, if properly dried. They are heaped systematically under the shed or in open areas if proper arrangement is done for resisting water to get into it. Any residual pulse crop (*bhusha*) can be stored after drying the materials properly. This is a good feed for cattle; but with moisture it is spoilt and the cattle (if fed to the cattle) is in danger.

Practical Class : To look into the arrangements for storage of the produce in godown/cold storage and record them individually from time-to-time

Lesson 156 & 157 : Visit the (a) godowns, (b) cold storages and (c) silage pits (silopit), for observing the storage of produce of (i) Cereals, (ii) Pulses, (iii) Oilseeds, (iv) Sugars, (v) Fibres and (vi) Potato (Take visual observation records) (Teachers should accompany the learners)

(a) Visiting the godown for observing storage of cereals, pulses, oil seeds, sugars, fibres and recording observations/procedures to be completed.

(b) Visiting the cold storage for storage of potato and other crops and recording observations therein.

(c) Visit for studying the silo-pits for making silage and record the process for *silageing*

[**N.B.** The teacher should accompany the learners and the learners should take note-book, pen/pencil and make to notes of the studies as per station officers/demonstrators].

Lesson 158 : Visit the dairy farm, for looking into feeding techniques of fodders along with the home system and also visit '*silopit*' areas in the same dairy farm and record the procedures for '*silage*' preparation

The learners should visit the dairy farm and will write the notes as follows: (teachers should accompany)

(a) Mechanised system for feeding animals.

(b) Ordinary system for feeding animals.

(c) Home system for feeding animals.

Then submit a full note on these systems to the class teacher.

CHAPTER 14

VITAL INFORMATION FOR PRACTICAL APPROACH

14.1 CLASSIFIED CROPS IN SEASONS (with scientific, English and local names, along with the different families)

Major field crops of India are arranged in different seasons under different economic/agronomic classifications, for better understanding of the students in agriculture. Therefore, the classification has been done accordingly along with scientific English- and local names, including families, under 3 different seasons, such as *rabi*, pre-*kharif* and *kharif*.

14.1.1 *Rabi* season :

[Cereals, pulses oilseeds, fibres, tubers, vegetables, fruits/spices, fod-ders/hays & miscellaneous crops (weeds-broad and narrow-leaved)]

In all the seasons (*rabi*/pre-*kharif* and *kharif*) the different crops, grown in that season, are arranged under 'group' cereals, pulses, oilseeds, fibres, tubers, vegetables, fruits/spices, fodders/hays and miscellaneous crops (weeds broad and narrow-leaved and green manure crops).

	Classification & crops (English)	Scientific name	Family	Local name (Bengali)
(i)	**Cereals :**			
1.	Wheat	*Triticum aestivum* (L.) emend Thell	Gramineae	Gam
2.	Maize	*Zea mays* L.	Gramineae	Bhutta
3.	Oat	*Avena sativa* L.	Gramineae	Jai
4.	Barley	*Hordeum vulgare* L.	Gramineae	Jab
5.	Rice (*boro*)	*Oryza sativa* L.	Gramineae	Borodhan
(ii)	**Pulses :**			
1.	Grasspea / *khesari*	*Lathyrus sativus* L.	Papilionaceae	Khesari

2.	Lentil	*Lens culinaris* Medik	Papilionaceae	Masur
3.	Pea	*Pisum sativum* L.	Papilionaceae	Matar
4.	Gram/ chickpea	*Cicer arietinum* L.	Papilionaceae	Chhola
5.	*Rabi arahar*	*Cajanus cajan* Millop	Papilionaceae	Rabi arahar
6.	*Rajmah*/ French bean	*Phaseolus vulgaris* L.	Papilionaceae	Rajma
(iii)	**Oilseeds :**			
1.	Mustard	*Brassica campestris* L.	Cruciferae	Sarisa
2.	Linseed	*Linum usitatissimum* L.	Linaceae	Tisi
3.	Rape	*Brassica juncia* L.	Cruciferae	Sarisa/sarse
4.	Sunflower	*Helianthus annuus* L.	Compositeae	Surjamukhi
5.	Safflower	*Carthamus tinctorius* L.	Compositeae	Kusum
(iv)	**Fibres :**			
1.	Flax	*Linum usitatissimum* L.	Linaceae	Phlaks
2.	Sunnhemp (seed)	*Crotolaria juncea*	Papilionaceae	Shan
(v)	**Tubers :**			
1.	Potato	*Solanum tuberosum* L.	Solanaeceae	Alu
(vi)	**Vegetables/fruits/spices :**			
1.	Spinach	*Spinacia oleracia* L.	Chinopodiaceae	Palang
2.	Coriander	*Coriandrum sativum* L.	Umbellifereae	Dhane
3.	Fenugreek	*Trigonella foenum-graecum* L.	Papilionaceae	Methi
4.	Fennel	*Foeniculum valgare Mill*	Umbellifereae	Mauri
5.	Black cumin	*Nigella sativa* L.	Umbellifereae	Kalojira
6.	Cucumber	*Cucumis sativa* L.	Cucurbitaceae	Shasha
7.	Cabbage	*Brassica oleracea* var. *capitata*	Crucifereae	Badhakapi
8.	Cauliflower	*Brassica oleracea* var. *bortrytis*	Crucifereae	Fulkapi
9.	Radish	*Raphanus* sativus L.	Crucifereae	Mula
(vii)	**Fodder :**			
1.	Grasspea	*Lathyrus sativus* L.	Papilionaceae	Khesari
2.	Oat	*Avena sativa* L.	Gramineae	Jai
3.	Senji	*Melilotus alba / indica* L.	Papilionaceae	Senjimethi
4.	Tiny vetch	*Vicia hirsuta* L.	Papilionaceae	Hetka
(viii)	**Miscellaneous (weeds-broad and narrow-leaved) :**			
I.	**Broad-leaved :**			
1.	Senji	*Melilotus parviflora* L.	Papilionaceae	Senjimethi
2.	Pig weed	*Chenopodium album* L.	Chenopo-diaceae	Bathua

3.	Sunberry	*Physalis minima* L.	Solanaceae	Bantepari
4.	Croton	*Croton sparsiflora* L.	Euphorbiaceae	Banmarich
5.	Anagallis	*Anagalis arvensis* L.	-	Krishnanil
6.	*Melilotus*	*Melilotus alba / indica* L.	Papilionaceae	Senjimethi
7.	*Vicia*	*Vicia hirsuta* L.	Papilionaceae	Hetka
II.	**Narrow-leaved**			
1.	Bermuda grass	*Cynodon dactylon* Pers.	Gramineae	Durba
2.	Nut grass	*Cypernus rotundus* L.	Cyperaceae	Mutha
3.	Holon weed	*Asphodelus tenuifolia* L.	Cyperaceae	Piaji
4.	Mutha weed	*Cyperus iria* L.	Cyperaceae	Jalmutha
5.	Wild oat	*Avena fatua* L.	Gramineae	Banjai
6.	Canary grass	*Phalaris minor* Rez	Gramineae	Phalaghash
7.	Rye	*Secale cereale* L.	Gramineae	Raighash

14.1.2 Pre-*kharif* season :

[Cereals, pulses, oilseeds, fibres, tubers/rhizomes, vegetables, fruit/ spices, fodders/hays and miscellaneous crops (weed-broad and narrow-leaved)]

Classification & crop (English)		Scientific name	Family	*Local name (Bengali)*
(i)	**Cereals :**			
1.	*Aus* rice	*Oryza sativa* L.	Gramineae	*Aus* dhan
2.	Maize	*Zea mays* L.	Gramineae	Bhutta
3.	Cat-tail millet	*Setaria italica* Beauv.	Gramineae	Kaon
4.	Finger millet	*Eleusine coracana* Gaetn	Gramineae	Ragi
5.	Sorghum	*Sorghum vulgare* Pers.	Gramineae	Jowar
6.	Pearl millet	*Pennisetum typhoide* Stapf.	Gramineae	Bajra
(ii)	**Pulses :**			
1.	Green Gram	*Vigna radiata* Wilczek.	Papilionaceae	Mug
2.	Cowpea	*Vigna unguiculata* Walp.	Papilionaceae	Barbati
3.	Horsegram	*Dolichos biflorus* L.	Papilionaceae	Kultikalai
4.	Ricebean	*Vigna umbellata* Ohwi & Ohashi	Papilionaceae	Gaimug
5.	Mothbean	*Vigna aconitifolia* Marechal	Papilionaceae	Kulatra kalai
(iii)	**Oilseeds :**			
1.	Sesame	*Sesamum indicum* L.	Pedaliaceae	Til

2.	Groundnut	*Arachis hypogea* L.	Papilionaceae	Badam
3.	Castor seed	*Ricinus communis* L.	Euphorbiaceae	Reri
(iv)	**Fibres :**			
1.	Jute	*Corchorus capsularis / oiltorius* L.	Tiliaceae	Pat
2.	Mesta	*Hibiscus sabdariffa* L.	Malvaceae	Mestapat
3.	Sunnhemp	*Crotolaria juncea* L.	Papilionaceae	Shan
(v)	**Tuber/Rhizomes :**			
1.	Arum/taro	*Colocasia esculenta* Scott.	Araceae	Kachu
2.	*Elephant-foot -yam*	*Amorphophallus campanulatus* Roxb.	Araceae	Ol
3.	Elephant-ear	*Colocasia antiquorun* Schott.	Araceae	Jalkachu
(vi)	**Vegetables/fruits/spices :**			
1.	Cucumber	*Cucumis sativa* L.	Cucurbitaceae	Shasha
2.	Musk melon	*Cucumis melo* Naud.	Cucurbitaceae	Tarmuj
3.	Lady's finger	*Abelmoschus esculentus* L.	Malvaceae	Dheras
4.	Chinese spinach	*Amaranthus tricolor* Moench.	Amarantaceae	Data
5.	Indian dill	*Anethum sowa* Kurz.	Umbelliferae	Saj
6.	Red gourd	*Cucurbita maxima* Duch	Cucurbitaceae	Lalkumra
7.	Bitter gourd	*Momordica charantia* L.	Cucurbitaceae	Karala
8.	Pumkin	*Cucurbita pepo* L.	Cucurbitaceae	Lau
9.	Ridge gourd	*Luffa acutangula* Roxb.	Cucurbitaceae	Jhinga
(vii)	**Fodders / hays :**			
1.	Thinnapier	*Pennisetum pedicellatum* Trin.	Gramineae	Dinanath ghash
2.	Elephant grass	*Pennisetum purpureum* Schum.	Gramineae	Napier ghash
3.	Blackgram	*Vigna mungo* Hepper.	Papilionaceae	Mashkalai
4.	Teosinte	*Euchlaena mexicana* Schrad.	Gramineae	Tiosinti
5.	Cowpea	*Vigna unguiculata* Walp.	Papilionaceae	Barbati
6.	Ricebean	*Vigna umbellata* Ohwi & Ohashi	Papilionaceae	Gaimug
7.	Moth bean	*Vigna aconitifolia* Marechal.	Papilionaceae	Kulatrakalai
(viii)	**Miscellaneous (weeds broad and narrow-leaved/green manure crops)**			
I.	**Broad-leaved weeds :**			
1.	Sunberry	*Physalis minima* L.	Solanaceae	Bantepari

2.	Croton	*Croton sparciflora* L.	Euphorbiaceae	Banmarich
3.	Michenia	*Michenia* spp.	-	Jhiarilata
II.	**Narrow-leaved weeds :**			
1.	Barn-yard grass	*Echinochloa crusgalli* Beauv.	Gramineae	Shamaghash
2.	Bermuda grass	*Cynodon dactylon* Pers	Gramineae	Durba
3.	Nut grass	*Cyperus rotundus* L.	Cyperaceae	Mutha
4.	Little millet	*Panicum sunatrense* Roth.	Gramineae	Chhotodana
5.	Gamba grass	*Andropogon gayanus* Kunth.	Gramineae	Batui
III.	**Green manure crops :**			
1.	Dhaincha	*Sesbania aculeata* Poir.	Papilionaceae	Dhaincha
2.	Sunnhemp	*Crotolaria juncea* L.	Papilionaceae	Shan
3.	Blackgram	*Vigna mungo* Hepper.	Papilionaceae	Mashkalai
4.	Green- gram	*Vigna radiata* Wilczek	Papilionaceae	Mug
5.	Cowpea	*Vigna unguiculata* walp.	Papilionaceae	Barbati
6.	Ricebean	*Vigna umbellata* Ohwi & Ohashi.	Papilionaceae	Gaimug
7.	Bagha-dhaincha	*Sesbania rostrata*	Papilionaceae	Bagha-dhaincha

Notes : *Aus* = direct-seeded rice.

14.1.3 *Kharif* season :

[Cereals, pulses oilseeds, fibres, tubers/rhizomes, vegetables, fruit/ spices, fodders/hays, and miscellaneous crops (weeds-broad and narrow-leaved)]

Classification & crops (English)		Scientific name	Family	*Local name (Bengali)*
(i)	**Cereals :**			
1.	Rice (*aman*)	*Oryza sativa* L.	Gramineae	Dhan
2.	Maize	*Zea mays* L.	Gramineae	Bhutta
3.	Sorghum	*Sorghum vulgare* Pers.	Gramineae	Jowar
4.	Pearl millet	*Pennisetum typhoides* Stapf.	Gramineae	Bajra
5.	Finger millet	*Eleusine coracana* Gaertn.	Gramineae	Ragi
(ii)	**Pulses :**			
1.	Greengram	*Vigna radiata* Wilczek.	Papilionaceae	Mug
2.	Cowpea	*Vigna unguiculata* Walp.	Papilionaceae	Barbati
3.	Blackgram	*Vigna mungo* Hepper.	Papilionaceae	Mashkalai
4.	Pigeon pea	*Cajanus cajan* Millsp.	Papilionaceae	Arahar

(iii)	**Oilseeds**			
1.	Sesame	*Sesamum indicum* L.	Pedaliaceae	Til
2.	Peanut/G. nut	*Arachis hypogaea* L.	Papilionaceae	Badam
3.	Castor-seed	*Ricinus communis* L.	Euphorbiaceae	Reri
4.	Jatropha	*Jatropha curcas* L.	Euphorbiaceae	Banbherenda
(iv)	**Fibre crops :**			
1.	Jute (seed)	*Corchorus capsularis / olitorius* L.	Tiliaceae	Pat
2.	Mesta (seed)	*Hibiscus sabdariffa* L.	Malvaceae	Mesta pat
3.	Sunnhemp	*Crotalaria juncea* L.	Papilionaceae	Shan
4.	Ramie	*Boehmeria nivea*(L.) Gaud.	Urticaceae	Remi
5.	Agave	*Agave sisalana* Perr.	Agavaceae	Sisal
(v)	**Tubers/rhizome :**			
1.	Arum/Taro	*Colocasia esculenta* Schott.	Araceae	Kachu
2.	*Elephant-foot-yam*	*Amorphophallus campanulatus* Roxb.	Araceae	Ol
(vi)	**Vegetables/fruits/spices :**			
1.	Cucumber	*Cucumis sativa* L.	Cucurbitaceae	Shasha
2.	Musk melon	*Cucumis melo* Naud.	Cucurbitaceae	Tarmuj
3.	Lady's finger	*Abelmoschus esculentus* Moench.	Malvaceae	Dheras
4.	Indian dill	*Anethum sowa* Kurz.	Umbellifereae	Saj
5.	Pepper/ chillis	*Capsicum annum* L.	Solanaceae	Lanka
6.	Ginger	*Zingiber officinale* Rosc.	Zingiberaceae	Ada
(vii)	**Fodders / hays :**			
1.	Elephant grass	*Pennisetum purpureum* Schum.	Gramineae	Napier grass
2.	Thin napier	*Pennisetum pedicellatum* Trin.	Gramineae	Dinanath grass
3.	Blackgram	*Vigna mungo* Hepper.	Papilionaceae	Mash- kalai
4.	Teosinte	*Euchlaena mexicana* Schrad.	Gramineae	Tiosinti
5.	Cowpea	*Vigna unguiculata* Walp.	Papilionaceae	Barbati
6.	Greengram	*Vigna radiata* Wilczek.	Papilionaceae	Mug
7.	Ricebean	*Vigna umbellata* Ohwi & Ohashi	Papilionacea	Gaimug
(viii)	**Miscellaneous (weed-broad & narrow-leaved and green manure crops)**			
I.	**Broad-leaved weeds :**			
1.	Sun berry	*Physalis minima* L.	Solanaceae	Bantepari
2.	Michenia	*Michenia* spp.	-	Jhiarilata

3.	Thorn weed	*Solanum khasianum*	Solanaceae	Banbegun
4.	Malva	*Malva* spp.	Malvaceae	Malva
II.	**Narrow-leaved weeds :**			
1.	Bermuda grass	*Cynodon dactylon* Pers.	Gramineae	Durba
2.	Iria cyperus	*Cyprus iria* L.	Cyperaceae	Jalmutha
3.	Asphodelus	*Asphodelus tenuifolia*	Cyperaceae	Chechur/ Piaji
III.	**Green manure crops :**			
1.	Dhaincha	*Sesbania aculeata* Poir.	Papilionaceae	Dhaincha
2.	Sunnhemp	*Crotalaria juncea* L.	Papilionaceae	Shan
3.	Blackgram	*Vigna mungo* Hepper.	Papilionaceae	Mash kalai
4.	Cowpea	*Vigna unguiculata* Walp.	Papilionaceae	Barbati
5.	Groundnut (haulm)	*Arachis hypogaea* L.	Papilionaceae	Badam
6.	Baghadhaincha	*Sesbania rostrata*	Papilionaceae	Baghadhaincha
7.	Cluster bean	*Cyamopsis teragonoloba* L.	Papilionaceae	Guar

14.2 SOIL ANALYTICAL DATA OF FARM AND MOISTURE STATUS IN RICE-FALLOWS IN DRY MONTHS

14.2.1 Soil analytical data of the farm under (a) Physical and (b) Chemical conditions along with (c) Land situations (BCKV Farm, Kalyani, W.B.)

(a) Physical properties of soil

Particle size distribution of soils (%)				
Coarse sand	Fine sand	Silt	Clay	Remarks
4.1	31.9	36.0	28.0	Clayey loam (alluvial)

Note : (i) Water-retention capacity – medium; (ii) Land situation – medium, having clayey alluvial soils

(b) Chemical properties of soils

Organic C %	Total N %	Av. P_2O_5 (kg/ha)	Av. K_2O (kg/ha)	pH	Remarks
0.69	0.07	18.5	168.0	7.1	Average soil fertility

(c) Land situation : Medium land

[**N.B.** (a) Physical and (b) Chemical properties of soils along with land situations above, indicated that this medium land is fit for dry and wet crops, as the soil fertility is average in rank. Data are from Kalyani BCKV Farm, W. Bengal.

14.2.2 Soil moisture status (%) in rice-fallows in dry months (December to March) at different soil depths of soil of Kalyani Farm (BCKV), West Bengal

Months & depth (cm)	Fallow (No-crop)	Cropped (with *khesari*)	Mean (depth)	Mean (month)
(a) December :				
0-10	11.4	14.3	12.8	-
40-50	23.7	24.0	23.9	21.3
90-100	25.4	27.4	27.0	-
Mean	20.5	22.0	-	-
(b) January :				
0-10	10.3	11.1	10.7	-
40-50	20.8	21.6	21.2	19.2
90-100	25.5	25.8	25.7	-
Mean	18.9	19.5	-	-
(c) February :				
0-10	10.0	12.5	11.3	-
40-50	15.5	14.1	14.8	16.2
90-100	24.4	22.5	22.5	-
Mean	16.0	16.9	-	-
(d) March :				
0-10	2.3	5.3	3.8	-
40-50	9.11	13.6	11.4	11.5
90-100	16.6	13.6	19.2	-
Mean	9.3	10.6	-	-
Grand mean (crop)	16.3	17.9	-	-

From the above data, it is observed that :

(i) Fallow (No-crop) lands remain normally fallow in *rabi* season, for lack of irrigation facilities, after the harvest of transplanted *kharif* rainfed rice and the *rabi*/winter (November to March) season is dry.

(ii) Cropped with grasspea (*Lathyrus sativus* L.) (i.e. *khesari* in Bengali) as *paira/utera* (no-tillage relay) crop, was sown on November 4 and the preceding crop (*kharif* transplanted rice) was harvested on November 5 and no-tillage was given.

(iii) At the time of sowing (broadcasting) *paira/utera* crop of grasspea, both under fallow or cropped-lands, soils were in muddy conditions, at full-water saturated point.

(iv) Moisture content of soils by the end of November either in fallow or cropped areas, was 32.5%, at 0-10 cm depth, but no individual records were taken depth-wise, because of wetness due to water saturation.

(v) Observations were recorded at the end of each month (December to March) from 26 to 29.

(vi) Rainfed winter *paira/utera* grasspea (cv. Nirmal 1) yielded 12.11 q/ha of seeds and 47.12 q/ha of stover yield. Preceding *kharif* transplanted rice (cv. IR 36) yielded 25.11 q/ha of grain and 56.14 q/ha of straw.

The findings from the above are as follows :

(a) Improvement of soil moisture status was noted in cropped plots, over no-crop.

(b) *Lathyrus sativus* (*khesari*) increased soil moisture in rice-fallows.

(c) Soil moisture in rice-fallows increased with depth.

(d) Soil moisture in the rice-fallows decreased with the advancement of time or month (in March, soil moisture was minimum, whereas it was the highest in December).

14.3 DATA ON RAINY/CLOUDY DAYS AND BRIGHT SUNSHINE HOURS, ALL-THE- YEAR-ROUND, AT BCKV (Kalyani Farms, W.B.)

Season & month	Rainy days no./ fortnight in month			Cloudy days no./ fortnight in month			Bright sunshine hr./day		
	1	2	Total	1	2	Total	1	2	Mean
(a) Pre-*kharif* (summer) season :									
April	2	4	6	6	7	13	10	8	9
May	4	4	8	7	7	14	10	8	9
June	7	10	17	9	14	23	9	7	8
July	9	8	17	10	14	24	9	7	8
Total/mean (sun)			48			64			9
(b) *Kharif* (rainy) season :									
July	9	8	17	10	14	24	9	7	8
August	8	9	17	9	10	19	7	7	7
September	5	10	15	5	10	15	9	9	9
October	3	3	6	3	3	6	9	8	9
November	0	2	2	3	2	5	9	9	9
Total/mean (sun)			57			62			8
(c) *Rabi* (winter) season :									
November	0	2	2	3	2	5	0	9	9
December	0	0	0	1	1	2	9	10	10
January	0	0	0	0	0	0	9	9	9
February	0	1	1	1	1	2	10	10	10
March	1	0	1	1	1	2	10	11	11
Total/mean (sun)			4			11			10

Note : 1 & 2 = First & Second fortnight, no. = number, hr. = hour, sun = bright sun shine.

From the above, it is observed that :

(i) Rainy days are maximum in rainy season, followed by summer and minimum in winter season.

(ii) Cloudy days are maximum in rainy and summer seasons, whereas it is minimum in winter.

(iii) Bright sunshine hours per day is maximum in winter season, followed by summer, whereas it is minimum in rainy season.

14.4 PRODUCTION ECONOMICS OF CROPS

Production economics of crop is very important for the growers, as it is now considered as a business. Naturally, if cost of production of any crop cultivation is minimised and total gross return is increased, due to higher crop production, then automatically net profit will be high. Hence, the grower will be benefited with crop cultivation.

14.4.1 Cost of cultivation of crops (CCC) (Rs./ha)

Costs of cultivation of crops (CCC) differ from place to place. The cost of cultivation, of course, may be in the following broad classifications like (a) cost of materials, (b) cost of tillage/irrigation arrangement (if irrigated) and (c) cost of laboures at different stages, for different functions.

14.4.1.1 *Cost of materials for cultivation*

This includes fertilizers, seed materials etc. All the prices of inputs, of course, change from place-to-place and time-to-time.

14.4.1.2 *Cost of tillage/ploughing/inter-tillage and irrigation*

This includes land preparation, inter-tillage operations, irrigation costs etc. The costs also differ from place to place and time-to-time.

14.4.1.3 *Cost of labourers at different stages for different works*

At different stages of crop cultivation, different functions (hoeing/ weeding etc.) are done. For that purpose some labourers are needed and there is involvement of cost. Of course they also vary from place-to-place and time-to-time.

14.4.2 Gross returns (GR) for cultivation (Rs/ha)

For gross returns (GR) in any cultivation, it is dependent on economic yields and by-products (yield), along with their market prices.

14.4.2.1 ***Produce of main/economic product (q/ha) and price (Rs/ quintal) for it and part gross return (Rs/ha)***

(A) Part gross return is like this :

(i) Grain/economic yield q/ha-

(ii) Price of grain/economic yield - Rs/q-

Therefore, part gross returns is (i) × (ii) = - Rs/ha.

14.4.2.2 ***Price of by-product***

(Rs/q ×total by – product = Rs/ha)

(B) Part gross return is like this –

(i) By-product (q/ha) -

(ii) Price of by-product (Rs./q),

Hence part gross return is (i) × (ii) = - Rs./ha. Add (A) and (B) together to get total 'gross return' (GR) total in Rs/ha.

14.4.3 Net returns (NR) (Rs/ha)

[Net returns (Rs./ha) = Gross returns (total) (Rs./ha) – Cost of cultivation of crops (Rs./ha)].

To get net return (NR) of crop cultivation, cost of cultivation, (13.4.1) should be subtracted from gross return (13.4.2).

[(i.e. Net returns (Rs./ha) = Gross retention (total) (Rs./ha) - cost of cultivation (CCC) (Rs/ha)].

In the similar way, net returns is calculated, in multiple cropping programmes, by adding the total yields (main and by-product) and total price of the crops and also by substracting the total cost of cultivation of crops.

Practical Class : Individual plots

Lessons 159 to 168: For management, '*individual plot*', to be distri-buted to all students in B.Sc. (Ag.) Hons class, with specific crop to each of them in different seasons (the class teacher will allot the plots and assess the work). Rough break-up of practical classes are as follows :

Management of individual plot, with specific crop, under irrigated condition, by the students of B.Sc. (Ag.) Hons.

Break-up of the lessons for individual plot

Lesson 159. Arrangement of inputs like seeds, manures, fertilizers, tillage implements, irrigation, appliances, etc.

Lesson 160. Tillage operation for the specific crop cultivation.

Lesson 161. Sowing the specific crop, with variety, seed rate etc.

Lesson 162. Fertilization in crop : Basal and top-dressing in crop.

Lesson 163. Irrigation in specific crop.

Lesson 164. Weeds and weed-management in crop (like weeding etc.).

Lesson 165. Intercultural operations (like hoeing/stirring, earthing-up etc.) in specific crop.

Lesson 166. Harvesting and processing of specific crop.

Lesson 167. Recording all sorts of data, regarding all operations for crop cultivation in individual plot.

Lesson 168. Production economics of crop/crops.

Practical Class: Calculation of production economics of crop cultivation

(A) For single crop programme:

Lesson 169: Calculate the net-income from the following data, in rupees per hectare

(a) Cost of cultivation of 'A' crop – Rs. 12,000,

(b) Grain and straw yields – 20 and 30 quintal per hectare respectively,

(c) Price of grain and straw are Rs. 1200 and Rs. 100 per quintal respectively.

Net income (Rs./ha) from the farm is, therefore,

Gross income – Cost of cultivation.

Gross income of that crop = Grain (q/ha) × Price of grain Rs/q + straw yield (q/ha) × price of straw rupees/q

= 20 × 1200 + 30 × 100 (Rs./ha) = 24,000 + 3000 (Rs./ha) = 27,000 (Rs/ha)

Hence, gross income of that crop is Rs. 27,000/ha.

Therefore, Net income = Rs. 27,000 – Rs.12,000 per hectare

= Rs. 15,000 per hectare.

(B) For multiple cropping programme:

Lesson 170 : Calculate the net income of crops from one hectare of land (in Rs/ha) under irrigated conditions, in '3–crop system' (with crops A, B and C) from the following data :

Crop (A) :

(i) Cost of cultivation – Rs. 20,000 per ha, (ii) Produce and by-product – 250 and 700 q/ ha respectively, and (iii) Selling price of produce and by-product Rs. 500 and 50/q respectively

Crop (B) :

(i) Cost of cultivation – Rs. 8,000 per ha, (ii) Produce and by-product - 22 and 35 q/ha respectively, and (iii) Selling price of produce and by-product – Rs. 1200 and Rs. 100 q respectively, and

Crop (C) :

(i) Cost of cultivation – Rs. 7000 per ha, (ii) Produce and by-product – 22 and 30 q/ha, respectively, and (iii) Selling price of produce and by-product – Rs. 1200 and Rs. 100 q, respectively.

From Crop A :

(i) Cost of cultivation – Rs. 20,000 per hectare,

(ii) Gross income = 250 q × Rs. 500 q + 700 q × Rs. 50 q = Rs. 1,25,000 + Rs. 35,000 = Rs. 1,60,000.

From Crop B :

(i) Cost of cultivation – Rs.8,000 per hectare,

(ii) Gross income = 22 quintal × Rs. 12,00 per quintal + 35 quinal × Rs.100 per quintal = Rs. 26,400 + Rs. 3,500 = Rs. 29,900.

From Crop C :

(i) Cost of cultivation – Rs. 7,000

(ii) Gross income = 22 q × Rs. 1200 q + 30 q × Rs. 100 q = Rs. 26,400 + Rs. 3000 = Rs. 29,400.

Therefore, cost of cultivation of all crops in multiple cropping programme

= A + B + C = Rs. 20,000 + Rs. 8,000 + Rs. 7,000= Rs. 35,000 for all crops

∴ Cost of cultivation of all crops (A + B + C) = Rs. 35,000 ha.

Gross income of all crops (A + B + C)

= Crop A Rs. 1,60,000 + Crop B Rs. 29,900 + Crop C Rs. 29,400

= Rs. 2,19,300 ha

∴**Net income =** Total Gross income (all crops) – total cost of cultivation (all crops)

= Rs. 2,19,300 – Rs. 35,000 = Rs. 1,84,300 ha.

Therefore, Net income for the '3 – crop system' is Rs. 1,84,300 ha annually.

14.5 SOIL FERTILITY RANKING / GRADING

Soil fertility of any land, is graded/ranked in terms of low, medium and high amount of total nitrogen, available phosphate, available potash, organic carbon contents of soils, along with the soil pH. These are as follows :

14.5.1 Total Nitrogen content in soils % (Average N kg/ha)

Low	Below 0.06	(280)
Medium	0.061 to 0.070	(280 – 450)
High	Above 0.070	(450)

14.5.2 Available phosphate content in soils (kg/ha)

Low	Below 15.0	—	(poor soil)
Medium	15.1 to 20.0	—	(good soil)
High	Above 20.1	—	(good soil)

14.5.3 Available potash content in soils (kg/ha)

Low	Below	150.0	— (poor soil)
Medium	150.1	340	— (good soil)
High	Above	340.1	— (good soil)

14.5.4 Soil pH

Low	3.5 to 5.5	Bad for most of the crops,
Medium	5.6 to 7.5	Good for most of the crops,
High	7.6 to 9.5	Bad for most of the crops.

14.5.5 Organic carbon content of soil (%)

Low	Below 0.05	— (poor soil)
Medium	0.50 – 0.75	— (good soil)
High	Above 0.75	— (good soil with high soil fertility)

14.6 REQUIREMENT OF NPK-FERTILIZERS FOR UNIT AREA OF CROP CULTIVATION FOR HIGHER PRODUCTION

For higher crop production, nutrients from fertilizers are needed very much. Different crops require different amount of fertilizers under relevant situations. NPK-fertilizers are discussed here.

14.6.1 Nitrogenous (N) fertilizers

All crops need nitrogenous fertilizer, though amounts vary from crop-to-crop. Generally cereal crops require high amount of N-fertilizers, followed by oilseed crops, fibre crops etc. but least requirement of nitrogenous fertilizers is for leguminous crop which itself fixes nitrogen from the soils. Wheat and rice the cereal crops, require nitrogen @ 100 kg/ha each, whereas in legumes its requirement is only 40 kg/ha. For vegetative growth of all crops, nitrogen is needed.

14.6.2 Phosphatic (P_2O_5) fertilizers

Though all crops need phosphate, but its requirement is high in legumes. For any reproductive growth or for any flower, development etc. phosphate is needed very much. Less amount of phosphate is required for the growth of cereals (wheat, rice etc.). Greengram (pulse i.e. legume) requires 60-80 kg phosphate. Some tuber crops or oilseed crops require high amount (50-60 kg P_2O_5/ha) of fertilizers.

14.6.3 Potassic (K_2O) fertilizers

Tuber crops/rhizomes/fibre crops etc. require, in general, high amount of potassic fertilizers, whereas in wheat, its requirement is the lowest. Requirement of potash in potato is 80-90 kg/ha, whereas its requirement in wheat is only 40 kg/ha. The requirement of potash is 60-70 in jute. Fibre crops, in general, require high amount of potash.

14.7 YIELD ESTIMATION OF SOME CROPS

Sometimes, it is necessary to estimate yield of crop before its harvest. Sometimes, yield can be estimated with the yield-components before harvesting of the crops. In some cases (jute), yield can be estimated at any time with its biometric observations even in standing crop.

14.7.1 Cereals crop (wheat)

Yields of cereal grains are dependent on the following yield-components : (i) Number of effective tiller (ear head) per unit area (say, per square metre), (ii) Number of grains per ear head and (iii) Grain weight (1000-seed weight in small grains (i.e. wheat) and 100-seed weight in big grains (i.e. maize) of cereals (in gm). Therefore, the grain yield of any cereal crop may be calculated as per the following formula :

$$Y = \frac{h \times gn \times gw}{1,000} \times \frac{10,000}{100 \times 1,000} = \frac{h \times gn \times gw}{10,000}$$

Where, Y = Grain yield of any cereal crop in q/ha,

h = Earhead (effective tillers) number m^2,

gn = Grain number per earhead (i.e. effective tillers), and

gw = Grain weight (g).

If normal wheat is considered, the average effective tillers/m^2, grain number/earhead and grain weight (1000-seeds) will be 400,25 and 40 respectively. Then the grain yield of wheat per hectare will be as follows :

$$Y = \frac{h \times gn \times gw}{10,000} \times \frac{400 \times 25 \times 40}{10,000} \text{ q/ha} = 40 \text{ q/ha (Ans.)}$$

Therefore, in one hectare of land wheat estimated production is 40 q/ha.

14.7.2 Fibre crop (jute for fibre and seed yield)

(a) For fibre yield of jute :

Jute is a bast fibre crop, hence for fibre it can be harvested at any time of its duration. Its fibre production is dependent on (a) age of the crop, (b) height of the crop, (c) basal diameter (i.e. girth) of the crop, (d) population m^2, (e) weather conditions etc. The farmers are sometimes interested to estimate the dry fibre yields of jute before the actual harvest, for catching the market price or for transplantation of the next transplanted *kharif* rice (in West Bengal). The estimation of dry fibre yield of jute in standing crop, is done by the Das formula :

$$Y = 2\pi r \times (h - 20) \times P / 20,$$

Where, Y = Dry fibre yield of jute (in kg/ha),

$2\pi r$ = Basal girth of jute (cm) stem at lower portion but above the soil,

h = Plant height upto smallest leaf seen, from the soil level (cm)

and p = Plant population m^2.

[Ref. Das, N.R. (1968)].

If the basal girth (cm), height (cm) and plant population (per m^2) of jute are 6,300 and 40 respectively, the dry fibre yield of jute in that field will be : $6 \times (300 - 20 \text{ cm}) \times 40/20 = 6 \times 280 \times 2 = 3360$ kg/ha,

i.e 33.60 quintal/ha.

Ans : Therefore, dry fibre yield of jute is 33.60 q/ha.

In the similar way, the fibre yield of mesta, sunnhemp etc. could be estimated.

The stick-yield of jute can also be estimated by multiplying the fibre yield by 2 i.e. jute stick will be, in this case : $33.60 \times 2 = 67.20$ q/ha.

Therefore, the stick yield of jute is 67.20 q/h. In the similar way, the stick yield of sunnhemp and mesta could also be estimated.

(b) For seed-yield of jute

Jute seed-yield can be obtained from the following formula, if (a) plant population/m^2, (b) number of pod/plant, (c) number of seed/pod and (d) 1000- seed weight (g), are known from the normal grown crop. If these are (a) 40, (b) 40, (c) 12 and (d) 5 respectively, the jute seed-yield will be obtained from the formula as follows :

Y = population/m^2 × pod number/plant × seed number/pod × 1000 – seed weight (g) × 10,000 m^2 (1 ha = 10,000 square metre)/1000 x 1000 x 100.

$$= \frac{40 \times 40 \times 12 \times 5 \times 10{,}000}{1000 \times 1000 \times 100} = \frac{96{,}000 \times 10{,}000}{1{,}000 \times 1{,}000 \times 100} = 96/10 = 9.6 \text{ q/ha}$$

Therefore, the seed yield of jute of that field will be 9.6 quintal per hectare.

14.7.3 Yields of oilseeds and pulses

In India, oilseeds and pulses are very important crops. Hence, the yields may be estimated, before harvest, by the formula roughly. In oilseeds and pulses, (a) plant population, (b) number of pods/plant, (c) number of seeds/pod and (d) 1000-seed weigh are very important yield-components. So, for estimating seed-yield, the following formula may be used :

(a) Oilseeds :

$$Y = \frac{p \times pd \times sn \times sw}{10{,}000} \text{ q/ha}$$

Where, Y = Seed-yield (q/ha), p = Plant population/m^2, pd = Pod number/ plant, sn = Seed number/pod, and sw = 1000-seed weight (g).

Under the normal conditions, oilseed crop mustard variety *seeta* (B-85) showed, (i) plant population/m^2, (ii) pod number/plant, (iii) seed number/pod, (iv) 1000-seed weight (g) as 15, 120, 8 and 5 respectively. Therefore the seed- yield of mustard will be:

$p \times pd \times sn \times sw/10{,}000$ q/ha, $= (15 \times 120 \times 8 \times 5)/10{,}000$ q/ha,

$= 72{,}000/10{,}000$ q/ha, 7.2 q/ha.

Therefore, the yield of oilseed crop in that field will be 7.2 q/ha.

(b) Pulses :

Under the normal conditions, the pulse green gram (cv. Pusabaisakhi), gave (i) plant population/m^2, (ii) pod number/plant, (iii) seed number/pod and (iv) 1000-seed weight (g), as 20, 60, 6 and 20 respectively.

Therefore, the seed yield of mungbean "Pusabaisakhi" will be :

$= p \times pd \times sn \times sw/10{,}000$ q/ha, $= (20 \times 60 \times 6 \times 20)/10{,}000$ q/ha,

$= 144{,}000/10{,}000$ q/ha $= 14.4$ q/ha.

Therefore, yield of Pusabaisakhi will be 14.4 q/ha.

14.7.4 Fodder crops [Non-legume (thinnapier) and legume (grasspea)]

(a) Napier grass : In non-legume fodder crop, the entire thin napier, crop at early/growing stage, is harvested green above the ground level. So total plants per square metre and total weight of green matter of that area, is very important.

In thin napier grass, if plant population is 100 per square metre area and the individual plant weight is 200 g., the fodder production per hectare will be :

$$= \frac{\text{Plant population}/m^2 \times \text{plant weight (g)} \times 10{,}000}{10{,}000}$$

$$= \frac{100 \times 200 \times 10{,}000}{1000 \times 100 \times 10} \text{ t/ha} = 200 \text{ tonnes/ha}$$

Therefore, the amount of green fodder of thinnapier grass in that field per hectare area, will be 200 tonnes/ha.

(b) Grasspea (*khesari*) :

In grasspea, if (i) plant population/m^2 (P) is 100, (ii) branch number/ plant, b is 3 and (iii) weight of branch (bw) is 10 g, then green fodder yield of grasspea per hectare will be as follows :

$$Y = \frac{P \times b \times bw \times 10{,}000}{1{,}000 \times 100 \times 10} \text{ t/ha} = \frac{100 \times 3 \times 10 \times 10{,}000}{1000 \times 100 \times 10} \text{ t/ha} = 30 \text{ t/ha}$$

Therefore, green yield of grasspea will be 30 t/ha.

14.8 NO-TILLAGE RELAY (*PAIRA* / *UTERA*) CROPS UNDER RAINFED CONDITIONS IN DIFFERENT SEASONS

The crop, which is sown by broadcasting seeds, over the matured prece-ding crop, under marshy/muddy conditions without tillage, and the said crop which is raised with the residual soil moisture/rainfall, is called the *paira/ utera* (no-tillage relay) crop. These types of crops are grown under rainfed conditions in different season, mostly in those areas where rainfall is high (mostly in the eastern parts of the country).

14.8.1 *Rabi* season (in weedy fields, rice-fallows and riverbeds)

Paira/utera crops may be raised in the rice-fallows, weedy fields and, riverbeds areas, where marshy/muddy conditions prevail. If slight water is there in the field in marshy/muddy conditions, grasspea (*khesari*) seeds could be broadcast over the matured transplanted *kharif* rice, and the said *khesari* could be raised with the soil residual moisture, after the harvest of rice. The *khesari* crop is the *paira* crop and the cropping system with rice, is called the *paira* cropping system. This *paira* crop also could be raised/grown in mesta crop grown for seeds. It could be grown in different season, under different conditions.

In rice-fallows, *paira* crops, could be raised with grasspea, lentil, oat, linseed, mustard, sunnhemp (for seed), coriander, black-cumin etc. in *rabi/* winter season, provided muddy/marshy conditions prevail in the lands where soil should be alluvial in nature.

If some fields are in weedy conditions, sunnhemp (for seeds), *khesari*, pea etc. could be grown as 'no-tillage crop' like *paira* crop in winter, provided land is in marshy/muddy conditions.

Sometimes, after the rainy season, water from the riverbanks recedes and marshy and muddy conditions prevail. Under the conditions, *khesari*, lentil, mustard, sunn-hemp, oat, coriander, black-cumin etc. could be grown in winter (If water recedes earlier and temperature is high, blackgram may be raised for good yields).

14.8.2 Pre-*kharif* season (in field)

Paira / utera crops may also be sown at the time of harvest of irrigated crops of wheat, mustard, potato, *boro* paddy etc. and they may be raised as rainfed crops in eastern parts of the country particularly in West Bengal, as it will be raining in pre-*kharif* season frequently, but with small amount. Blackgram may also be grown as *paira* crop in pre-*kharif* seasons, if its seeds are broadcast over matured *aus* rice.

Broadcasting of 60 kg seeds of green gram/blackgram (for fodder), over the matured *rabi* crops of irrigated wheat, mustard etc. before harvest, followed by harvest of mustard or wheat and 2 ladderings only, give good yields of crops. After the harvest of green gram in pre-*kharif* season, transplanted *kharif* rice may be raised in the same land, paving the 3-crops systems.

14.8.3 *Kharif* season (in fields, home stead & orchard)

In *kharif* season, rainfed ricebean, blackgram, cowpea etc. could be raised as *paira* crop, after pre-*kharif* crops like maize *aus* rice etc., in high rainfall areas, where frequent rainfall will take place. Seeds of *paira* crops, [(double the amount of seed rate) of normal sowing conditions] should be broadcast over the crops just before the harvest of the preceding crops. In home stead and orchard areas, blackgram may be grown.

14.9 CLIMATOLOGICAL DATA AND THE RAINFED CROPS

It is essential to go through the climatological data, for all the seasons of the year, in particular areas, for any crop-production developmental progra-mmes. The main weather parameters are (i) Rainfall, (ii) Temperature and (iii) Relative humidity. There are other minor parameters of weather also that play the vital role of crop production.

14.9.1 Weather parameters [Rainfall, temperature and relative humi-dity of BCKV Farm (Kalyani, WB)] in different seasons

Season & month	Rainfall (mm)	Temperature °C		Relative humidity (%)	
		Maximum	Minimum	Maximum	Minimum
(a) Summer/ pre-*kharif* season :					
April	62.3	36.8	23.9	88.3	43.3
May	102.6	36.4	25.1	87.9	58.0
June	271.3	36.4	25.0	89.5	76.1
July	282.6	32.3	26.7	92.2	81.1
Total	724	-	-	-	-
(b) Rainy / *kharif* season :					
July	282.6	32.3	26.7	92.2	81.1
August	291.1	32.6	24.7	93.2	63.7
September	267.2	32.7	25.5	92.9	76.6
October	114.7	30.2	23.8	90.0	60.7
November	17.2	29.4	16.1	90.8	49.1
Total	975.8	-	-	-	-

(c)	Winter/*rabi* season :				
November	17.2	29.4	16.1	90.8	49.1
December	3.2	25.9	11.2	93.9	52.1
January	7.6	25.4	9.9	63.7	59.1
February	7.9	27.8	13.2	94.0	51.2
March	23.4	32.9	19.9	86.7	41.1
Total	59.3	-	-	-	-
Annual rainfall (mm)	1457	-	-	-	-

From the sub-chapter 14.9.1, it is obvious that summer is not so severe hot. Total rainfall is 724 mm during the period from April to July. During this period maximum temperature is about 37°C, whereas minimum temperature is 24°C and the relative humidity is maximum (92%) in July and minimum (43%) in April. Rainfall, though not so high but is frequent.

Rainy or *kharif* season starts from July and continues upto November. Rainfall is maximum i.e. 976 mm, during the rainy season and maximum temperature (37°C) and minimum temperature (16°C) are not so high, though relative humidity is high (maximum 93% and minimum 49%).

During winter season (November to March), rainfall is very scanty (59 mm). Maximum temperature is 33°C and minimum 9.9°C with low humidity (41.59%).

14.9.2 Rainfed crops in different season

In arid sub-tropics and semi-arid tropics, it is not possible to grow rainfed crop, but it is possible to grow in other 3 zones. [There are 5 ecosystem zones and these are (i) Arid zone, (ii) Semi-arid zone, (iii) Sub-humid zone, (iv) Humid zone and (v) Coastal zone], where Humid zone is the best for rainfed crops in different seasons.

(a) *Rabi* season :

The criteria for rainfed crops should be : (i) Shorter in duration, (ii) Lower in water requirement, (iii) Capable to grow in low moisture regime, (iv) With narrow and small leaves, (v) Responsive to fertilizer with low moisture, (vi) Less sensitivity to tillage, (vii) Less foliage, (viii) Synchronizing flowering system, (ix) Deep roots-system for water extraction, (x) Capable to extract moisture from lower depths, (xi) Harder but robust root-system, (xii) Narrow straw grain ratio and (xiii) High bulking-rate (from flowering to maturity). The following crops could be grown in *rabi* season in 'dry sub-humid', wet sub-humid' and wet-humid regions :

(1) Mustard, (2) Oat, (3) Barley, (4) Linseed, (5) Lentil, (6) Grasspea, (7) Pea, (8) Gram, (9) Safflower, (10) Niger, (11) Wheat, (12) Rape, (13) *Rajma*, (14) Sesame, (15) Coriander, (16) Fenugreek, (17) Blackgram, (18) Sunflower etc.

(b) Pre-*kharif* season :

Wherever pre-*kharif* season exists (mainly in eastern part of the country – wet humid zone), the following rainfed crops could be grown (in pre-*kharif* season) :

(1) Jute, (2) *Aus* rice (direct-seeded), (3) Green gram, (4) Sesame, (5) Cowpea, (6) *Dhaincha* (for green manure), (7) Blackgram (for fodder or green manure), (8) *Amaranthus (data)*, (9) *Baghadhaincha* (for green manure), (10) *Kulatra / Kulthikalai*, (11) Maize (for green-cobs), (12) Napier grass, (13) Dinanath grass (for fodder), (14) *Kaon* (*Setaria italica*), (15) Ricebean, (16) Groundnut, (17) Mesta, (18) Sunnhemp, (19) *Ragi*, (20) Musk melon, (21) Cucumber, (22) *Lady's finger*, (23) Gourd, (24) Sunflower etc.

(c) *Kharif* season :

In most of the areas of the country, *kharif* season prevails, having more rainfall, medium temperature and high relative humidity. Under these conditions the following rainfed crops may be grown :

(1) *Kharif* transplanted rice (*aman*), (2) Direct-seeded rice (*aus*), (3) Jute (for seed), (4) Ricebean (for fodder and seeds), (5) Blackgram, (6) Green gram, (7) Groundnut, (8) Maize (for green-cobs), (9) Cowpea, (10) Sesame, (11) *Dhaincha* (for seeds), (12) *Baghadhaincha* (for seeds), (13) *Kulatra/ Kulthikalai*, (14) Napier grass, (15) Dinanath grass (for fodder), (16) Sunnhemp (for fibre), (17) *Arahar* / Redgram, (18) Gourd, (19) Cucumber, (20) Sunflower, (21) *Jowar*, (22) *Bajra*, (23) Pearl millet, (24) Finger millet etc.

14.10 PREPARATION OF HERBARIUM

Herbarium, the classified collection of preserved plants, is very important in respect of teaching, research and other activities. The different small crops/plants are withered and attached on the special papers with some writings. These are as follows :

14.10.1 Field crops

For big crop plants, the parts like root, leaf, stem, inflorescence, seed etc. and for small crops plants as such, should be withered in the sun and pasted on the special type of paper, along with the writing to be placed in the corner of the album. For field crops, the following should be written on piece of paper, in the right corner below :

(1) Common name of the crops, (2) Scientific name of crops, (3) Date of collection, (4) Place of collection, (5) Types of crop, with the family and (6) Use of crop, if known and (6) Other information (if any).

14.10.2 Weeds of different seasons

Like field crops, weeds of big size, only the root, leaf, stem and inflorescence should be preserved in the album with paste, after partial withering in the sun for some days. Small weeds, of course, may be kept on the album, as such. In the right hand (below) corner of the album, in a piece of paper the following should be written :

(1) Common name of weeds, (2) Scientific names, (3) Date of collection, (4) Place of collection, (5) Family (if known) and (6) Other information (if any). The writing should be like this :

1.	*Common name*	:	5.	Name of family	:
2.	Scientific name	:	6.	Other information (if any)	:
3.	Date of collection	:	7.	Use (if any)	:
4.	Place of collection	:			

14.10.3 Fodder crops (legumes and non-legumes)

Like field crops, the fodder crops (both legumes and non-legumes), also the roots, leaves, stems and inflorescences should be preserved in the album with paste after withering in the sun for some days. In the corner (right hand corner, below) of the album the following should be written on a piece of paper and pasted :

(1) Common names of the fodders, (2) Scientific names of the fodders, (3) Date of collection, (4) Place of collection, (5) Types of crop (legume/non legume), (6) Whether annual/perennial, (7) Cutting or non-cutting and (8) Their families (if known). Besides these, the feeding technique (to the cattle) to be written also.

14.11 MAINTENANCE OF SMALL GLASS JARS

Maintaining the glass/plastic jars in the laboratories/ field, for keeping (i) different samples of seeds, (ii) fertilizers/manures, (iii) soil samples and (iv) other purposes (like plant and grain ground samples) for showing to the learners or for practical analysis of plants, grain, soil and fertilizer manures etc.

14.11.1. Keeping different types of seeds

The big or small jars (mainly glass/plastic or ebonite etc.) are generally maintained in the agronomy laboratories or in the field-office of the farms, for keeping some seed samples of different types of crops. These seed-samples are shown to the learners, in their practical classes.

14.11.2 Keeping manures and fertilizers

In the similar way (as above), big or small plastic, glass or ebonite jars are maintained in agronomy laboratories or in the field office of the farms, for keeping some manures and fertilizers of different nutrients, to show the learners/students in their practical classes. Of course, for fertilizers, 'glass jars' are better, because of chemical changes.

14.11.3 For keeping different types and categories of soils

Different types of soils, collected from different places, are preserved in small or big containers/jars, mostly made of glass, with fully airtight lids, either to show to the learners or to analyze, in future, for nitrogen, phosphate, potash, soil pH etc.

14.11.4 For other purposes (research materials)

Sometimes, plant materials like grains, straws, total plant materials, soils etc. are preserved in the glass jars, after putting them into paper packets, giving treatment number, by putting them with the airtight lids, for research work.

14.12 METEOROLOGICAL OBSERVATORY EQUIPMENTS (ORDINARY OBSERVATORY)

The following are the required instruments, for ordinary observatory

(a) Essential Instruments

(1) Dry bulb thermometer, (2) Wet bulb thermometer, (3) Maximum thermometer, (4) Minimum thermometer, (5) Single stevension screen, (6) Raingauge, (7) Soil thermometer, (8) Wind vane, (9) Anemometer, (10) Class A open-pan-evaporimeter, (11) Assmann psychrometer, (12) Soil moisture equipment.

(b) Optional Instruments

(1) Sunshine recorder, (2) Dewgauge, (3) Self recording raingauge, (4) Thermograph, (5) Air hygrograph, (6) Double stevension screen, (7) Gross minimum thermometer, (8) Ground surface thermometer, (9) Rapid drying infrared moisture balance meter, (10) Net Radio-meter, (11) Pyrheliometer, (12) Pyrgeometer.

14.13 VISIT TO DIFFERENT FARMS

14.13.1 State Govt. Farms

State Govt. have different farms, starting from the district levels to sub-divisional level, having different areas and different facilities. The sub-divisional farms has only 25 acre of land only. Maximum facilities are available in state level farms, having big areas. State Govt. farms are maintained for production of seeds and general cultivations of crops is done for some varieties.

14.13.2 University teaching farms

The University farms are mainly for teaching and research activities, hence facilities are more, having all types of implements and arrangements. All the big areas are well laid out; growing and experimenting all types of crops grown in the localities, besides museum plots.

14.13.3 ICAR Research Institutes (of the states concern) farms

The farms under ICAR research institutes, are fully equipped with modern facilities and implements and meant for research, based on problems in the localities. More emphasis in given on local crop-developments.

14.13.4 Farmers' fields

The farmers fields/farms, are mostly devoid of modern equipments/ implements. Crops are cultivated with old methods in most of the localities. In some areas, of course, good implements are visible. Now-a-days the farmers, of course, grow crops with scientific methods.

14.14 ALLOTMENTS OF INDIVIDUAL PLOTS FOR THE STUDENTS/ LEARNERS

14.14.1 Distribution of plots

The individual plots are distributed to the students of B.Sc. (Ag.) 2nd or 3rd year class, for raising particular crop in the season. The plots should have all facilities. The plot should not be too small and too big. The teachers should guide them for raising crops scientifically.

14.14.2 Cultivation of specified crops in different seasons

The crops for individual plots, should be based on localities under different seasons. All the selections should be done by the concerned teachers, assigned to individual plots.

14.14.3 Keeping records of seeds, tillage, inter-tillage operations, irrigation, fertilization, harvesting etc.

The students should keep all the following records in respect of individual crop-plot. It will change according to distribution of crops :

(1) Seed, (2) Tillage, (3) Intercultural-operation, (4) Irrigation, (5) Fertilization, (6) Harvesting, (7) Threshing, (8) Other processing etc.

14.14.4 Record maintenance for grain and fibre or other yields and straw/stover yields of the crops cultivated in individual plots

Maintenance of records, on yields (grain, fibre or straw or stick) are as follows :

Crop (s) (Name)	Yields (q/ha)	
	Grain or fibre	Straw or stick

14.14.5 Market value of produce and by-products

The maintenance of records for market value of produce (grain or fibre) and by-products (straw or stick) will be like this :

Crop grown (Name)	Market price of produce and by-product	
	Grain or fibre –(Rs./q)	Straw or stick (Rs./q)

14.15 PRODUCTION ECONOMICS OF INDIVIDUAL CROP CULTIVATION (individual crops and cropping systems)

For production economics of crops or cropping systems should be maintained, based on the following aspects:

14.15.1 Cost of cultivation

Cost of cultivation should be maintained for individual plot, for calculation of production economics.

14.15.2 Gross returns for single and multiple crop(s)

Gross returns for crop(s) should be calculated according to market price of the locality.

14.15.3 Net returns for crop(s) of individual plot

Net returns should be calculated as follows :

Net returns (Rs/ha) = Gross return (for crop) – Cost of cultivation (of crop).

LIST OF SKETCHES OF (A) TILLAGE IMPLEMENTS, (B) CROPS, (C) WEEDS AND (D) OTHERS

A. Tillage implements :

Figure No. 1. *Deshi* plough and yoke; 2. Mouldboard plough; 3. Jute seeddrill; 4. Seedrill (3-row); 5. Wheel hoe; 6. Paddy weeder; 7. *Khurpi*; 8. Sickle; 9. *Hasua*; 10. *Da*; 11. *Nirani*; 12. Spade; 13. Tine; 14. Fork; 15. *Khanta*; 16. Rake; 17. Tractor; 18. Small plough (for potato harvest); 19. Ladder; 20. Wooden hammer (*itamugur*).

B. Crops :

Figure No. 1. Maize; 2. Wheat; 3. Rice; 4. Sesame; 5. Groundnut; 6. Mustard; 7. Linseed; 8. Sunflower; 9. Safflower; 10. Gram; 11. Lentil; 12. *Pigeon Pea*; 13. Pea; 14. Soybean; 15. Cowpea; 16. *Khesari*; 17. Blackgram; 18. Greengram; 19. Jute–*Capsularis;* 20. Jute – *Olitorius*; 21. Potato; 22. Sugarcane; 23. Paragass; 24. Oat; 25. Lucern; 26. Teosinte; 27. Berseem.

C. Weeds :

Figure No. 1. *Cynodon dactylon*; 2. *Ipomoea disitata*; 3. *Mikania micrantha*; 4. *Centella asiatica;* 5. *Solanum nigram*; 6. *Melilotus indica;* 7. *Physalis minima*; 8. *Phalaris minor;* 9. *Lolium temulentum;* 10. *Vicia hirsuta;* 11. *Vicia sativa;* 12. *Asphodelus tenuifolia;* 13. *Lathyrus aphaca;* 14. *Echinochloa crusgalli;* 15. *Eclipta alba;* 16. *Euphorbia hirta;* 17. *Lantana camera;* 18. *Cassia tora;* 19. *Convolvulus arvensis;* 20. *Phyllanthus niruri;* 21. *Imperata cylindrical;* 22. *Eleusine indica;* 23. *Leucas aspera;* 24. *Cyperus rotundus;* 25. *Alternanthera sessilis;* 26. *Amaranthus spinosus;* 27. *Sporobolus diander;* 28. *Commelina benghalensis;* 29. *Croton sparsiflorus;* 30. *Argemone mexicana;* 31. *Fussiaea repens;* 32. *Fussiaea suffruticosa;* 33. *Portulaca oleracea;* 34. *Mimosa pudica;* 35. *Chenopodium album;* 36. *Parthenium hysterophorus*.

D. Others :

Figure No.1. Spring balance; 2. Basket; 3. Bowl; 4. Sack; 5. Peg; 6. Scale; 7. Tape; 8. Sprayer, 9. Measuring cylinder; 10. Note book; 11. Pen; 12. Pencil; 13. Album; 14. Gum; 15. Scissor; 16. Correction pen; 17. Quadrant; 18. Crop-beating stick; 19. Puddling; 20. Soil aggregates; 21. Rope/string; 22. Setts; 23. Cuttings; 24. Propagules; 25. Raising seedlings; 26. Gypsum blocks; 27. Rubber.

IMPORTANT INFORMATION

I. ABILITY AND DISABILITY OF GUIDING FERTILIZER MIXTURES

FERTILIZER NUMBERS

FERTILIZER NUMBERS	1	2	3	4	5	6	7	8	9	10	11	12	13	14	15	16	17
1	□	▥	▥	▥	▥	▥	▥	■	▥	▥	■	■	▥	▥	■	▥	▥
2	▥	□	▥	□	□	□	▥	▥	□	□	▥	▥	□	□	□	□	▥
3	▥	▥	□	▥	□	▥	▥	■	■	▥	■	■	■	■	□	▥	□
4	▥	□	▥	□	▥	□	▥	■	■	□	▥	▥	■	■	▥	□	▥
5	▥	□	□	▥	□	▥	▥	□	■	□	▥	▥	■	■	□	□	▥
6	▥	□	▥	□	▥	□	□	▥	■	□	□	□	■	■	□	□	□
7	▥	▥	▥	▥	▥	□	□	▥	■	▥	□	□	■	■	▥	□	□
8	■	▥	■	■	■	▥	▥	□	□	▥	■	■	▥	▥	▥	□	□
9	▥	□	■	■	■	■	■	□	□	■	■	■	□	□	▥	□	□
10	▥	□	□	□	□	□	▥	▥	■	□	□	□	■	■	□	□	▥
11	■	▥	■	▥	▥	□	□	■	■	□	□	□	■	■	□	□	□
12	■	▥	■	▥	▥	□	□	■	■	□	□	□	■	■	□	□	□
13	▥	□	■	■	■	■	■	▥	□	■	■	■	□	□	□	□	□
14	▥	□	■	■	■	■	■	▥	□	■	■	■	□	□	▥	▥	▥
15	■	□	□	▥	□	□	▥	▥	▥	□	□	□	□	▥	□	□	□
16	▥	□	▥	□	□	□	□	□	□	□	□	□	□	▥	□	□	□
17	▥	▥	▥	▥	▥	□	□	□	□	▥	□	□	□	▥	□	□	□

Note : (i) □ = Fertilizers which can be mixed together & stored; (ii) ▥ = Fertilizers which can be mixed but not stored longer than 2-3 days; (iii) ■ = Fertilizers which can not be mixed together at all, because of chemical reaction.

LEGEND (Fertilizer numbers) : (1) Calcium nitrate, (2) Chilean nitrate, (3) Calcium ammonium nitrate (Lime stone amm. nitrate), (4) Ammonium sulphate nitrate, (5) Nitropotash (Potash) ammonium nitrate, (6) Sulphate of ammonia, (7) Nitrogen magnesia, (8) Urea, (9) Calcium cynamide, (10) Diammonium phosphate, (11) Superphosphate, (12) Triple superphosphate, (13) Basic slag, (14) Rock phosphate, (15) Muriate of potash, (16) Sulphate of potash, (17) Sulphate of potash – magnesia.

II. MORPHOLOGICAL CHARACTERISTICS OF SOME MAIN CROPS

	Crops (1)	Roots (2)	Stems (3)	Leaves (4)	Flowers (5)	Tallness (6)
(a)	**Cereal crops** (Rice, wheat, maize and barley)					
(i)	**Rice** (all seasons)	Fibrous	Hollow	Rough, long leaves	Panicle/ spikelets	Short
(ii)	**Wheat** (winter)	Fibrous	Hollow	Rough long leaves	Earhead, spikelets	Short
(iii)	**Maize** (all seasons)	Fibrous	Solid	Rough, long wide leaves	Cobs/silks	Tall
(iv)	**Barley** (winter)	Fibrous	Hollow	Rough, long leaves	Earhead/spikelets	Short
(b)	**Pulse crops** (Pea, lentil, *Pigeon pea*, gram and *Green gram*)					
(i)	**Pea** (winter)	Tap root	Solid	Lobed	Pods	Trailing
(ii)	**Lentil** (winter)	Tap root	Solid	Small lobes	Pods	Trailing
(iii)	***Pigeon pea*** (winter)	Tap root	Solid	Lobed	Pods	Tall
(iv)	**Gram** (winter)	Tap root	Solid	Multi lobed	Pods	Straight
(v)	***Greengram*** (pre-*kharif*)	Tap root	Solid	Trilobed	Pods	Small
(c)	**Oilseed crops** (Mustard, sunflower, groundnut, and sesame)					
(i)	**Mustard** (winter)	Tap root	Solid /pith	Serrated leaves	Siliqua	Tall/short
(ii)	**Sunflower** (all season)	Tap root	Solid	Soft leaves	Head	Tall/short
(iii)	**Groundnut** (*kharif*)	Tap root	Solid	Lobed	Pod (in soil)	Vine
(iv)	**Sesame** (pre-*kharif*)	Tap root	Solid	Wide leaves	Pods	Tall/short
(d)	**Fibre crops** [Jute (2), mesta and sunnhemp]					
(i)	**Titapat** (pre-*kharif*) (Bitter jute, *Capsularis* jute)	Tap root	Longstem with fibre	Serrated leaves	Pod (capsule)	Tall
(ii)	**Mithapat** (pre-*kharif*) (sweet jute, *Olitorius* jute)	Tap root	Longstem with fibre	Serrated leaves	Long pod	Tall
(iii)	**Mesta** (pre-*kharif*)	Tap root	Solid stem with fibre	Serrated, wide big leaves	Capsules	Tall
(iv)	**Sunnhemp** (pre-*kharif*)	Tap root	Solid (fibre)	Lobed leaves	Pods	Tall
(e)	**Sugar crops** (Sugarcane and sugar beet)					
(i)	**Sugarcane** (all seasons)	Fibrous roots	Solid stem	Rough, long leaves	Inflorescence	Tall plant
(ii)	**Sugar beet** (winter)	Tap root	No stem, but soft mass	Big broad leaves	Long inflorescence	Small height
(f)	**Tuber crops** (Potato and *amorphophallus / ol*) :					
(i)	**Potato** (winter)	Tap root	Solid but soft	Lobed leaves	Big flower, small seeds	Vine
(ii)	Amorphophallus/ol	Root from stem	Pseudo Stem (corm) soft	Leaves big variegated with long stalk	Inflorescence	Medium high

III. SKETCHES OF (A) TILLAGE IMPLEMENTS, (B) CROPS, (C) WEEDS AND (D) OTHERS.

A. Tillage implements:

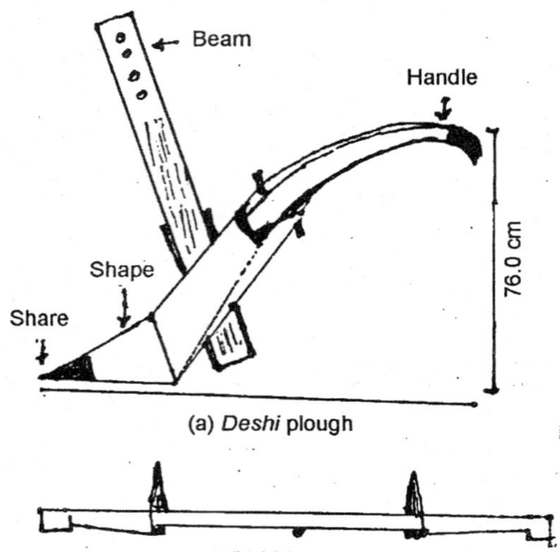

(a) *Deshi* plough

(b) Yoke

Fig. 1. *Deshi* plough & yoke

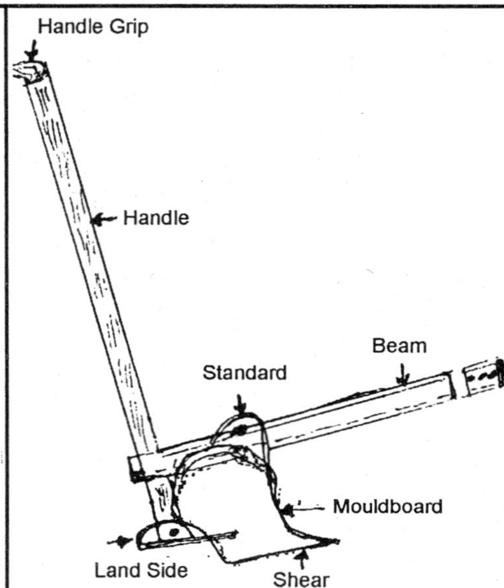

2. Mouldboard plough

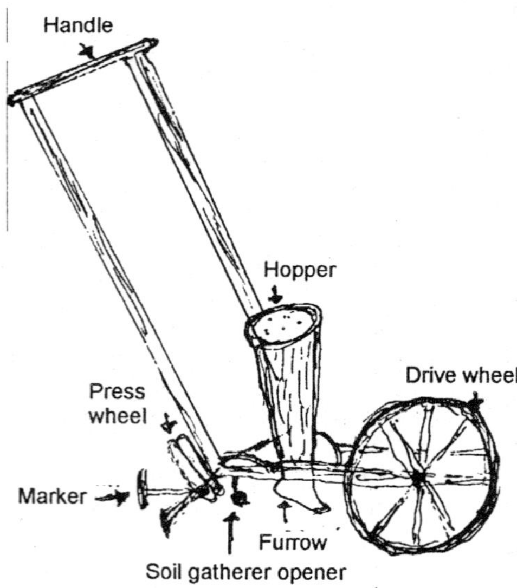

3. Jute seed drill

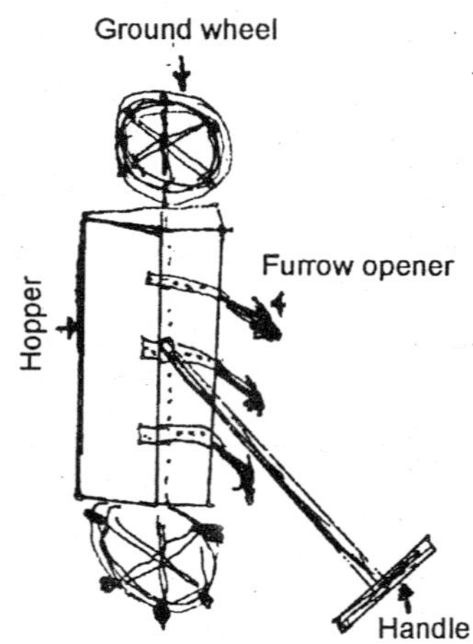

4. Seed drill (3-row)

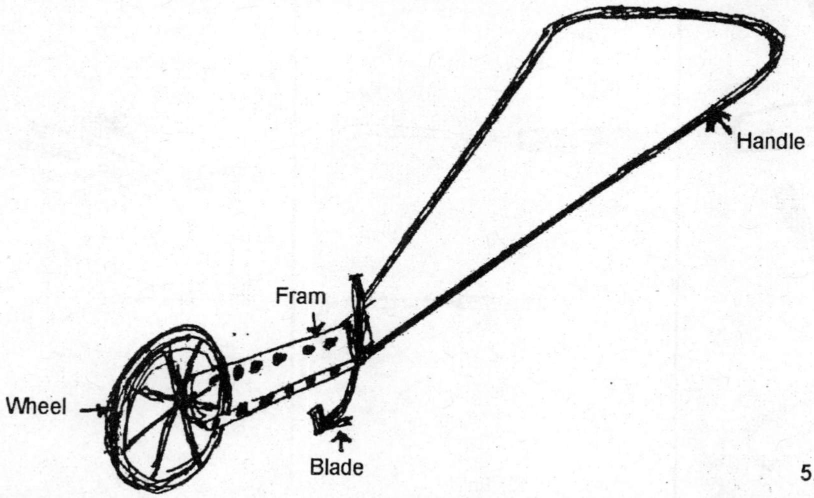

5. Wheel hoe

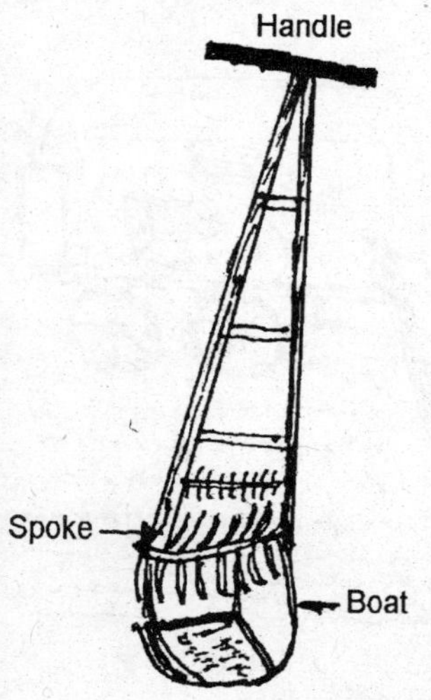

6. Paddy weeder

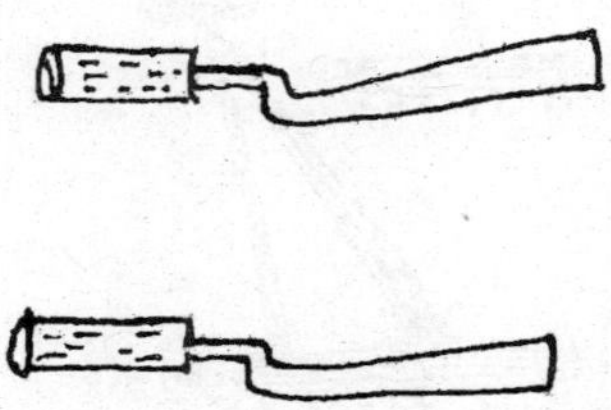

7. *Khurpi*

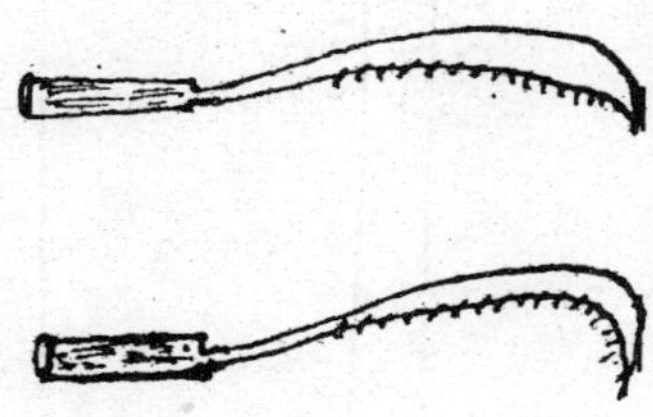

8. Sickle

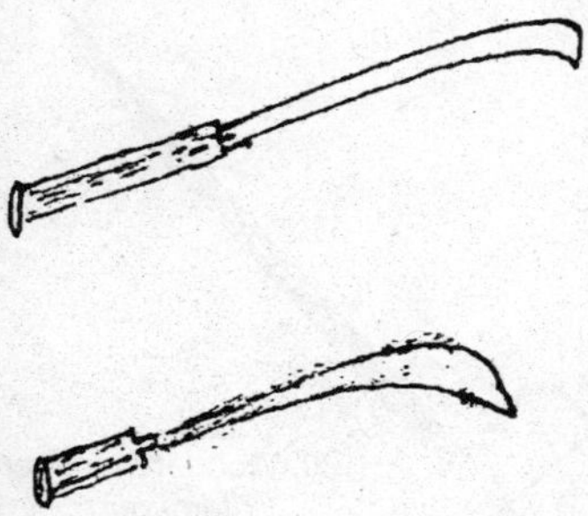

9. *Hasua*

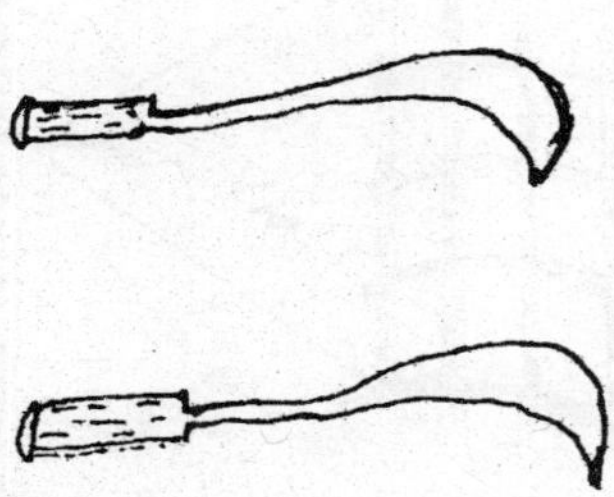

10. *Da*

11. *Nirani*

12. Spade

13. Tine

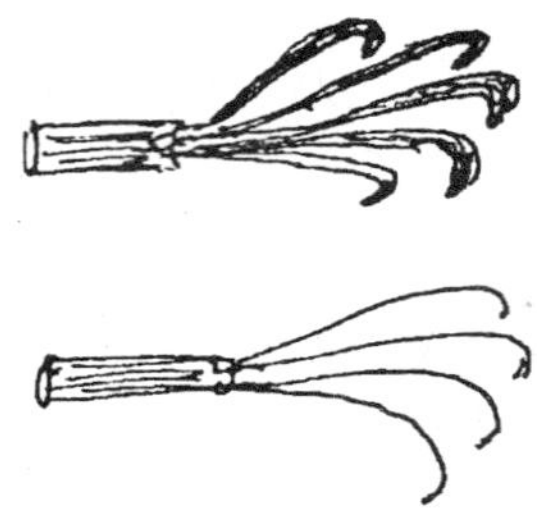

14. Fork

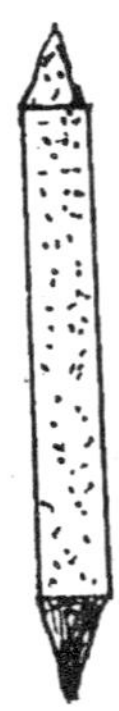

15. *Khanta*

16. Rake

17. Tractor

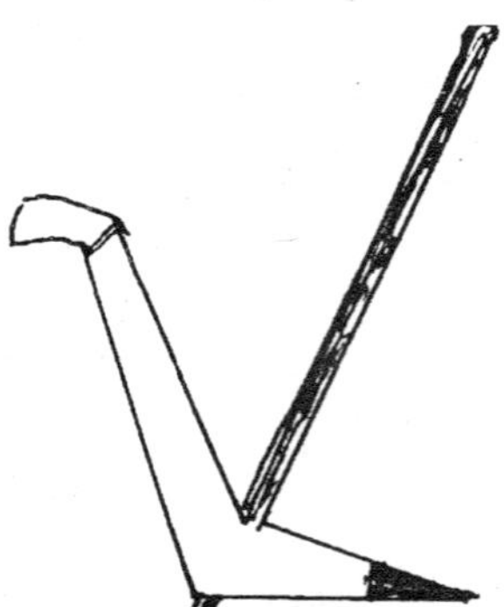

18. Small plough

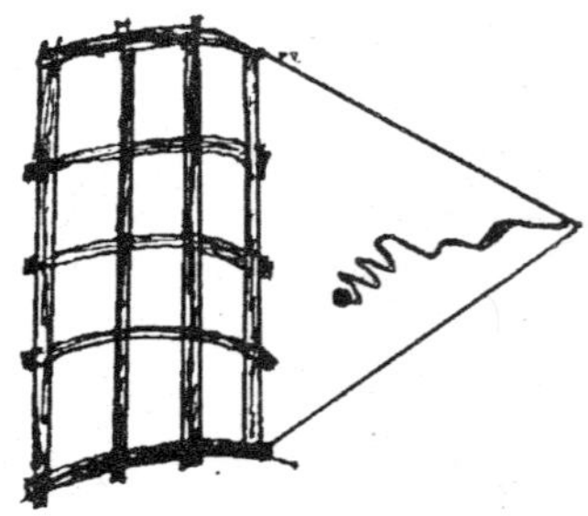

19. Ladder

20. Wooden hammer *(Itamugur)*

B. Crops

Fig. 1. Maize

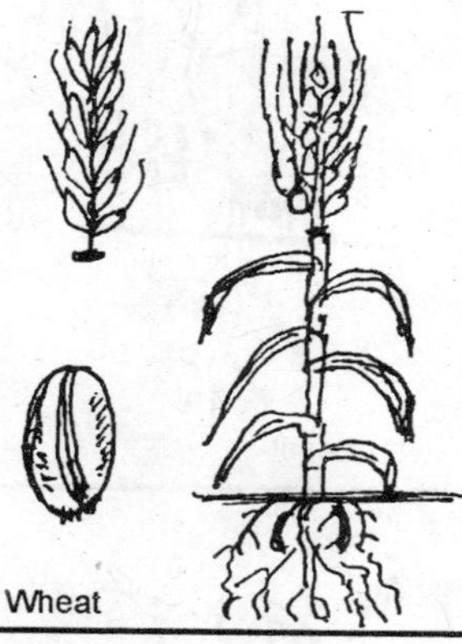
2. Wheat

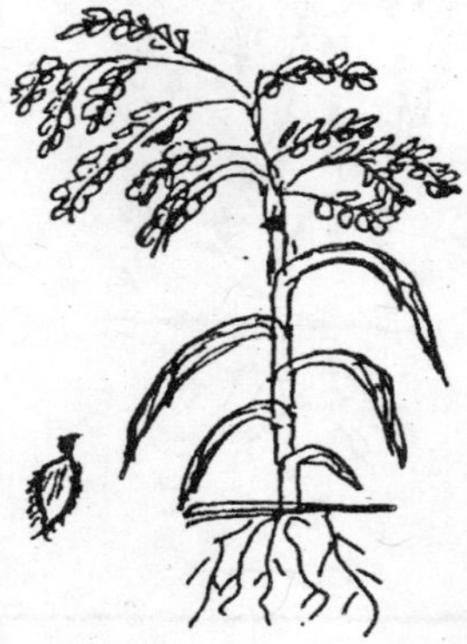
3. Rice

4. Sesame

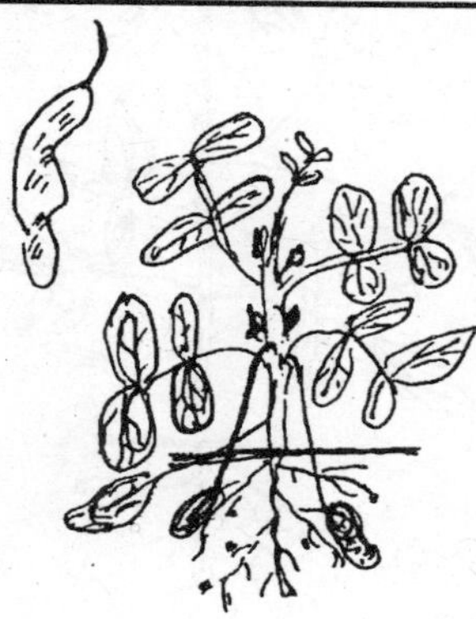
5. Groundnut

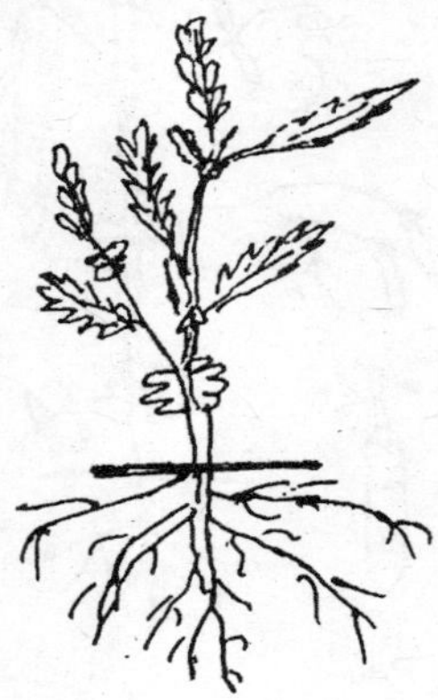
6. Mustard

7. Linseed

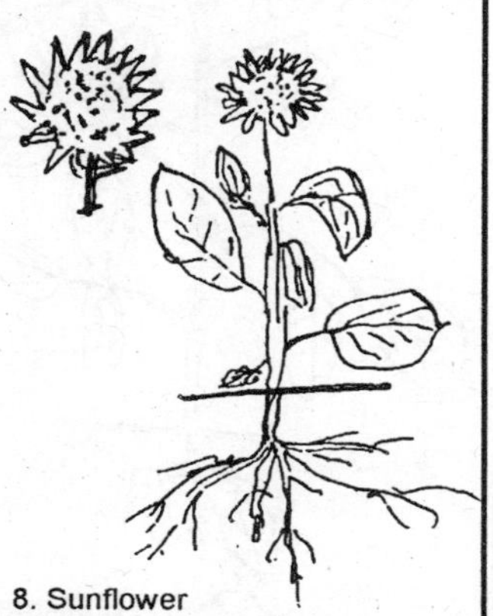
8. Sunflower

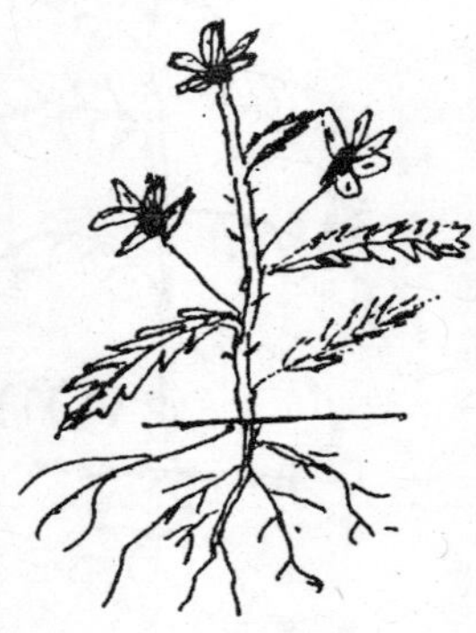
9. Safflower

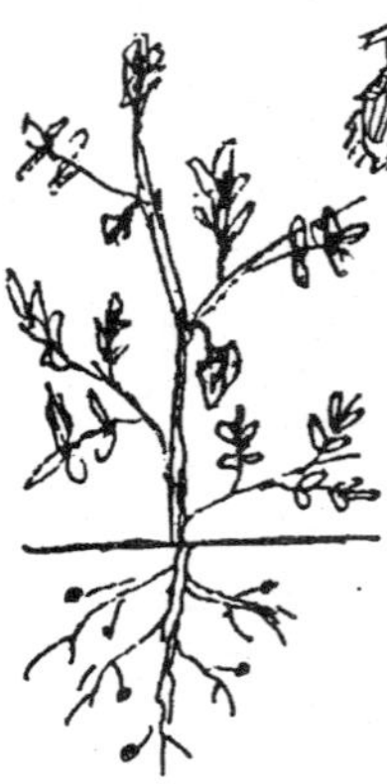

10. Gram

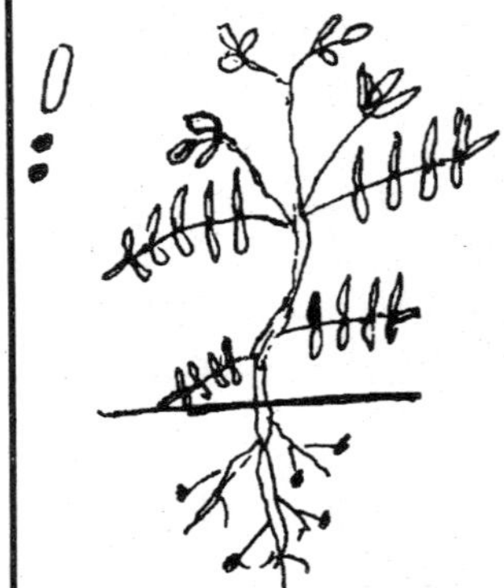

11. Lentil

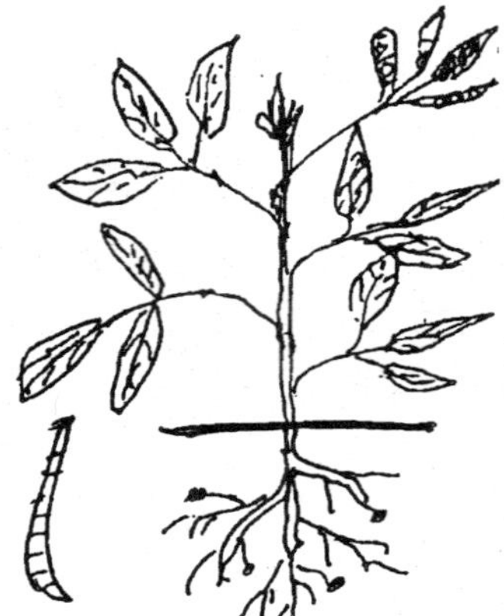

12. *Arhar*

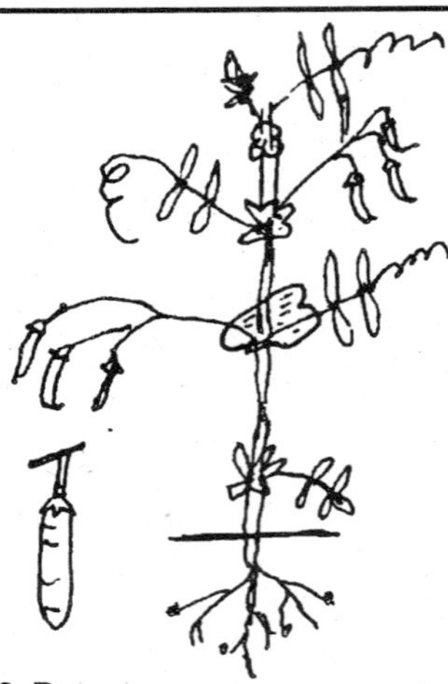

13. Pea

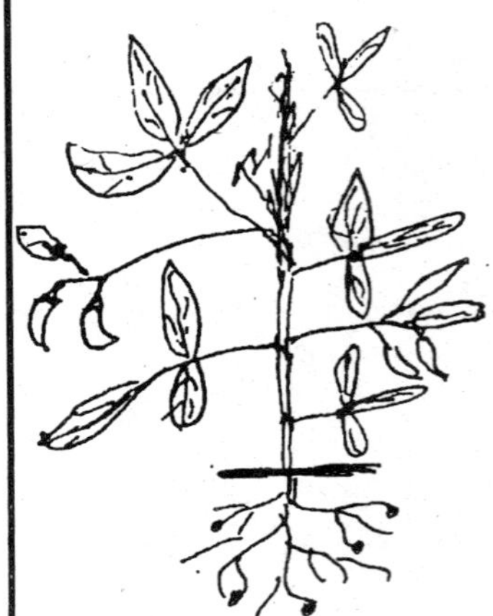

14. Soybean

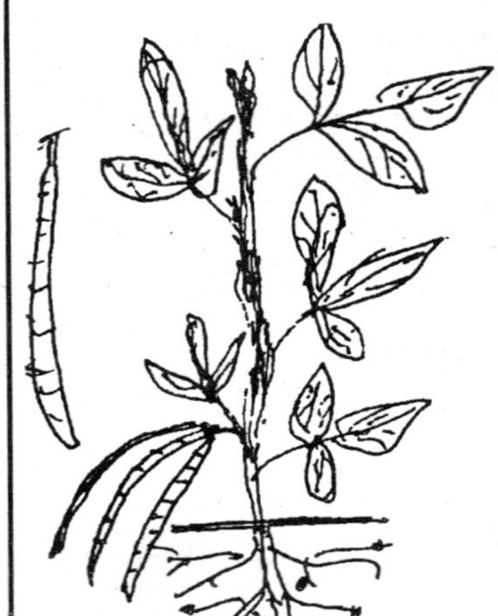

15. Cowpea

16. *Khesari*

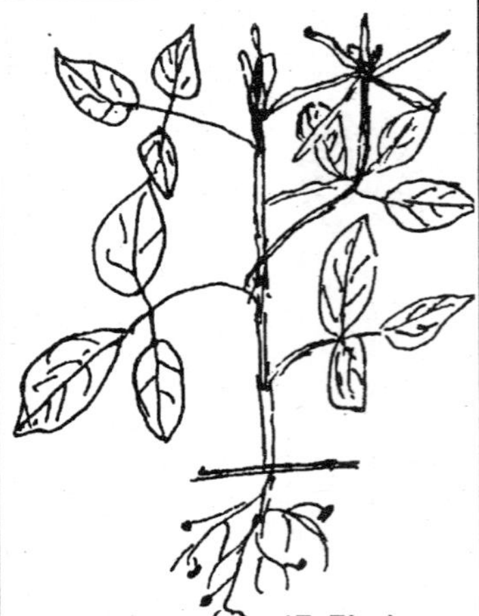

17. Blackgram

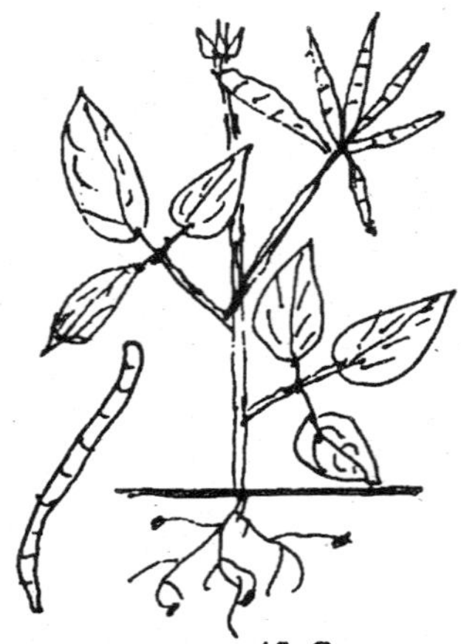

18. Green gram

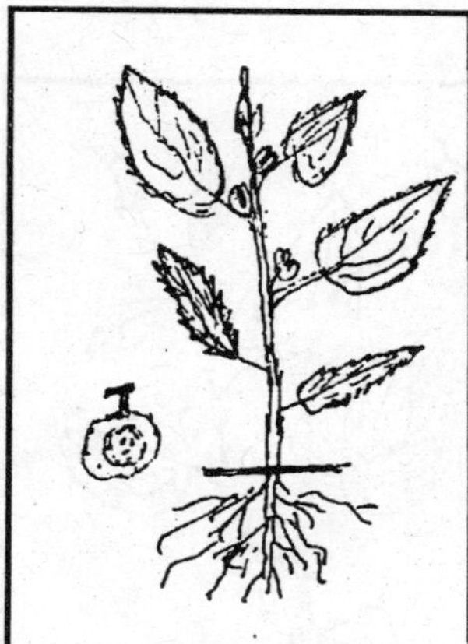

19. Jute- *Capsularis*

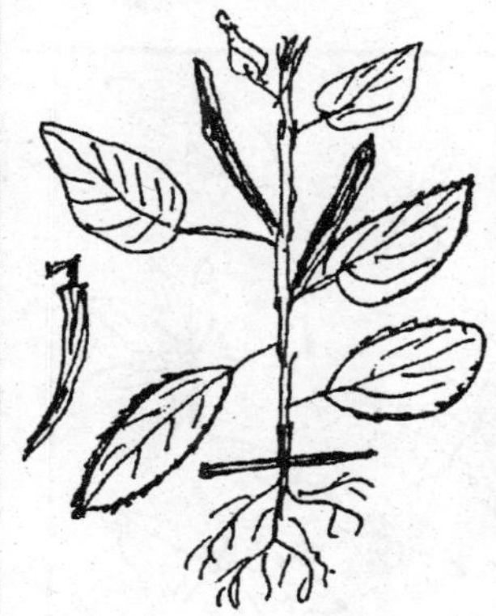

20. Jute - *Olitorius*

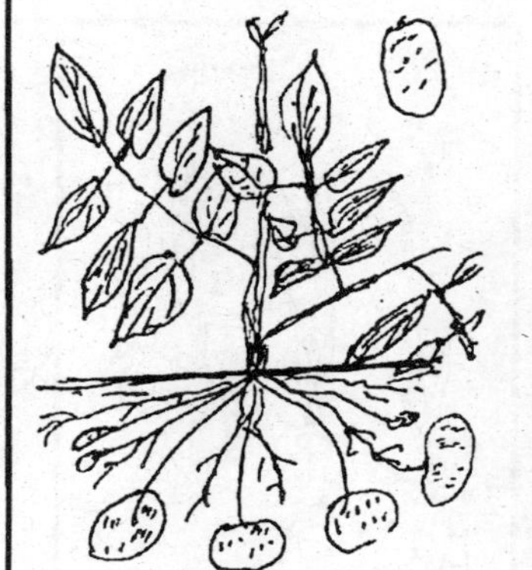

21. Potato

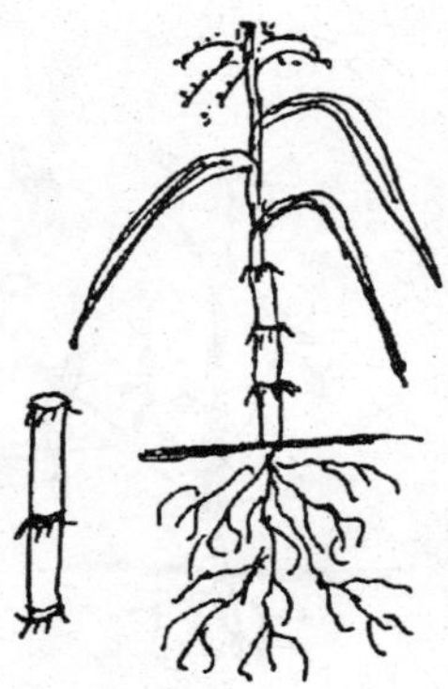

22. Sugarcane

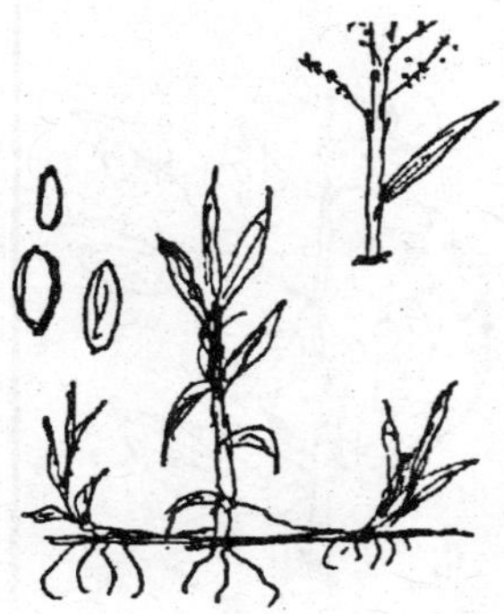

23. Para grass

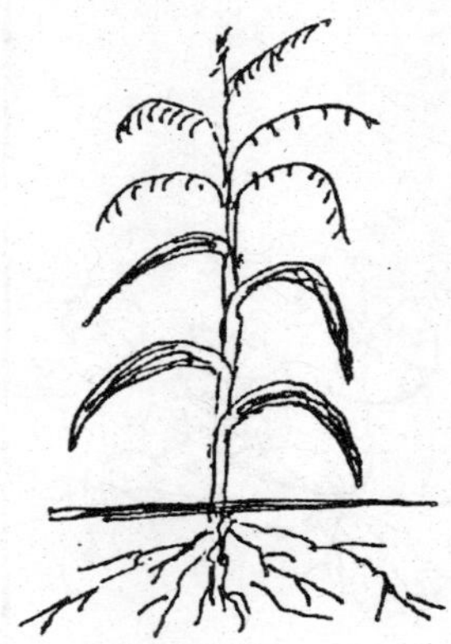

24. Oat

25. Luncern

26. Teosinte

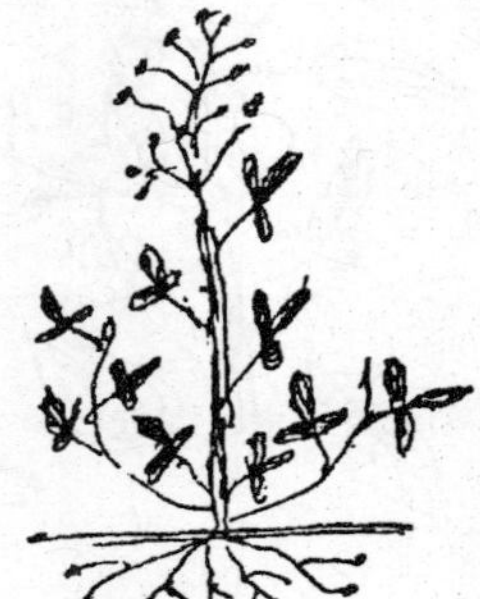

27. Berseem

C. Weeds

Fig. 1. *Cynodon dactylon*

2. *Ipomoea disitata*

3. *Mikania micrantha*

4. *Centella asiatica*

5. *Solanum nigram*

6. *Melilotus indica*

7. *Physalis minima*

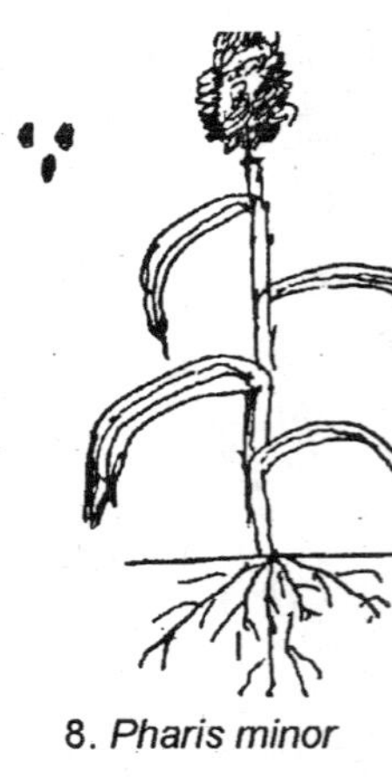

8. *Pharis minor*

9. *Lolium temulentum*

10. *Vicia hirsuta*

11. *Vicia sativa*

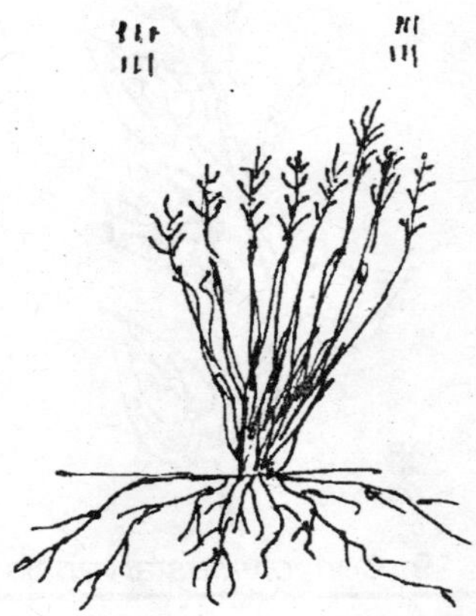

12. *Asphodelus tenuifolia*

13. *Lathyrus aphaca*

14. *Echinochloa crusgalli*

15. *Eclipta alba alba*

16. *Euphorbia hirta*

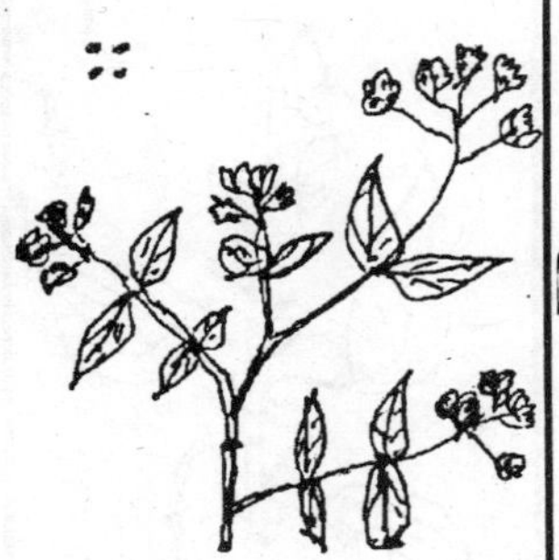

17. *Lantana camera*

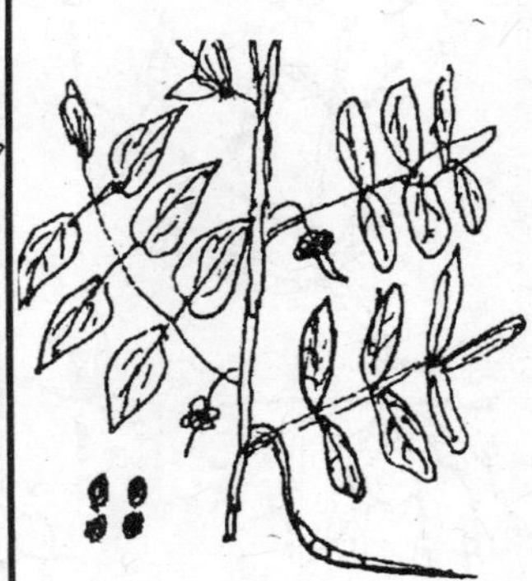

18. *Cassia tora*

19. *Convolvulus arvensis*

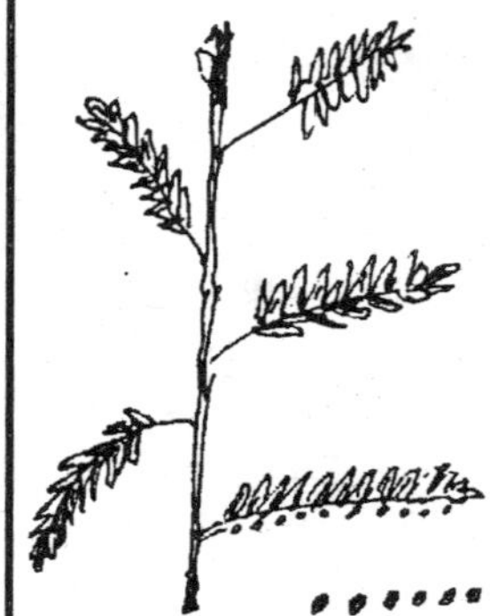

20. *Phyllanthus niruri*

21. *Imperata cylindrica*

22. *Eleusine indica*

23. *Leucas aspera*

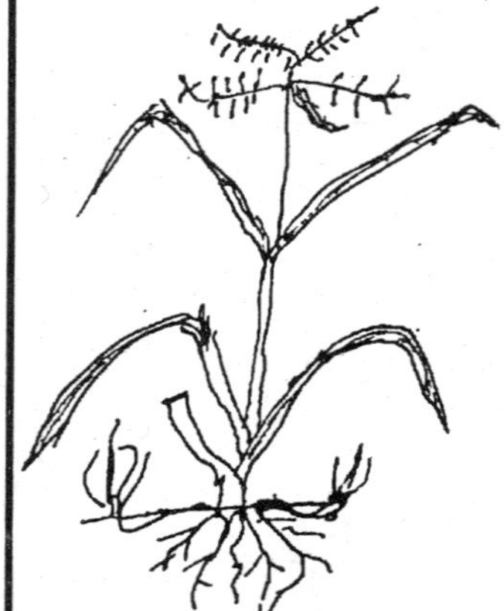

24. *Cyperus rotundus*

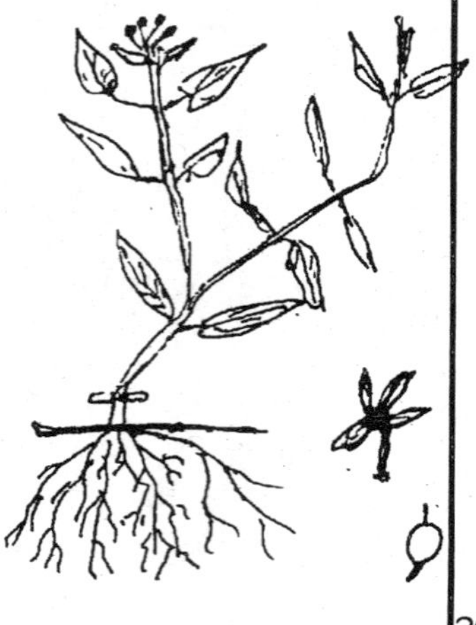

25. *Alternanthera sessilis*

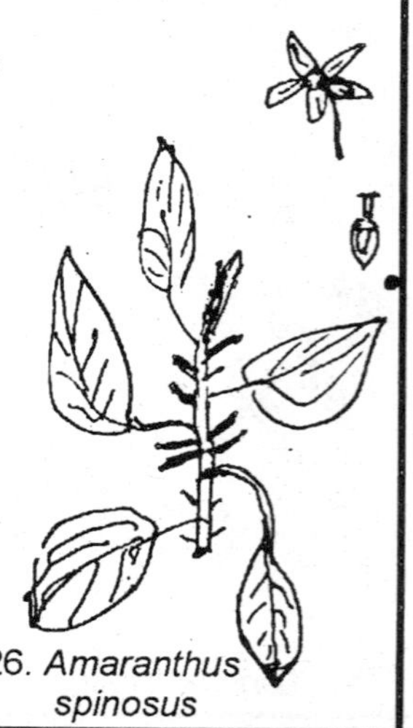

26. *Amaranthus spinosus*

27. *Sporobolus diander*

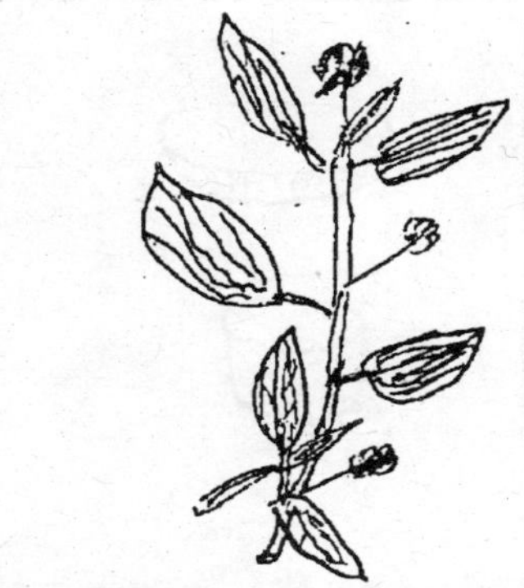

28. *Commelina benghalensis*

29. *Croton sparsiflorus*

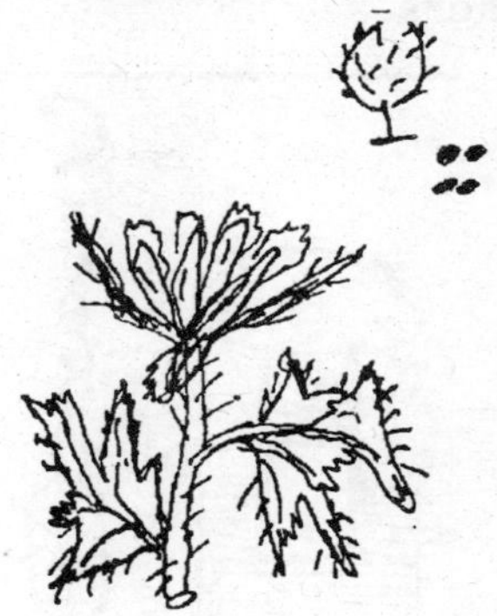

30. *Argemone mexicana*

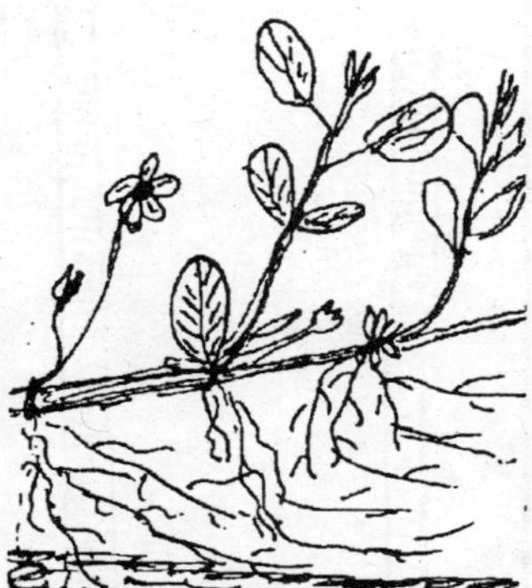

31. *Fussiaea repens*

32. *Fussiaea suffruticosa*

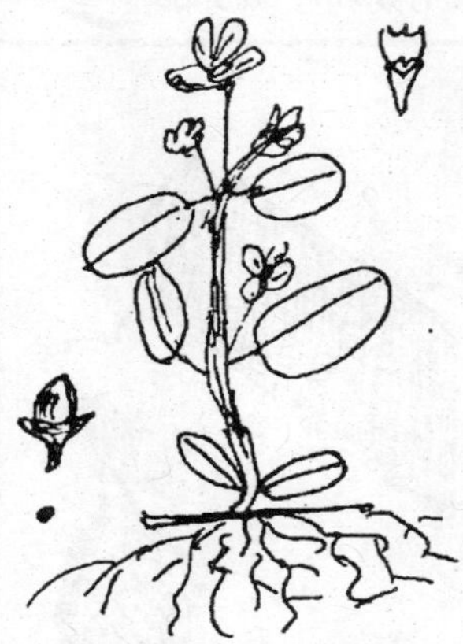

33. *Portulaca oleracea*

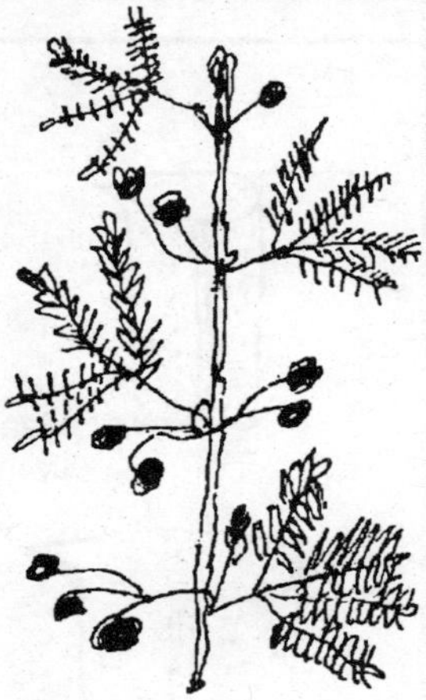

34. *Mimosa pudica*

35. *Chenopodium album*

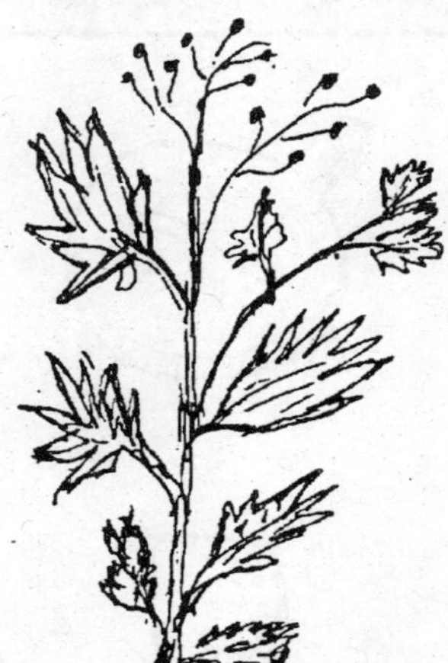

36. *Parthenium hysterophorus*

D. Others

Fig. 1. Spring balance

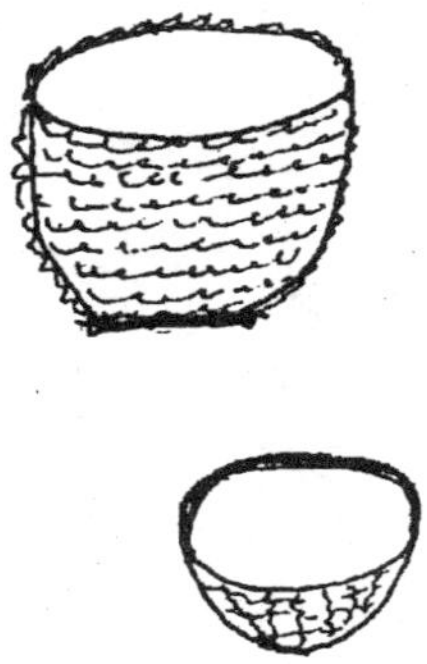

2. Basket

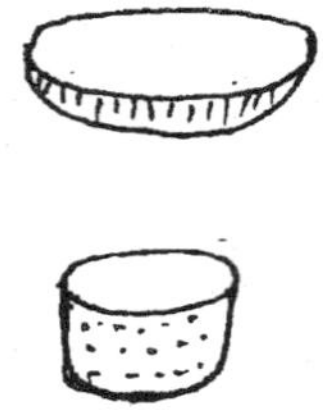

3. Bowl

4. Sack

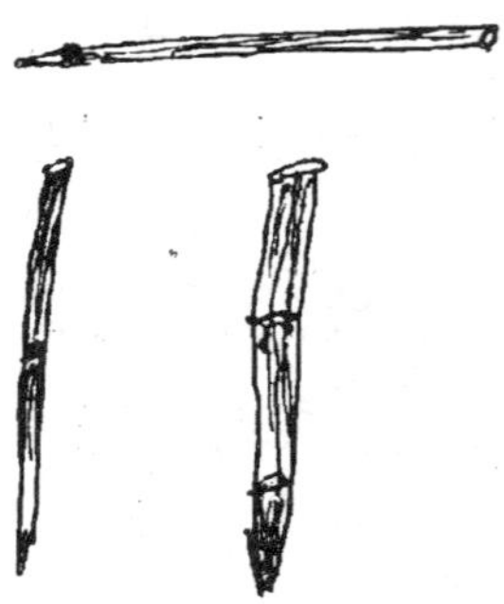

5. Peg

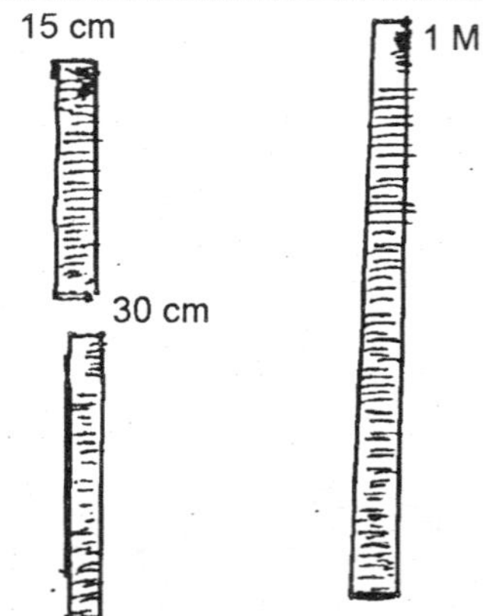

6. Scale

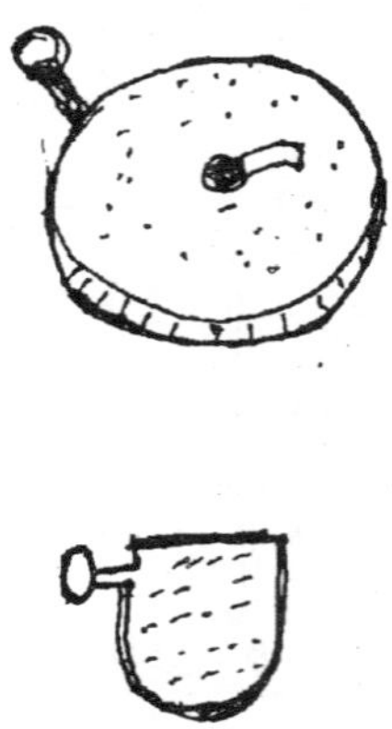

7. Tape

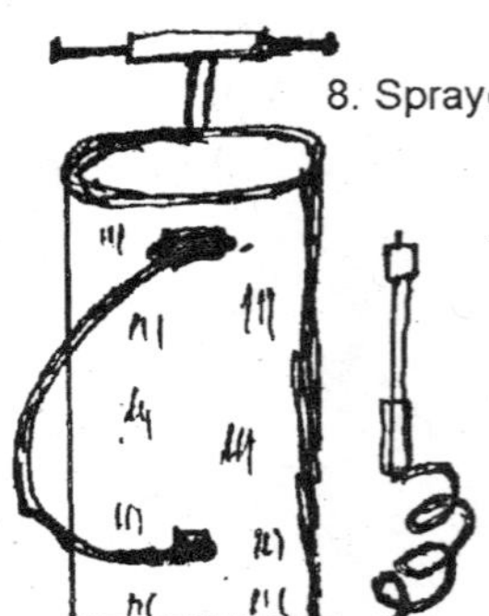

8. Sprayer

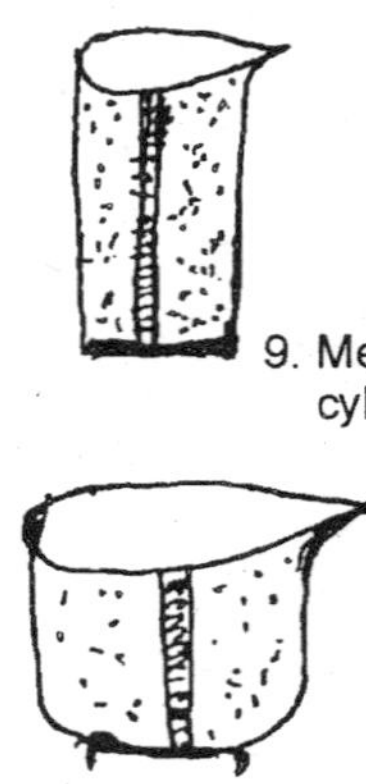

9. Measuring cylinder

10. Note Book

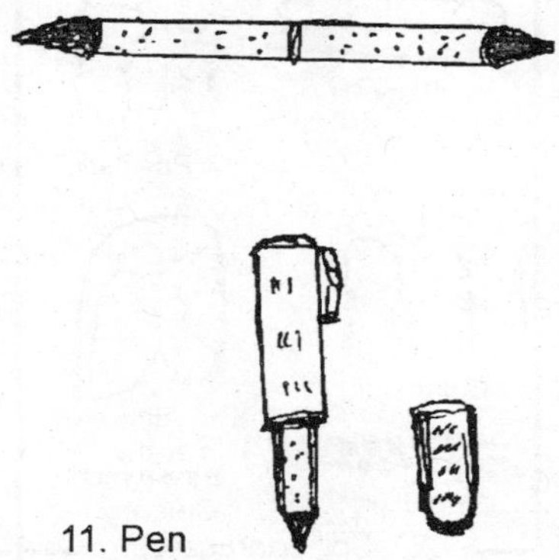
11. Pen

12. Pencil

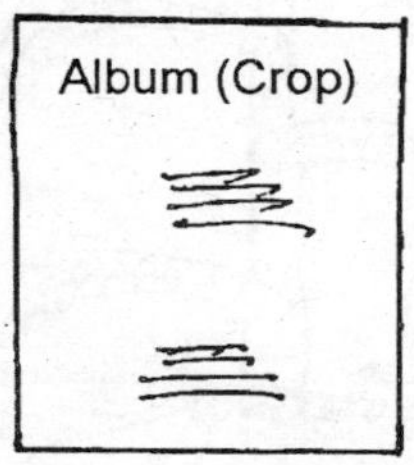

13. Album

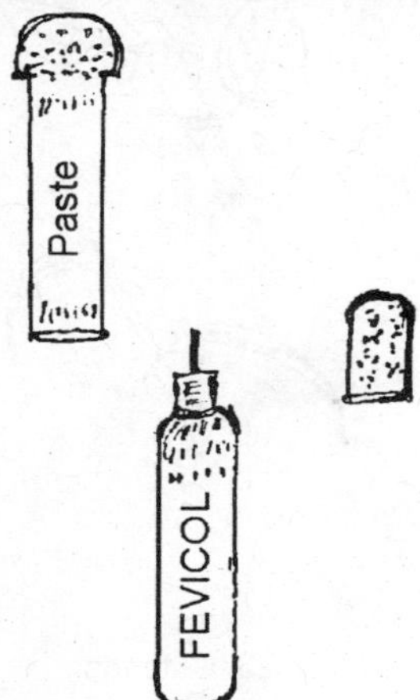

14. Gum

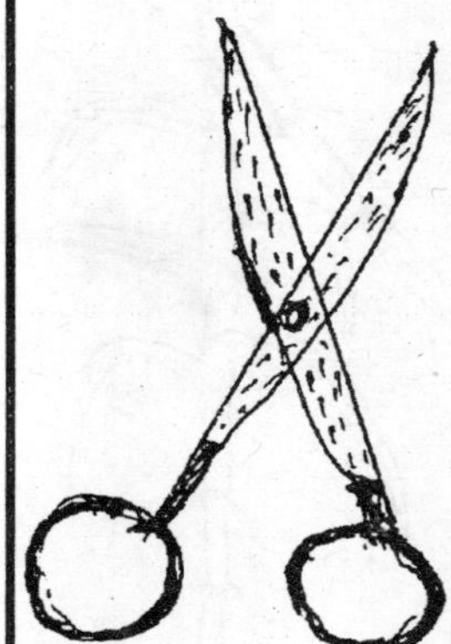
15. Scissor

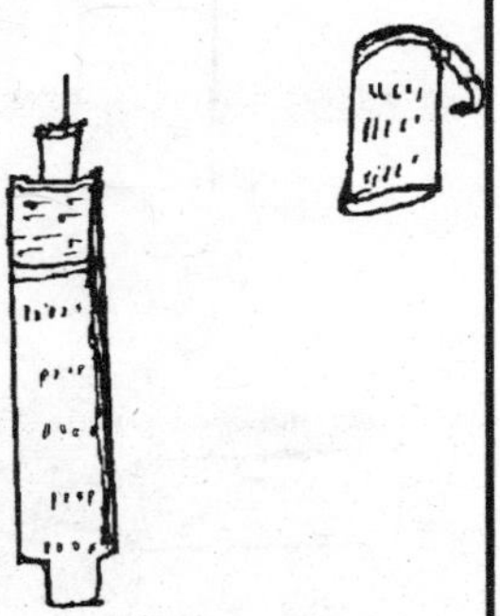
16. Correction pen

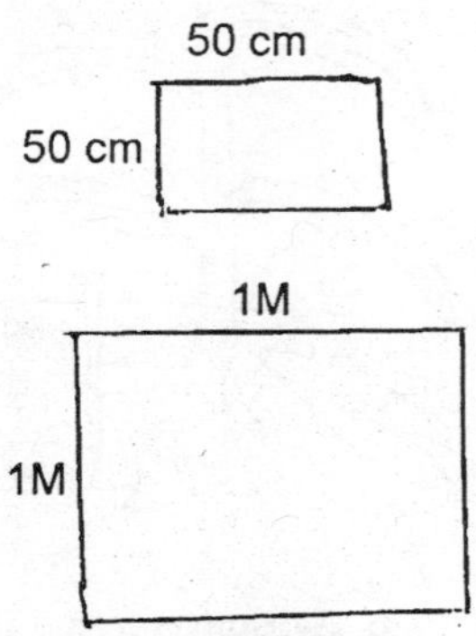

17. Quadrant

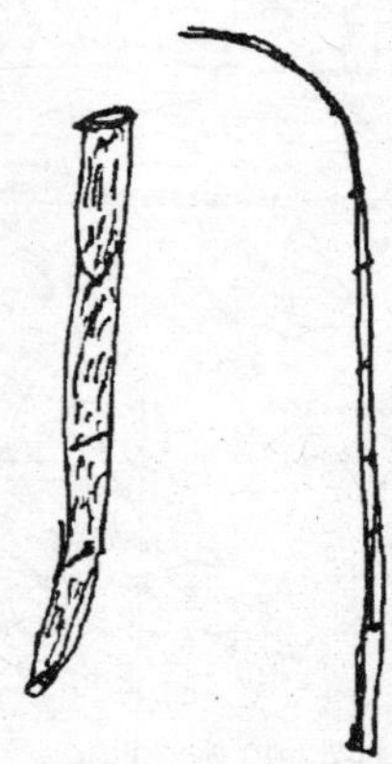
18. Crop-beating stick

19. Pudding

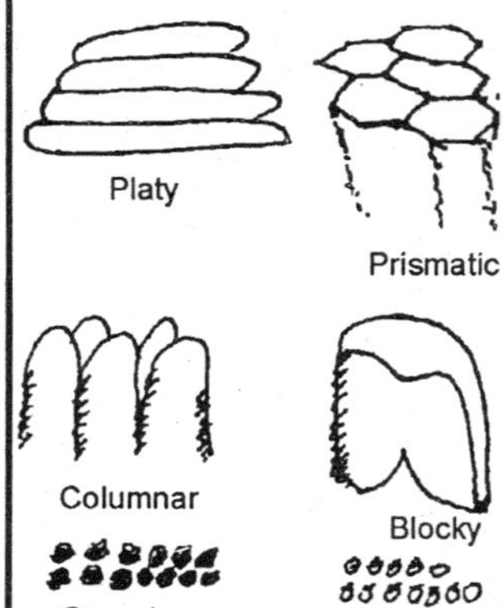

20. Soil aggregates

21. Rope/string

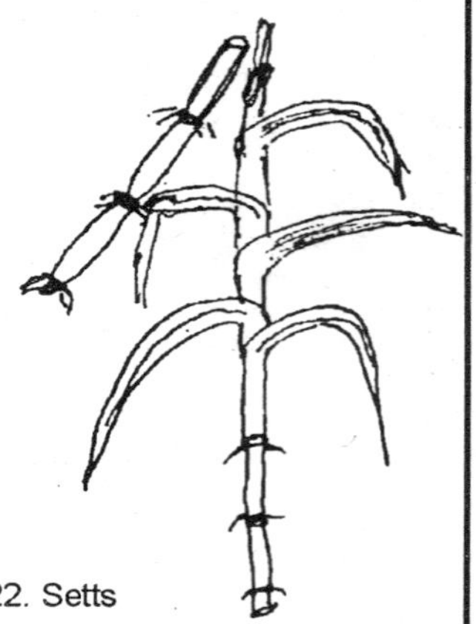
22. Setts

23. Cuttings

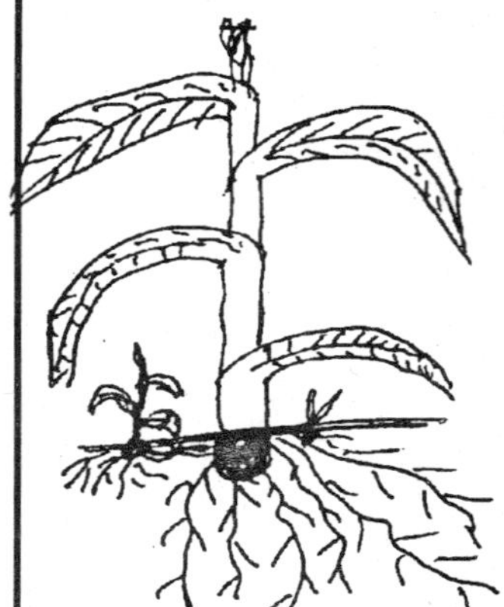
24. Propagules

25. Raising seedlings

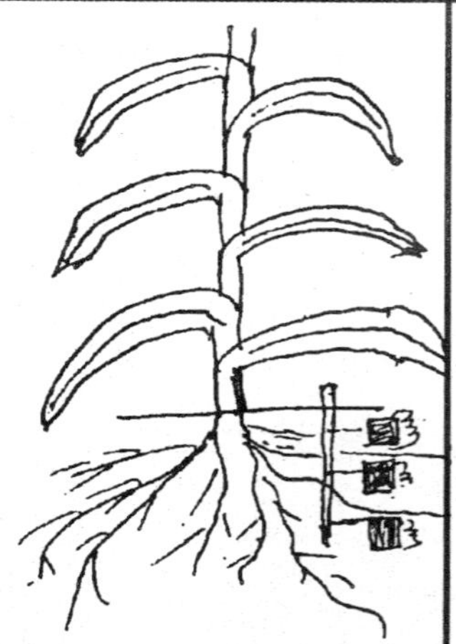
26. Gypsum blocks

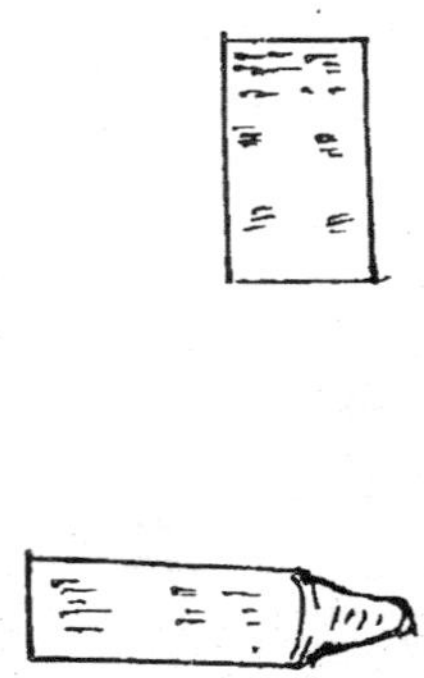
27. Rubber

REFERENCES

Chatterjee, B.N. and Maity, S. (1984). *Cropping System* (Theory & Practice). Oxford & IBH Publishing Co. Pvt. Ltd., Calcutta, First Edition.

Chatterjee, B.N. and Bhattacharyya, K.K. (1986). *Principles and Practices of grain legumes production*. Oxford & IBH Publishing Co. Pvt. Ltd., New Delhi, pp.1-492.

Chatterjee, B.N. and Das, P.K. (1989). *Forage crop production. Principles and practices*, Oxford & IBH Publishing Co. Pvt. Ltd., New Delhi, pp. 1-484.

Chatterjee, B.N. and Maiti, S. (1979). *Rice Production Technology Manual*. Oxford & IBH Publishing Co. Pvt. Ltd., New Delhi, pp. 1-139.

Das, N.R. (2006). *Rainfed paira/utera crops in India*, Kalyani Publishers, Ludhiana, First Edition, pp. 1-182.

Das, N.R. (2007). *Introduction to Crops of India*, Scientific Publishers (India), Jodhpur, First Edition, pp. 1-334.

Das, N.R. (2008). *Tillage and Crop production*, Scientific Publishers (India), Jodhpur, First Edition, pp. 1-292.

Das, N.R. (2008). *Agronomic Research Management*, Agrotech Publishing Academy, Udaipur, First Edition, pp. 1-328.

Das, N.R. (2008). *Wheat Crop Management,* Scientific Publishers (India), Jodhpur, First Edition, pp. 1-316.

De, G.C. (1989). *Fundamentals of Agronomy*. Oxford & IBH Publishing Co. Pvt. Ltd., Calcutta, First Edition, pp. 1-401.

Datta, A.C. (1971). *A Class Book of Botany*, Oxford University Press, Ely House, London, W.1. Fourteenth Edition.

ICAR (1970). *Pulse crops in India*, Indian Council of Agricultural Research, New Delhi-110012.

ICAR (1975). *Agronomy of dwarf wheat.* Indian Council of Agricultural Research, New Delhi.

ICAR (2006). *Hand Book of Agriculture,* Fifth Edition, ICAR, New Delhi.

ICAR (1985). *Rice Research in India,* Publication and Information Division, Indian Council of Agricultural Research, New Delhi-110012.

ISA (1989). *Agronomic Terminology*. Indian Society of Agronomy, Division of Agronomy, IARI, New Delhi-110 012, Second Edition.

Kundu, B.C., Basak, K.C. and Sarkar, P.B. (1959). *Jute in India*. ICJC, Calcutta, pp. 1-395.

Lal, R. (1997). *Tillage System in the Tropics : Management options and Sustainability implication*. Published jointly by Food and Agriculture Organization (FAO) of United Nations (UN) and Oxford & IBH Publishing Co. Pvt. Ltd., Calcutta, First Indian Edition.

Menon, P.A. (1973). Our Weather. National Book Trust, India, Second Print, New Delhi.

Mengel, K. and Kirkby, E.A. (1978). *Principles of Plant Nutrition*, International Potash Institute, P.O. Box CH-3048, Worblanfen – Bern, Switzerland.

Mudaliar, V.T., Subbiah (1969). *Principles of Agronomy*. The Bangalore Printing and Publishing Co. Ltd., Bangalore, p. 1-499.

Pal, Mahendra (2002). *Basics of Agriculture,* Jain Brothers, New delhi-110 005.

Palaniappan, S.P. and Annadurai, K. (1999). *Organic Farming (Theory & Practice)*. Scientific Publishers (India). Jodhpur, First Edition.

Reddy, T.Y. and Reddi, G.H.S. (1992). *Principles of Agronomy*, Kalyani Publishers, Calcutta, First Edition.

Relwani, L.L. (1979). *Fodder crops and grasses, ICAR Publication Series No. 5*. Indian Council of Agricultural Research, Delhi.

Singh, S.S. (1998). *Crop Management*, Kalyani Publishers, Calcutta, Third Revised Edition.

Singh, S.S. (1998). *Principles and Practices of Agronomy*, Kalyani Publishers, pp.1-324.

Singh, U., Wadhwani, A.M. and Johri, B.N. (1996). *Dictionary of Economic Plants in India*. Indian Council of Agricultural Research, New Delhi-110 012, Fourth Printing.

Venkataraman, S. and Krishnan, A. (1992). *Crops and Weather*, Publication and Information Division, Indian Council of Agricultural Research, Krishi Anusandhan Bhavan, IARI, New Delhi-110 012.

■■■■